# Introduction to Digital Electronics

# Introduction to Digital Electronics

**John Crowe and Barrie Hayes-Gill**
Both Lecturers in the
Department of Electrical and Electronic Engineering
University of Nottingham

OXFORD AMSTERDAM BOSTON LONDON NEW YORK PARIS
SAN DIEGO SAN FRANCISCO SINGAPORE SYDNEY TOKYO

Newnes
An imprint of Elsevier Science
Linacre House, Jordan Hill, Oxford OX2 8DP
200 Wheeler Road, Burlington, MA 01803

First published by Arnold 1998
Reprinted 2001, 2002, 2003



**British Library Cataloguing in Publication Data**
A catalogue record for this book is available from the British Library

**Library of Congress Cataloguing in Publication Data**
A catalogue record for this book is available from the Library of Congress

ISBN 0 340 64570 9

For information on all Newnes publications
visit our website at www.newnespress.com

Typeset by AFS Image Setters Ltd, Glasgow
Printed and bound in Great Britain by JW Arrowsmith Ltd, Bristol

# Contents

# Series preface

In recent years there have been many changes in the structure of undergraduate courses in engineering and the process is continuing. With the advent of modularisation, semesterisation and the move towards student-centred learning as class contact time is reduced, students and teachers alike are having to adjust to new methods of learning and teaching.

*Essential Electronics* is a series of textbooks intended for use by students on degree and diploma level courses in electrical and electronic engineering and related courses such as manufacturing, mechanical, civil and general engineering. Each text is complete in itself and is complementary to other books in the series.

A feature of these books is the acknowledgement of the new culture outlined above and the fact that students entering higher education are now, through no fault of their own, less well equipped in mathematics and physics than students of ten or even five years ago. With numerous worked examples throughout, and further problems with answers at the end of each chapter, the texts are ideal for directed and independent learning.

The early books in the series cover topics normally found in the first and second year curricula and assume virtually no previous knowledge, with mathematics being kept to a minimum. Later ones are intended for study at final year level.

The authors are highly qualified chartered engineers with wide experience in higher education and in industry.

R G Powell
Nottingham Trent University

# Preface

This book covers the material normally found in first and second year modules on digital electronics. It is intended for use by degree, diploma and TEC students in electronic engineering, computer science and physics departments. It is also aimed at those professionals who are revisiting the subject as a means of updating themselves on recent developments.

The philopsophy of this book is to cover the basics in detail and to give a taste of advanced topics such as asynchronous and synchronous circuit design, and semi-custom IC design. In other words we have adopted a broad brush approach to the subject. This will provide the reader with an introduction to these advanced techniques which are covered in more specialist and comprehensive texts. The book is sprinkled with numerous practical examples indicating good and poor design techniques and tips regarding circuit pitfalls. To provide you with confidence in your understanding each chapter is concluded with both short and long questions and follows the same format as the other texts in this series.

## Acknowledgements

Thanks are due to David Ross for his patience and understanding when every deadline was set and missed, and for offering the opportunity for us to be involved with this series. Finally, and not least, we would like to thank Robert, Emily, Charlotte, Jessica and Nicki (who also assisted in the proof reading) for their patience.

Solutions to the self-assessment and other exercises are available from the publishers to lecturers only. Please e-mail nicki.dennis@hodder.co.uk

# 1 Fundamentals

## 1.1 INTRODUCTION

This chapter introduces the essential information required for the rest of the book. This includes a description of Boolean algebra, the mathematical language of digital electronics, and the logic gates used to implement Boolean functions. Also covered are the 'tools' of digital electronics such as truth tables, timing diagrams and circuit diagrams. Finally, certain concepts such as duality, positive and negative assertion level logic and universal gates, that will be used in later chapters, are introduced.

## 1.2 BASIC PRINCIPLES

### 1.2.1 Boolean algebra – an introduction

The *algebra* of a number system basically describes how to perform arithmetic using the *operators* of the system acting upon the system's *variables* which can take any of the allowed values within that system. Boolean algebra describes the arithmetic of a two-state system and is therefore the mathematical language of digital electronics. The variables in Boolean algebra are represented as symbols (e.g. *A, B, C, X, Y* etc.) which indicate the state (e.g. voltage in a circuit). In this book this state will be either 0 or 1.[1] Boolean algebra has only three operators: NOT, AND and OR. The symbols representing these operations, their usage and how they are used verbally are all shown in Table 1.1. Note that whereas the AND[2] and OR operators operate on two or more variables the NOT operator works on a single variable.

Table 1.1 Boolean variables and operators

| Operator | Symbol | Usage | Spoken as |
|---|---|---|---|
| NOT | – | $\bar{A}$ | not A; or A bar |
| AND | · | A · B | A and B |
| OR | + | A+B | A or B |

[1]In other textbooks, and occasionally later on in this one, you may see these states referred to as HIGH and LOW or ON and OFF.

[2]Sometimes the AND symbol, $A \cdot B$, is omitted and the variables to be AND'd are just placed together as $AB$. This notation will be adopted in later chapters.

**Example 1.1**

A circuit contains two variables (i.e. signals), $X$ and $Y$, which must be OR'd together. How would this operation be shown using Boolean algebra, and how would you describe it verbally?

***Solution***

The operation would be spoken as *X or Y* and written as $X+Y$.

**Example 1.2**

The output $Y$ of a logic circuit with two inputs, $A$ and $B$, is given by the Boolean arithmetic expression, $Y=A\cdot\bar{B}$. How would this be described verbally?

***Solution***

This would be spoken as either *Y equals A and B bar*, or alternatively *Y equals A and not B*.

## 1.2.2 The three Boolean operators

The basic gates (i.e. circuit elements) available in digital electronics perform the three Boolean algebraic operations of NOT, AND and OR. The symbols for these gates are shown in Fig. 1.1. In order to both design and analyse circuits it is necessary to know the output of these gates for any given inputs.

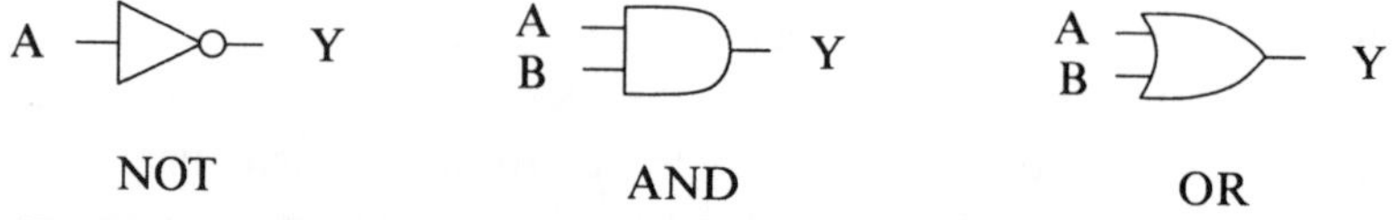


Fig. 1.1 The three basic Boolean operators

**The NOT operator**

Since any Boolean variable can only be either 0 or 1 (Boolean algebra is a two- state system) then if it is 0 its complement is 1 and vice versa. The NOT gate performs this operation (of producing the complement of a variable) on a logic signal, so if $A$ is input to the NOT gate then the output is represented by $Y=\bar{A}$. Therefore if $A=0$ then $Y=1$, or else $A=1$ and $Y=0$ (there are only two possibilities).

The *truth table* of a logic system (e.g. digital electronic circuit) describes the output(s) of the system for given input(s). The input(s) and output(s) are used to label the columns of a truth table, with the rows representing all possible inputs to the circuit and the corresponding outputs. For the NOT gate there is only one input (hence one input column, $A$), which can only have two possible values (0 and 1), so there are only two rows.[3] As there is only one output, $Y$, there is only

[3]The number of possible inputs, and hence rows, is given by $2^i$ (where $i$ is the number of inputs) since each of the $i$ inputs can only take one of two possible values (0 and 1).

one output column. The truth table for the NOT gate in Table 1.2 shows that $Y=1$ if $A=0$, and $Y=0$ if $A=1$. So $Y=\overline{A}$, the complement of $A$. The NOT gate is also sometimes referred to as an *inverter* due to the fact that it complements (inverts) its input.

Table 1.2 Truth tables for the three basic logic gates

| | $A$ | $Y$ |
|---|---|---|
| row 1 | 0 | 1 |
| row 2 | 1 | 0 |

NOT
$Y=\overline{A}$

| | $A$ | $B$ | $Y$ |
|---|---|---|---|
| row 1 | 0 | 0 | 0 |
| row 2 | 0 | 1 | 0 |
| row 3 | 1 | 0 | 0 |
| row 4 | 1 | 1 | 1 |

AND
$Y=A \cdot B$

| $A$ | $B$ | $Y$ |
|---|---|---|
| 0 | 0 | 0 |
| 0 | 1 | 1 |
| 1 | 0 | 1 |
| 1 | 1 | 1 |

OR
$Y=A+B$

**The AND operator**

The AND operator takes a number of variables as its input and produces one output whose value is 1 if and only if *all* of the inputs are 1. That is *the output is 1 if input 1* and *input 2* and *all the other inputs are 1*. Hence its name.

Considering a two-input (although it can be any number) AND gate its truth table will have two input columns, $A$ and $B$, and one output column, $Y$. With two inputs there are $2^2=4$ input combinations (since both $A$ and $B$ can be either 0 or 1) and so four rows. The output of the gate, $Y$, will be 0 unless all (i.e. both $A$ and $B$) inputs are 1, so only the last row when $A$ and $B$ are 1 gives an output of 1. The truth table (see Table 1.2) describes completely the output from an AND gate for any combination of inputs.

Alternative, but exactly equivalent, descriptions of this operation are given by use of either the circuit symbol or the Boolean equation, $Y=A \cdot B$. (This is true of all combinational logic circuits.)

**Example 1.3**

Consider a three-input AND gate. How many columns and rows would its truth table have? What would the Boolean expression describing its operation be? What would its truth table and circuit symbol be?

***Solution***

The truth table would have four columns; three for the inputs and one for the output. Since there are three inputs it would have $2^3=8$ rows corresponding to all possible input combinations. Its Boolean algebraic expression would be $Y=A \cdot B \cdot C$, assuming the inputs are $A$, $B$ and $C$. Its truth table and circuit symbol are shown in Fig. 1.2.

| $A$ | $B$ | $C$ | $A \cdot B \cdot C$ |
|---|---|---|---|
| 0 | 0 | 0 | 0 |
| 0 | 0 | 1 | 0 |
| 0 | 1 | 0 | 0 |
| 0 | 1 | 1 | 0 |
| 1 | 0 | 0 | 0 |
| 1 | 0 | 1 | 0 |
| 1 | 1 | 0 | 0 |
| 1 | 1 | 1 | 1 |

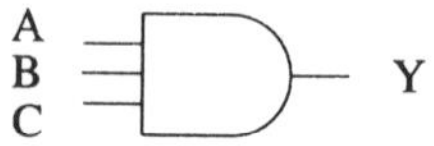

A
B
C
Y

Fig. 1.2 Truth table and symbol for a three-input AND gate as discussed in Example 1.3

**The OR operator**

The OR operator takes a number of variables as its input and produces an output of 1 if *any* of the inputs are 1. That is *the output is 1 if input 1* or *input 2* or *any input is 1.* The layout of the truth table for a two-input OR gate is the same as that for the two-input AND gate for the same reasons given above (since both have two inputs and one output). The entries in the output column are all that differ with $Y=1$ whenever any input, either $A$ or $B$, is 1.[4] Note that this includes an output of 1 if *both* inputs are 1.[5] The Boolean algebraic equation for this gate is $Y=A+B$.

**Example 1.4**

Draw the circuit symbol and truth table for a four-input OR gate.

***Solution***

These are shown in Fig. 1.3.

## 1.3 BOOLEAN ALGEBRA

Boolean algebra is the mathematical language of digital logic circuits, which are simply circuits built out of the three gates (operations) introduced above. It provides techniques for describing, analysing and designing their operation. Using the above descriptions of the operation of the three basic gates, and their Boolean descriptions, $Y=\bar{A}$, $Y=A \cdot B$ and $Y=A+B$, the additional rules and laws of Boolean logic which are needed will now be introduced.

### 1.3.1 Single-variable theorems

As the heading suggests, this section details those rules which describe the opera-

[4]Another way of looking at this is that the output is only 0 if both input $A$ and input $B$ are 0. This (negative logic) approach is discussed in more detail in Section 1.7.

[5]A gate which is similar to the OR gate in all but this aspect, the exclusive-OR (XOR) gate, will be considered in Section 1.4.

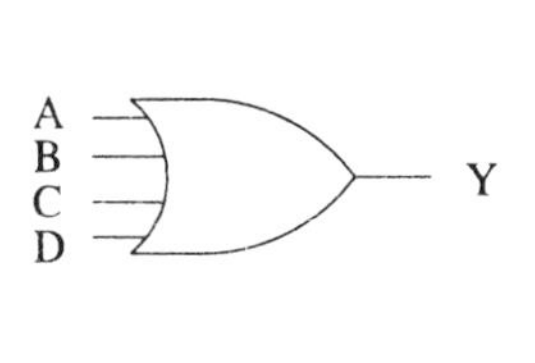


| $A$ | $B$ | $C$ | $D$ | $A+B+C+D$ |
|---|---|---|---|---|
| 0 | 0 | 0 | 0 | 0 |
| 0 | 0 | 0 | 1 | 1 |
| 0 | 0 | 1 | 0 | 1 |
| 0 | 0 | 1 | 1 | 1 |
| 0 | 1 | 0 | 0 | 1 |
| 0 | 1 | 0 | 1 | 1 |
| 0 | 1 | 1 | 0 | 1 |
| 0 | 1 | 1 | 1 | 1 |
| 1 | 0 | 0 | 0 | 1 |
| 1 | 0 | 0 | 1 | 1 |
| 1 | 0 | 1 | 0 | 1 |
| 1 | 0 | 1 | 1 | 1 |
| 1 | 1 | 0 | 0 | 1 |
| 1 | 1 | 0 | 1 | 1 |
| 1 | 1 | 1 | 0 | 1 |
| 1 | 1 | 1 | 1 | 1 |

Fig. 1.3 Truth table and symbol for a four-input OR gate as discussed in Example 1.4

tion of logic gates when only one variable is present. Note that these laws, given in Table 1.3, provide *exactly* the same information as the truth tables.

Table 1.3 Single-variable Boolean theorems

Idempotent laws: Rows 1 and 4 of the truth tables, demonstrate the effect of a variable operating upon itself:

$$A \cdot A = A \quad (1.1)$$

$$A + A = A \quad (1.2)$$

Property of inverse elements: Rows 2 and 3 of the truth tables show the effect of a variable operating on its complement:

$$A \cdot \bar{A} = 0 \quad (1.3)$$

$$A + \bar{A} = 1 \quad (1.4)$$

Involution (NOT) law:

$$\bar{\bar{A}} = A \quad (1.5)$$

Property of identity elements: The effect of operating on a variable with 0 or 1:

$$A \cdot 0 = 0 \quad (1.6)$$

$$A \cdot 1 = A \quad (1.7)$$

$$A + 0 = A \quad (1.8)$$

$$A + 1 = 1 \quad (1.9)$$

**Idempotent laws**

The idempotent[6] laws describe the effect of a variable operating upon itself (i.e. the same variable goes to all inputs). For the two-input AND gate this gives $Y = A \cdot A$ which will give 1 if $A = 1$, and 0 if $A = 0$; hence $Y = A \cdot A = A$.

[6] *idem* means 'same' in Latin.

The OR operator gives exactly the same result and these laws give the outputs in rows 1 and 4 (see Table 1.2) of the truth tables.

**Inverse elements**

The law of *inverse elements* describes the effect of operating on a variable, $A$, with its complement, $\overline{A}$. For the AND gate this gives $Y=A\cdot\overline{A}=0$, since $A$ and $\overline{A}$ must have complementary values and therefore $Y=0$.

For the OR gate $Y=A+\overline{A}=1$, since either $A$ or $\overline{A}$ must be 1. This law describes the operations in rows 2 and 3 of the truth tables.

**Involution law**

This describes the effect of operating on a variable twice with the NOT operator (i.e. passing a signal through two NOT gates). The effect of this is to return the variable to its original state. So $Y=\overline{\overline{A}}=A$.

Note that the truth tables could be derived from the above three laws as they give exactly the same information. It will become apparent that there is always more than one way of representing the information in a digital circuit, and that you must be able to choose the most suitable representation for any given situation, and also convert readily between them.

**Properties of identity elements**

The above laws give all of the information held in the truth tables. However another way of expressing this information is as the *properties of identity elements.* These just give the output of the AND and OR gates when a variable, $A$, is operated on by a constant (an identity element). (So for a two-input gate one of the inputs is held at either 0 or 1.) Obviously these laws, shown in Table 1.3, can also be used to completely determine the truth tables.

Note that Equation 1.6 in Table 1.3 states that AND'ing any variable (or Boolean expression) with 0 gives 0, whilst Equation 1.9 means that OR'ing any variable (or Boolean expression) with 1 gives 1. However AND'ing with 1 (Equation 1.7) or OR'ing with 0 (Equation 1.8) gives the Boolean value of the variable or expression used in the operation.

**Example 1.5**

What is the result of the operations $(X\cdot 0)$ and $((X\cdot Y)+1)$?

***Solution***

The output from $(X\cdot 0)$ will be 0, since anything AND'd with 0 gives a digital output of 0. The result of $((X\cdot Y)+1)$ will be 1, since any expression OR'd with 1 gives 1. Note that in the second example it is the Boolean expression $(X\cdot Y)$ (which must be either 0 or 1) that is OR'd with 1.

Table 1.4 Multivariable Boolean theorems

Commutative laws: Show that the order of operation under AND and OR is unimportant:

$$A \cdot B = B \cdot A \tag{1.10}$$
$$A + B = B + A \tag{1.11}$$

Associative laws: Show how variables are grouped together:

$$(A \cdot B) \cdot C = A \cdot B \cdot C = A \cdot (B \cdot C) \tag{1.12}$$
$$(A + B) + C = A + B + C = A + (B + C) \tag{1.13}$$

Distributive laws: Show how to expand equations out:

$$A \cdot (B + C) = A \cdot B + A \cdot C \tag{1.14}$$
$$A + (B \cdot C) = (A + B) \cdot (A + C) \tag{1.15}$$

De Morgan's theorem:

$$\overline{A + B} = \bar{A} \cdot \bar{B} \Rightarrow \overline{A + B + C + \ldots} = \bar{A} \cdot \bar{B} \cdot \bar{C} \ldots \tag{1.16}$$
$$\overline{A \cdot B} = \bar{A} + \bar{B} \Rightarrow \overline{A \cdot B \cdot C \ldots} = \bar{A} + \bar{B} + \bar{C} \ldots \tag{1.17}$$

Other laws which can be proved from the above are the:

Absorption laws:

$$A \cdot (A + B) = A \tag{1.18}$$
$$A + (A \cdot B) = A \tag{1.19}$$

and 'other identities':

$$A \cdot (\bar{A} + B) = A \cdot B \tag{1.20}$$
$$A + (\bar{A} \cdot B) = A + B \tag{1.21}$$

**Example 1.6**

What is the result of the operations $(Y \cdot 1)$ and $((X \cdot Y) + 0)$?

***Solution***

The outputs will be whatever the digital values of $Y$ and $(X \cdot Y)$ are, since anything AND'd with 1 or OR'd with 0 is unchanged.

### 1.3.2 Multivariable theorems

These rules describe the operations of Boolean algebra when more than one variable is present. This includes defining the equivalence of certain groups of operations (i.e. groups of gates forming a circuit). All of the multivariable theorems described below are given in Table 1.4.

**Commutative laws**

These simply state that it does not matter which way two variables are AND'd or OR'd together. So

$$Y = A \cdot B = B \cdot A \quad \text{and} \quad Y = A + B = B + A$$

This is the same as saying it does not matter which inputs of a two-input gate the two variables are connected to.

**Associative laws**

These show how operations can be associated with each other (grouped together). Essentially if three or more variables are to be AND'd or OR'd together it does not matter in which order it is done. This is relevant if three variables are to be operated upon and only gates with two inputs are available.

### Example 1.7

If only two input OR gates are available draw the circuit to implement the Boolean expression $Y = A + B + C$.

*Solution*

The circuit is shown in Fig. 1.4. Note that because of the associative law it does not matter which two of the three variables are OR'd together first.

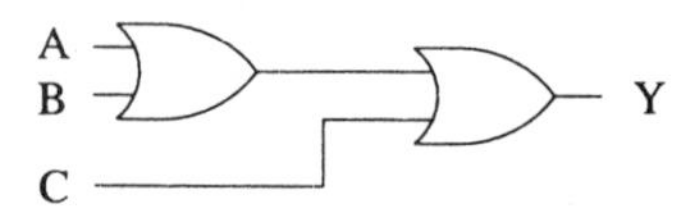


Fig. 1.4 Implementation of $Y = A + B + C$ using two-input OR gates as discussed in Example 1.7

**Distributive laws**

The rules given by the commutative and associative laws are intuitive. However, the remaining multivariable theorems require more thought and are less obvious. The distributive laws (Equations 1.14 and 1.15) show how to expand out Boolean expressions and are important because it is upon them that the factorisation, and hence simplification, of such expressions are based.

### Example 1.8

What does the expression $(A + B) \cdot (C + D)$ become when expanded out?

*Solution*

Doing this rigourously let us replace the Boolean expression $(A + B)$ with $X$. (This sort of substitution of one Boolean expression for another is perfectly legitimate.) We then have $X \cdot (C + D)$ to expand, which using the distributive law becomes

$$X \cdot C + X \cdot D = (A + B) \cdot C + (A + B) \cdot D$$

Using the commutative law to reverse these AND'd expressions and then the distributive law again gives the result of

$$(A + B) \cdot (C + D) = A \cdot C + B \cdot C + A \cdot D + B \cdot D$$

## De Morgan's theorem

De Morgan's theorem states (Equation 1.16) that complementing the result of OR'ing variables together is equivalent to AND'ing the complements of the individual variables. Also (Equation 1.17), complementing the result of AND'ing variables together is equivalent to OR'ing the complements of the individual variables.

## Example 1.9

Use Boolean algebra and de Morgan's theorem for two variables, $\overline{A+B}=\bar{A}\cdot\bar{B}$, to show that the form given in Equation 1.16 for three variables is also true.

### *Solution*

$$
\begin{aligned}
\overline{A+B+C} &= \overline{(A+B)+C} && \text{associative law}\\
&= \overline{(A+B)}\cdot\bar{C} && \text{De Morgan's theorem}\\
&= (\bar{A}\cdot\bar{B})\cdot\bar{C} && \text{De Morgan's theorem}\\
&= \bar{A}\cdot\bar{B}\cdot\bar{C} && \text{associative law}
\end{aligned}
$$

## Example 1.10

Draw the circuits that will perform the functions described by both sides of the first of De Morgan's theorems (Equation 1.16) given in Table 1.4, and also demonstrate the theorem is true using a truth table.

### *Solution*

The circuits and truth table are shown in Fig. 1.5.

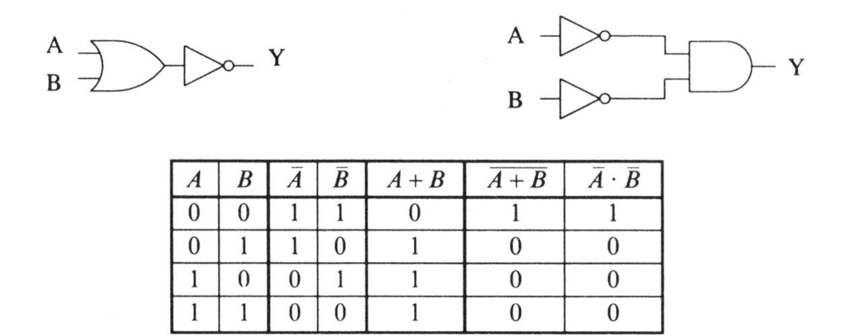


| $A$ | $B$ | $\bar{A}$ | $\bar{B}$ | $A+B$ | $\overline{A+B}$ | $\bar{A}\cdot\bar{B}$ |
|---|---|---|---|---|---|---|
| 0 | 0 | 1 | 1 | 0 | 1 | 1 |
| 0 | 1 | 1 | 0 | 1 | 0 | 0 |
| 1 | 0 | 0 | 1 | 1 | 0 | 0 |
| 1 | 1 | 0 | 0 | 1 | 0 | 0 |

Fig. 1.5 Solution to Example 1.10 regarding De Morgan's theorem

De Morgan's theorems prove very useful for simplifying Boolean logic expressions because of the way they can 'break' an inversion, which could be the complement of a complex Boolean expression.

**Example 1.11**

Use De Morgan's theorems to produce an expression which is equivalent to $Y=\overline{\overline{A}+\overline{B}\cdot C}$ but only requires a single inversion.

***Solution***

| | |
|---|---|
| $\overline{\overline{A}+(\overline{B}\cdot C)}=\overline{\overline{A}}\cdot(\overline{\overline{B}\cdot C})$ | de Morgan's theorem |
| $=A\cdot(\overline{\overline{B}}+\overline{C})$ | de Morgan's theorem |
| $=A\cdot(B+\overline{C})$ | |
| $=A\cdot B+A\cdot\overline{C}$ | distributive law |

De Morgan's theorems can also be used to express logic expressions not originally containing inversion terms in a different way. This can again prove useful when simplifying Boolean equations. When used in this way care must be taken not to 'forget' the final inversion, which is easily avoided by complementing both sides of the expression to be simplified before applying De Morgan's theorem, and then complementing again after simplification. The following example illustrates this point.

**Example 1.12**

Use De Morgan's theorem to express $Y=A+B$, the OR operation, in a different form.

***Solution***

The conversion could be performed directly but when used on more complicated expressions it is easy to 'forget' an inversion as mentioned above. We therefore firstly invert both sides of the expression giving $\overline{Y}=\overline{A+B}$. Applying De Morgan's theorem gives $\overline{Y}=\overline{A}\cdot\overline{B}$, with both sides again inverted to give the final expression $Y=\overline{\overline{A}\cdot\overline{B}}$.

Finally we note that one way of interpreting De Morgan's theorem is that *any* AND/OR operation can be considered as an OR/AND operation as long as NOT gates are used as well (see last example). This approach will be considered later on in this chapter when we look at the principle of duality in Section 1.6.

**Absorption laws**

Although these can be proved from the above laws, they nevertheless merit inclusion in their own right as they are often used to simplify Boolean expressions. Their value is clear since they take an expression with two variables and reduce it to a single variable. (For example $B$ is 'absorbed' in an expression containing $A$ and $B$ leaving only $A$.)

**Example 1.13**

Use Boolean algebra to rigorously prove the first absorption law (Equation 1.18)

$$A \cdot (A+B) = A$$

***Solution***

| | |
|---|---|
| $A \cdot (A+B) = A \cdot A + A \cdot B$ | distributive law, Equation 1.14 |
| $= A \cdot 1 + A \cdot B$ | Equations 1.1 and 1.7 |
| $= A \cdot (1+B)$ | distributive law |
| $= A \cdot 1$ | Equation 1.9 |
| $= A$ | Equation 1.7 again |

**'Other identities'**

The remaining identities are grouped together under this heading since, like the absorption laws, they can be proved from the earlier theorems, but nevertheless are not entirely obvious or expected. These identities are again valuable when trying to simplify complicated Boolean expressions.

**Example 1.14**

Use Boolean algebra and a truth table to rigorously prove the first 'other identity' (Equation 1.20)

$$A \cdot (\bar{A}+B) = A \cdot B$$

***Solution***

| | |
|---|---|
| $A \cdot (\bar{A}+B) = A \cdot \bar{A} + A \cdot B$ | distributive law, Equation 1.14 |
| $= 0 + A \cdot B$ | Equation 1.3 |
| $= A \cdot B$ | Equation 1.8 |

The truth table is shown in Table 1.5.

Table 1.5 Truth table for Equation 1.20 as discussed in Example 1.14

| $A$ | $B$ | $\bar{A}$ | $\bar{A}+B$ | $A \cdot (\bar{A}+B)$ | $A \cdot B$ |
|---|---|---|---|---|---|
| 0 | 0 | 1 | 1 | 0 | 0 |
| 0 | 1 | 1 | 1 | 0 | 0 |
| 1 | 0 | 0 | 0 | 0 | 0 |
| 1 | 1 | 0 | 1 | 1 | 1 |

**The similarity between Boolean and ordinary algebra**

You may have wondered why the AND and OR operations are indicated by the symbols for multiplication, ·, and addition, +. The reason is that many of the laws in Table 1.3 and 1.4 hold for both Boolean and ordinary algebra. Indeed, of all the

Boolean laws not involving complements (inversions) the only ones that are not true in ordinary algebra are 1.1, 1.2, 1.9 , 1.15 and the absorption laws.

It is for this reason that variables AND'd together are often referred to as *product* terms and variables OR'd together as *sum* terms. Hence the expression:

$$Y=A \cdot B+\bar{A} \cdot B$$

is a *sum of products* expression. The two product terms are $(A \cdot B)$ and $(\bar{A} \cdot B)$ which are then summed (OR'd) together. We will return to the use of this kind of terminology in Chapter 3.

A final point on this topic is that in the same way that multiplication takes precedence over addition so too does AND'ing over OR'ing. That is why when sum terms are to be AND'd together they are enclosed in brackets.

**Example 1.15**

What type of expression is $Y=(\bar{A}+B) \cdot (A+\bar{B})$, what sum terms are included in it, and what is its expanded form?

***Solution***

This is a *product of sums* expression, with two sum terms, $(\bar{A}+B)$ and $(A+\bar{B})$.

Using Boolean algebra:

$$\begin{aligned}(\bar{A}+B) \cdot (A+\bar{B}) &= (\bar{A}+B) \cdot A+(\bar{A}+B) \cdot \bar{B} && \text{distributive law}\\ &= \bar{A} \cdot A+B \cdot A+\bar{A} \cdot \bar{B}+B \cdot \bar{B} && \text{distributive law}\\ &= \bar{A} \cdot \bar{B}+A \cdot B && \text{Equation 1.3}\end{aligned}$$

(This is the Boolean expression for the exclusive-NOR gate discussed in the next section.)

## 1.4 LOGIC SYMBOLS AND TRUTH TABLES

Digital electronics is about designing and analysing circuits and although this could be done using only the mathematical language of Boolean algebra introduced above, it is often more convenient to use circuit diagrams to show how the logic gates are connected together. The logic symbols for the three basic Boolean operators have already been given in Fig. 1.1, and are included again in Fig. 1.6 which shows all of the logic gates that are commonly used together with their Boolean algebraic expressions, truth tables and the alternative IEEE/ANSI symbols for the gates.

The gates shown in Fig. 1.6 include the NAND and NOR gates which are the NOT'd versions of the AND and OR gates (i.e. NOT-AND and NOT-OR). This simply means that their outputs are inverted, which is indicated by the *bubbles* on

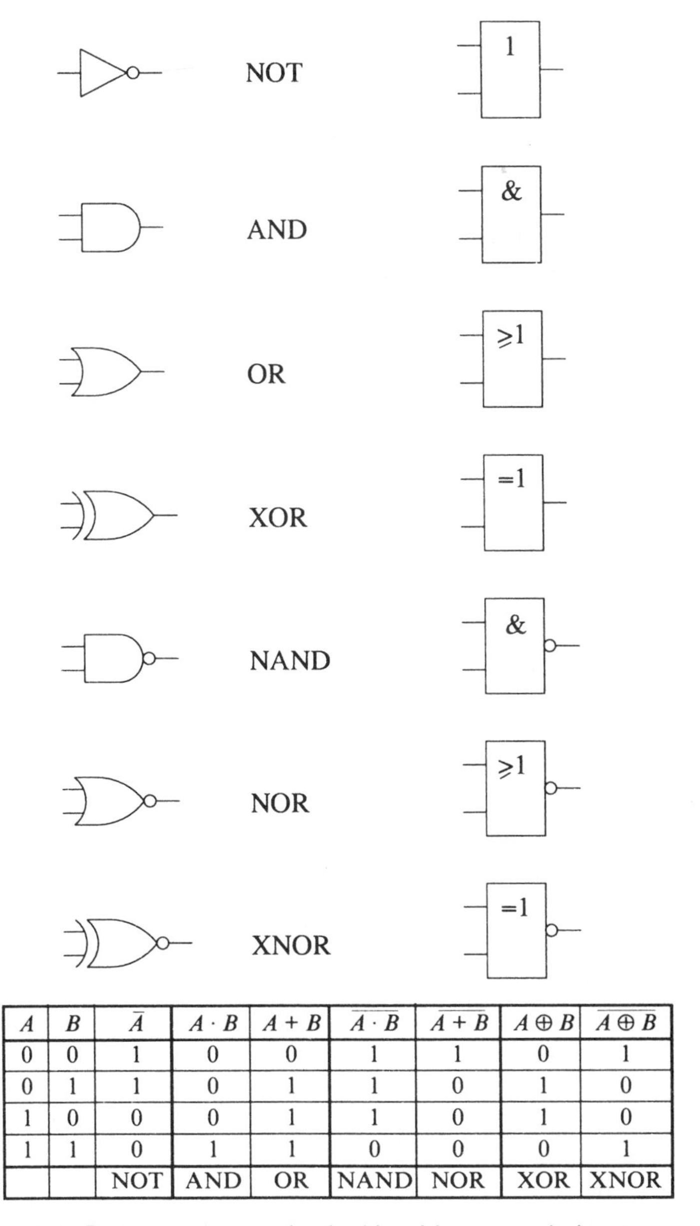


| $A$ | $B$ | $\bar{A}$ | $A \cdot B$ | $A + B$ | $\overline{A \cdot B}$ | $\overline{A + B}$ | $A \oplus B$ | $\overline{A \oplus B}$ |
|---|---|---|---|---|---|---|---|---|
| 0 | 0 | 1 | 0 | 0 | 1 | 1 | 0 | 1 |
| 0 | 1 | 1 | 0 | 1 | 1 | 0 | 1 | 0 |
| 1 | 0 | 0 | 0 | 1 | 1 | 0 | 1 | 0 |
| 1 | 1 | 0 | 1 | 1 | 0 | 0 | 0 | 1 |
| | | NOT | AND | OR | NAND | NOR | XOR | XNOR |

Fig. 1.6 Logic symbols, Boolean operators and truth tables of the common logic gates

the outputs.[7] They are equivalent to AND and OR gates whose outputs are then passed through an inverter (NOT gate).

**The exclusive-OR gate**

The other new gate introduced at this stage is the exclusive-OR (XOR) gate whose *output is 1* if and only if *an odd number of inputs are 1*. So a two-input XOR gate differs from the corresponding OR gate because $Y=0$ if both $A$ and $B$ are 1 since

[7]Bubbles are also sometimes used on the inputs to gates to indicate inversion of an input.

in this case an even number of inputs is 1. The Boolean expression for the output from a two-input XOR gate is:

$$Y = A \cdot \overline{B} + \overline{A} \cdot B$$

## Example 1.16

Write out the truth table for a three-input XOR gate and draw its circuit symbol.

### *Solution*

These are shown in Fig. 1.7

| $A$ | $B$ | $C$ | $A \oplus B \oplus C$ |
|---|---|---|---|
| 0 | 0 | 0 | 0 |
| 0 | 0 | 1 | 1 |
| 0 | 1 | 0 | 1 |
| 0 | 1 | 1 | 0 |
| 1 | 0 | 0 | 1 |
| 1 | 0 | 1 | 0 |
| 1 | 1 | 0 | 0 |
| 1 | 1 | 1 | 1 |

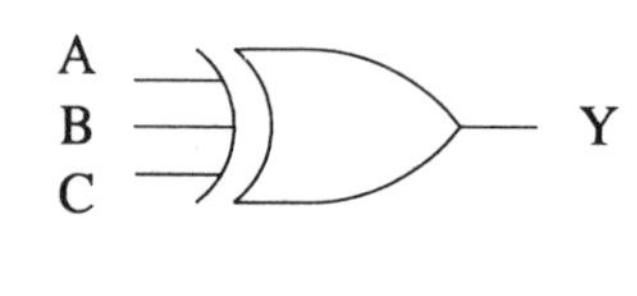


Fig. 1.7 Truth table and symbol for a three-input XOR gate (see Example 1.16)

In addition to the operation of logic circuits being described in terms of Boolean equations and circuit diagrams, remember that truth tables can also be used, as shown in Table 1.2. To recap, a truth table shows how the output(s) of a circuit (i.e. whether 0 or 1) depends upon the input(s). We now have three ways of representing the operation of a digital circuit: by a Boolean algebraic expression; a circuit diagram; or a truth table. Note that the rows of the truth table are ordered in binary code: i.e. 000, 001, 010, 011, etc. (for a table with three input variables).

## Example 1.17

Draw the circuit, and write out the truth table, for the Boolean expression $Y = (A + B) \cdot (\overline{A \cdot B})$ stating what single gate it is functionally equivalent to. Then prove this equivalence using Boolean algebra.

### *Solution*

The circuit and truth table are shown in Fig. 1.8. This complete circuit performs the function of an XOR gate.

$$\begin{aligned}(A + B) \cdot (\overline{A \cdot B}) &= (A + B) \cdot (\overline{A} + \overline{B}) && \text{De Morgan's theorem}\\ &= A \cdot \overline{A} + A \cdot \overline{B} + B \cdot \overline{A} + B \cdot \overline{B} && \text{distributive law}\\ &= A \cdot \overline{B} + \overline{A} \cdot B && \text{Equation 1.3}\end{aligned}$$

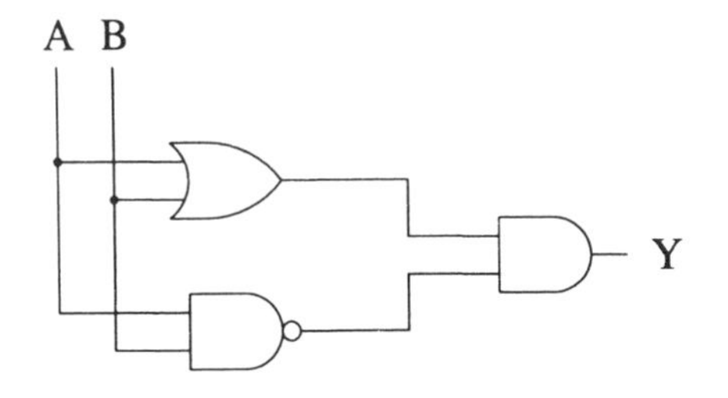


| $A$ | $B$ | $A+B$ | $\overline{(A \cdot B)}$ | $Y$ |
|---|---|---|---|---|
| 0 | 0 | 0 | 1 | 0 |
| 0 | 1 | 1 | 1 | 1 |
| 1 | 0 | 1 | 1 | 1 |
| 1 | 1 | 1 | 0 | 0 |

Fig. 1.8 Solution to Example 1.17, which produces the XOR function

## Example 1.18

Another common Boolean expression is the AND-OR-INVERT function, $Y=\overline{(A \cdot B)+(C \cdot D)}$. Draw out the circuit for this function together with its truth table.

### *Solution*

This is shown in Fig. 1.9. Note that this function has four variables and so there are $2^4=16$ rows in the truth table.

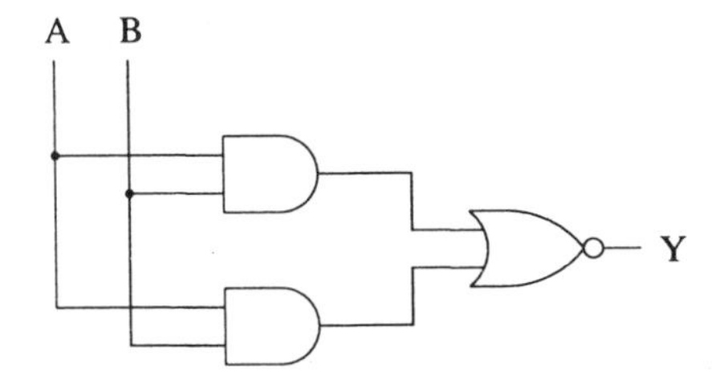


| $A$ | $B$ | $C$ | $D$ | $A \cdot B$ | $C \cdot D$ | $(A \cdot B)+(C \cdot D)$ | $\overline{Y}$ |
|---|---|---|---|---|---|---|---|
| 0 | 0 | 0 | 0 | 0 | 0 | 0 | 1 |
| 0 | 0 | 0 | 1 | 0 | 0 | 0 | 1 |
| 0 | 0 | 1 | 0 | 0 | 0 | 0 | 1 |
| 0 | 0 | 1 | 1 | 0 | 1 | 1 | 0 |
| 0 | 1 | 0 | 0 | 0 | 0 | 0 | 1 |
| 0 | 1 | 0 | 1 | 0 | 0 | 0 | 1 |
| 0 | 1 | 1 | 0 | 0 | 0 | 0 | 1 |
| 0 | 1 | 1 | 1 | 0 | 1 | 1 | 0 |
| 1 | 0 | 0 | 0 | 0 | 0 | 0 | 1 |
| 1 | 0 | 0 | 1 | 0 | 0 | 0 | 1 |
| 1 | 0 | 1 | 0 | 0 | 0 | 0 | 1 |
| 1 | 0 | 1 | 1 | 0 | 1 | 1 | 0 |
| 1 | 1 | 0 | 0 | 1 | 0 | 1 | 0 |
| 1 | 1 | 0 | 1 | 1 | 0 | 1 | 0 |
| 1 | 1 | 1 | 0 | 1 | 0 | 1 | 0 |
| 1 | 1 | 1 | 1 | 1 | 1 | 1 | 0 |

Fig. 1.9 Circuit and truth table for the AND-OR-INVERT function, as discussed in Example 1.18

## Example 1.19

Use De Morgan's theorem to convert the AND-OR-INVERT function into an alternative form using the four input variables in their complemented forms. Then draw the circuit to produce this function and the truth tables to prove it is the same as the previous circuit.

### *Solution*

$$\overline{(A \cdot B)+(C \cdot D)}=(\overline{A \cdot B}) \cdot (\overline{C \cdot D})$$
$$=(\bar{A}+\bar{B}) \cdot (\bar{C}+\bar{D})$$

The circuit and truth tables are shown in Fig. 1.10.

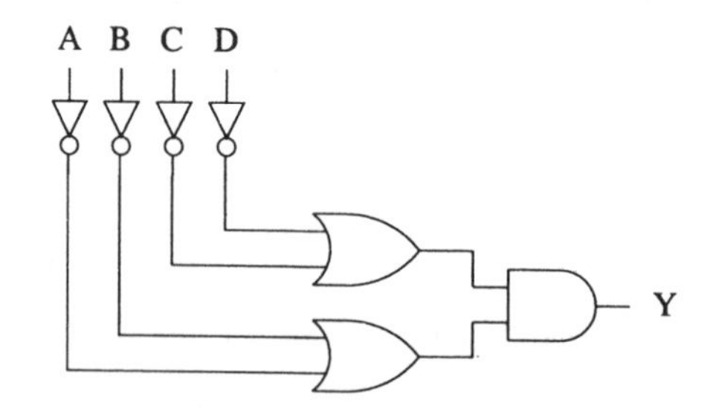


| $A$ | $B$ | $C$ | $D$ | $\bar{A}$ | $\bar{B}$ | $\bar{C}$ | $\bar{D}$ | $\bar{A}+\bar{B}$ | $\bar{C}+\bar{D}$ | $Y$ |
|---|---|---|---|---|---|---|---|---|---|---|
| 0 | 0 | 0 | 0 | 1 | 1 | 1 | 1 | 1 | 1 | 1 |
| 0 | 0 | 0 | 1 | 1 | 1 | 1 | 0 | 1 | 1 | 1 |
| 0 | 0 | 1 | 0 | 1 | 1 | 0 | 1 | 1 | 1 | 1 |
| 0 | 0 | 1 | 1 | 1 | 1 | 0 | 0 | 1 | 0 | 0 |
| 0 | 1 | 0 | 0 | 1 | 0 | 1 | 1 | 1 | 1 | 1 |
| 0 | 1 | 0 | 1 | 1 | 0 | 1 | 0 | 1 | 1 | 1 |
| 0 | 1 | 1 | 0 | 1 | 0 | 0 | 1 | 1 | 1 | 1 |
| 0 | 1 | 1 | 1 | 1 | 0 | 0 | 0 | 1 | 0 | 0 |
| 1 | 0 | 0 | 0 | 0 | 1 | 1 | 1 | 1 | 1 | 1 |
| 1 | 0 | 0 | 1 | 0 | 1 | 1 | 0 | 1 | 1 | 1 |
| 1 | 0 | 1 | 0 | 0 | 1 | 0 | 1 | 1 | 1 | 1 |
| 1 | 0 | 1 | 1 | 0 | 1 | 0 | 0 | 1 | 0 | 0 |
| 1 | 1 | 0 | 0 | 0 | 0 | 1 | 1 | 0 | 1 | 0 |
| 1 | 1 | 0 | 1 | 0 | 0 | 1 | 0 | 0 | 1 | 0 |
| 1 | 1 | 1 | 0 | 0 | 0 | 0 | 1 | 0 | 1 | 0 |
| 1 | 1 | 1 | 1 | 0 | 0 | 0 | 0 | 0 | 0 | 0 |

Fig. 1.10 Solution to Example 1.19 regarding an alternative form for the AND-OR-INVERT function

## 1.5 TIMING DIAGRAMS

Yet another way of demonstrating the operation of a logic circuit is with a timing diagram. This shows how the outputs of a circuit change in response to the inputs which are varying *as a function of time*. More complex versions of such diagrams than the ones considered here appear in the data sheets of components such as analogue-to-digital converters and solid state memory which give the timing information which is needed to design circuits containing them.

For simple circuits like those we have looked at so far the output waveform is simply obtained by using the Boolean expression or truth table.[8]

## Example 1.20

The inputs to a two-input AND gate have the following values for equal periods of time. $(A, B) = (0, 0), (1, 1), (0, 1), (0, 0), (1, 0), (1, 1), (0, 0)$. Draw the timing diagrams showing the waveforms for the inputs $A$ and $B$ and the output $Y$.

### *Solution*

The timing diagram is shown in Fig. 1.11.

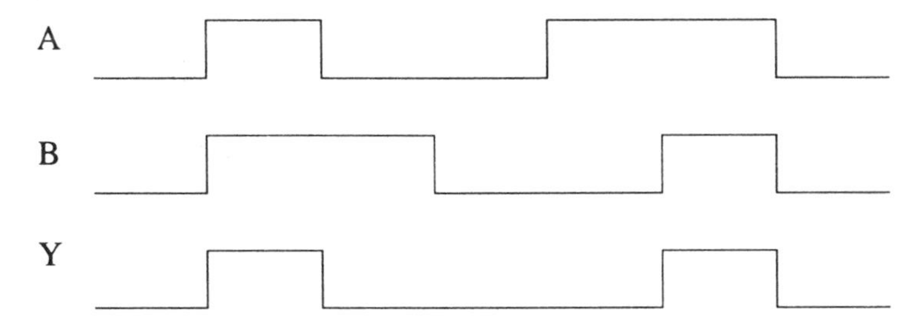


Fig, 1.11 Timing diagram for a two-input AND gate for the inputs given in Example 1.20

## Example 1.21

Given the timing diagram in Fig. 1.12, write out the truth table for the circuit responsible for it, the Boolean equation describing its operation and draw the actual circuit.

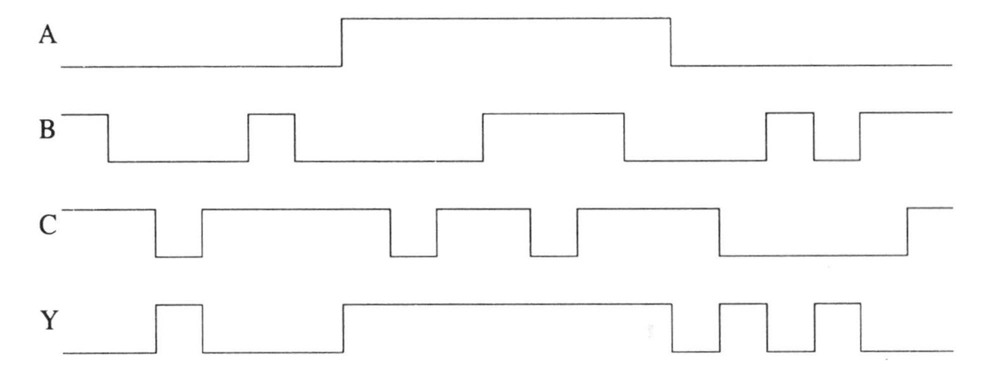


Fig. 1.12 Timing diagram for Example 1.21

[8]Note that in the timing diagrams shown here, the logic gates used to implement the Boolean functions have been considered ideal in that the signals are passed through the gates with no delay. The consequences of what happens in practice will be considered in Section 4.3.

***Solution***

From the timing diagram we can determine the values of the output, $Y$, for given input values of $A$, $B$ and $C$. These values can be used to produce the truth table in Fig. 1.13 together with the circuit.

| A | B | C | Y |
|---|---|---|---|
| 0 | 0 | 0 | 1 |
| 0 | 0 | 1 | 0 |
| 0 | 1 | 0 | 0 |
| 0 | 1 | 1 | 0 |
| 1 | 0 | 0 | 1 |
| 1 | 0 | 1 | 1 |
| 1 | 1 | 0 | 1 |
| 1 | 1 | 1 | 1 |

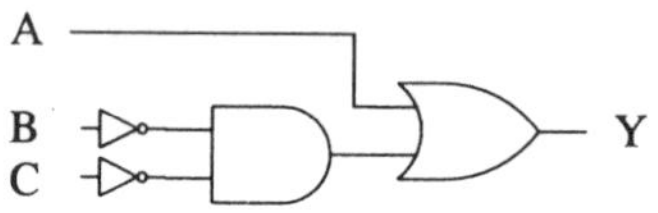

Fig. 1.13 Truth table and circuit produced from the timing diagram in Fig. 1.12 (Example 1.21)

Then, from the truth table we can see that the output is 1 if either $A=1$, irrespective of the values of $B$ and $C$ (i.e. the bottom four rows of the truth table), OR if $((B=0)$ AND $(C=0))$. So we can deduce that $Y=A+(\bar{B}\cdot\bar{C})$. (We will look at more rigorous ways of obtaining such information in later chapters.) Note that the truth table has been written in binary ordered fashion, as is usual, even though the values are not read off from the waveform in this order.

## 1.6 DUALITY AND GATE EQUIVALENCE

De Morgan's theorem (Section 1.3.2) indicates a certain equivalence between AND and OR gates since it states that the result of AND'ing/OR'ing two variables together is the same as OR'ing/AND'ing their complements. Consequently so long as we have inverters (NOT gates) available we can convert *any* circuit constructed from AND and OR gates to one composed of OR and AND gates.

This fact is generally referred to as the principle of 'duality' and arises because Boolean algebraic variables can only take one of two values and can only be operated upon (in combinations) by the two operations (AND and OR). (Its most trivial form is that if a variable is *not* 1 then it must be 0, and vice versa.) Duality has wide-ranging implications for digital electronics since it means that any circuit must have a 'dual'. That is, a replica of the circuit can be made by basically swapping bits of the circuit.

**Rules for dualling**

De Morgan's theorem tells us:

$$Y=\overline{A+B}=\bar{A}\cdot\bar{B}$$

which gives us an alternative way of constructing a NOR gate using an AND gate and two inverters. In addition:

$$Y=\overline{A\cdot B}=\bar{A}+\bar{B}$$

tells us a NAND gate can be made up of an OR gate and two inverters.

These equations, and corresponding circuits shown in Fig. 1.14, are duals and demonstrate how it is always possible to use AND/OR gates to perform any functions originally using OR/AND gates (as long as NOT gates are also available).

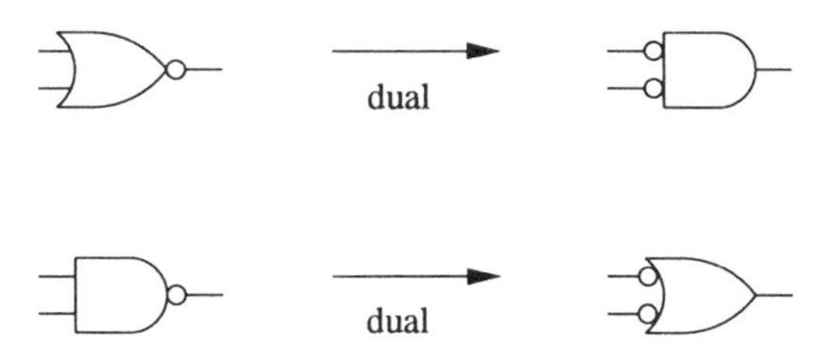


Fig. 1.14 The NOR and NAND gates and their duals

From these simple examples the 'rules' for dualling circuits can be seen to be that:

1. All input and output variables must be complemented ('bubbled') (e.g. $A$ becomes $\bar{A}$, and $\bar{B}$ becomes $B$). Note that a bubbled output feeding a bubbled input cancel each other due to Equation 1.5.
2. The type of operation must be swapped (i.e. OR gates replace AND gates, and AND gates replace OR gates).

This can be applied to any circuit no matter how complex. When using these rules to dual circuits, remember that: inverted (bubbled) inputs/outputs can rather be used to bubble preceding outputs/following inputs; and that an inverted output feeding an inverted input cancel each other.

## Example 1.22

Draw the circuit to implement $Y=(A+B)\cdot(\bar{A}+\bar{B})$ and its dual. Write out the Boolean expression for $Y$ directly from this dualled circuit, and then prove this is correct using Boolean algebra.

### *Solution*

The circuits are shown in Fig. 1.15. From the dualled circuit:

$$Y=\overline{(\bar{A}\cdot\bar{B})+(A\cdot B)}$$

Using Boolean algebra we first invert the whole equation as given:

$$\begin{aligned}\bar{Y}&=\overline{(A+B)\cdot(\bar{A}+\bar{B})}\\&=\overline{(A+B)}+\overline{(\bar{A}+\bar{B})} && \text{De Morgan's theorem}\\&=(\bar{A}\cdot\bar{B})+(A\cdot B) && \text{De Morgan's theorem}\end{aligned}$$

Hence $Y=\overline{(\bar{A}\cdot\bar{B})+(A\cdot B)}$ as above.

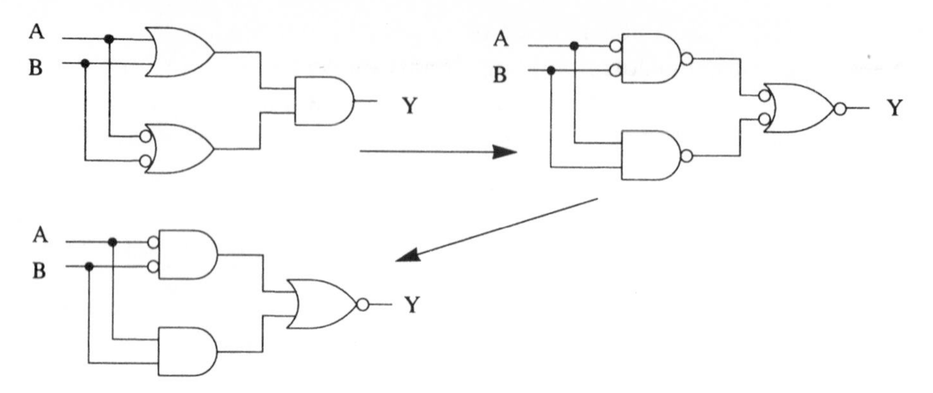


Fig. 1.15 Solution to Example 1.22

Earlier we discussed multivariable theorems which you may recall occurred in pairs. We can now see that the reason for this is due to duality, with the pairs of Boolean expressions for the associative and distributive laws and 'other identities' in Table 1.4 all being duals. (Note however that this does not mean they are equivalent Boolean expressions.)

**Example 1.23**

Show that the distributive laws:

$$A \cdot (B+C) = A \cdot B + A \cdot C$$

and

$$A + (B \cdot C) = (A+B) \cdot (A+C)$$

are duals.

***Solution***

Using the above 'rule' of complementing all variables and swapping operators the first equation becomes:

$$\bar{A} + (\bar{B} \cdot \bar{C}) = (\bar{A} + \bar{B}) \cdot (\bar{A} + \bar{C})$$

then letting $\bar{A} = X$, $\bar{B} = Y$, $\bar{C} = Z$ gives:

$$X + (Y \cdot Z) = (X + Y) \cdot (X + Z)$$

which has the same form as the second equation.

**Example 1.24**

What is the dual of $Y = A \cdot (\bar{A} + B)$, which is the left-hand side of Equation 1.20 in Table 1.4?

### *Solution*

The circuit to implement $Y$ and its dual are shown in Fig. 1.16. This gives

$$Y = \overline{\bar{A} + (A \cdot \bar{B})}$$

Using Boolean algebra:

$$\begin{aligned} \bar{Y} &= \overline{A \cdot (\bar{A} + B)} && \text{inverting both sides} \\ &= \bar{A} + \overline{(\bar{A} + B)} && \text{De Morgan's theorem} \\ &= \bar{A} + A \cdot \bar{B} && \text{De Morgan's theorem} \end{aligned}$$

Hence $Y = \overline{\bar{A} + A \cdot \bar{B}}$ as above.

Note that this has the same form as the left-hand side of Equation 1.21, as expected due to duality.

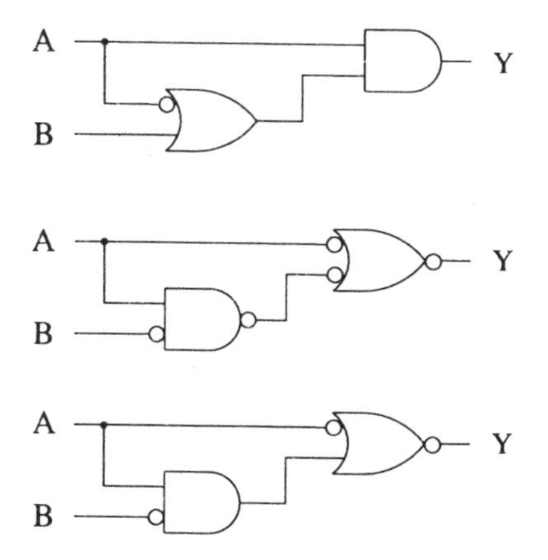


Fig. 1.16 Circuits relating to Example 1.24

## Example 1.25

Draw the circuit for $Y = A \cdot (\overline{B + C})$ and produce alternative representations of it using only a three-input AND and three-input OR gate (assuming NOT gates are also available). Also obtain the same expressions using Boolean algebra, and write out the truth table of these functions.

### *Solution*

The original circuit, and the effects of replacing the NOR and AND operators, respectively, are shown in Fig. 1.17. So:

$$Y = A \cdot \bar{B} \cdot \bar{C} = \overline{\bar{A} + B + C}$$

which can be implemented using three-input AND and OR gates respectively.

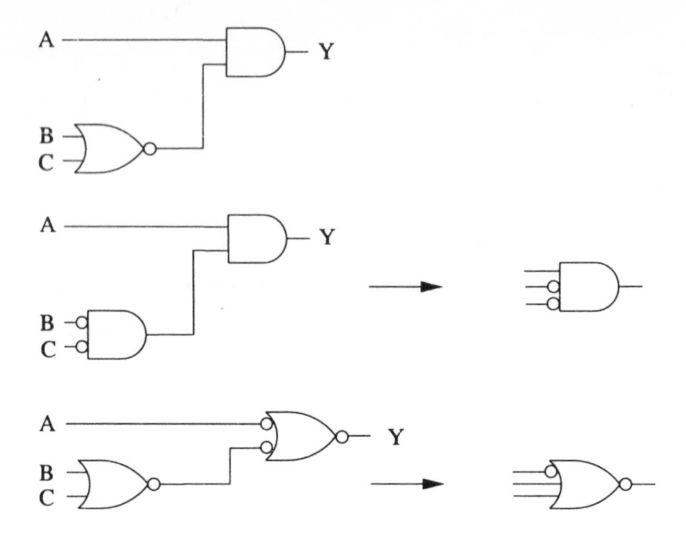


Fig. 1.17 Circuits relating to Example 1.25

Using Boolean algebra, since $(\overline{B+C})=(\bar{B}\cdot\bar{C})$ then $Y=A\cdot\bar{B}\cdot\bar{C}$, the AND gate implementation. To obtain the sum expression (OR gate implementation):

$$\begin{aligned}\bar{Y} &= \overline{A\cdot(\overline{B+C})} \quad \text{inverting both sides}\\ &= \bar{A}+(\overline{\overline{B+C}}) \text{ de Morgan's theorem}\\ &= \bar{A}+B+C\end{aligned}$$

Hence $Y=\overline{\bar{A}+B+C}$ as above.

The truth table for these is given in Table 1.6. Note that the AND gate produces a 1 when $A=1$ and $B=C=0$, which specifies a single row in the truth table. The OR operator can be considered in a slightly different way. This is that $Y=0$ (or alternatively $\bar{Y}=1$) when either ($A=0$) OR ($B=1$) OR ($C=1$). This is again a consequence of duality.

Table 1.6 Truth table relating to Example 1.25

| A | B | C | Y | $\bar{Y}$ |
|---|---|---|---|---|
| 0 | 0 | 0 | 0 | 1 |
| 0 | 0 | 1 | 0 | 1 |
| 0 | 1 | 0 | 0 | 1 |
| 0 | 1 | 1 | 0 | 1 |
| 1 | 0 | 0 | 1 | 0 |
| 1 | 0 | 1 | 0 | 1 |
| 1 | 1 | 0 | 0 | 1 |
| 1 | 1 | 1 | 0 | 1 |

# 1.7 POSITIVE AND NEGATIVE ASSERTION LEVEL LOGIC

The idea of positive and negative assertion level logic arises directly out of duality. It is again based upon the fact that because Boolean algebra describes a two-state system then specifying the input conditions for an output of 1 also gives the conditions (i.e. all other input combinations) for an output of 0.

For instance the NOR operator, $Y=\overline{A+B}$, tells us that $Y=1$ if the result of $(A+B)$ is 0 (since it is inverted to give $Y$). However, an alternative way of interpreting this operation is that $Y=0$ when $(A+B)$ is 1. Both views tell us all there is to know about how the circuit operates. The bubble on the output of the NOR gate indicates this second interpretation of the gate's operation since it signifies the output is 0 when either of the inputs is 1.

Regarding positive and negative assertion level logic, a non-bubbled input or output indicates a 1 being input or output (positive level logic) whilst a 0 indicates a 0 being input or output (negative level logic). In the case of the NOR operator such assertion level logic indicates that $Y$ is active-LOW (gives a 0) when either $A$ OR $B$ is active-HIGH (an input of 1). The dual of the NOR gate tells us that $Y$ is active HIGH if $A$ AND $B$ are active LOW (i.e. both 0).

The value of assertion level logic is that it is sometimes informative to consider the inputs and output(s) from logic circuits in terms of when they are 'active', which may be active-HIGH (an input or output of 1 being significant) or active-LOW (an input or output of 0 being significant). This is because it is useful to design circuits so that their function is as clear as possible from the circuit diagram.

Imagine an alarm is to be turned ON given a certain combination of variables whilst a light may have to be turned OFF. It could be helpful to think of the output from the corresponding circuits being 1 to turn ON the alarm (active-HIGH) and 0 (active-LOW) to turn OFF the light. In this case *assertion level* logic would be being used. Assertion level logic is also useful when interfacing to components such as microprocessors which often (because of the circuits from which they are constructed) initiate communication with other ICs by sending signals LOW.

Obviously because of duality we can always draw a circuit using the most appropriate assertion level logic. (Remember that dualling a circuit always inverts the output.) However, although a circuit may be drawn a certain way it may actually be implemented, for practical reasons, in its dualled form.

**Example 1.26**

Draw a NAND gate and its dual and describe their outputs in terms of assertion level logic.

### *Solution*

These are shown in Fig. 1.18. In the NAND form the output is active-LOW (because of the bubble) if both inputs are active-HIGH. In its dualled (OR) form the output is active-HIGH if either input is active-LOW (because of the bubbles). The two forms effectively describe the neccesary conditions for outputs of 0 (LOW) and 1 (HIGH) respectively from the circuit.

Fig. 1.18 A two-input NAND gate and its dual (see Example 1.26)

## Example 1.27

A circuit is needed to give an output of 1 when any of its three inputs are 0. Draw the truth table for this circuit and state what single gate could implement this circuit. Then derive its dual and state which gives the most appropriate desription of the circuit's operation.

### *Solution*

The truth table, which clearly describes the NAND function, is shown in Fig. 1.19 together with the single gate capable of implementing this function, and its dual.

The NAND based circuit shows the output is active-LOW if *all* of the inputs are active-HIGH, whereas the OR based circuit shows the output is active-HIGH if *any* of the inputs are active-LOW. The OR based circuit is the most appropriate given the stated circuit requirements.

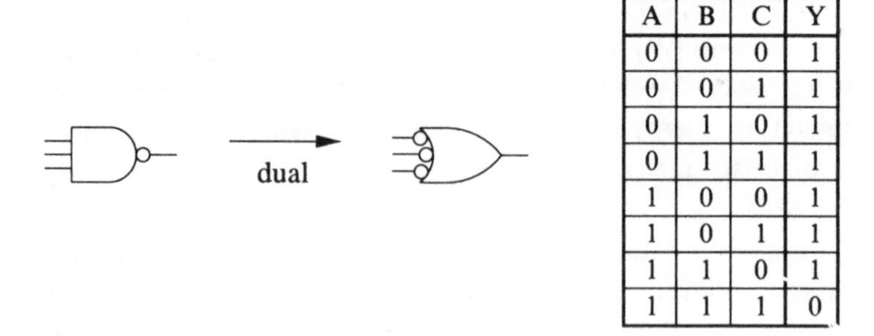


| A | B | C | Y |
|---|---|---|---|
| 0 | 0 | 0 | 1 |
| 0 | 0 | 1 | 1 |
| 0 | 1 | 0 | 1 |
| 0 | 1 | 1 | 1 |
| 1 | 0 | 0 | 1 |
| 1 | 0 | 1 | 1 |
| 1 | 1 | 0 | 1 |
| 1 | 1 | 1 | 0 |

Fig. 1.19 Truth table and gates relating to Example 1.27

## Example 1.28

The courtesy light of a car must go off when the door is closed and the light switch is off. What gate is required to implement this and what is the alternative way of looking at this circuit?

*Solution*

This function can be implemented by an AND gate with active-LOW inputs and outputs. Hence the output will go LOW (and the light OFF) when both inputs are LOW (door closed and switch off).

The alternative interpretation is that the light in on (active-HIGH) when either the door is open (active-HIGH) or the switch is on (active-HIGH). This would require an OR gate for its implementation.

## 1.8 UNIVERSAL GATES

Universal gates, as the name suggests, are gates from which any digital circuit can be built. There are two such gates, but far from being more complex than anything considered so far they are in fact the NAND and NOR gates.

The reason they are universal is that because any circuit can be dualled (by complementing all of the variables and swapping operators) then any gate capable of being used (either singly or in combinations of itself) to implement either the AND or OR operation *and* the NOT operator must be universal.

The NAND and NOR gates fulfil these requirements since tying their inputs together produces a NOT gate (rows 1 and 4 of the truth tables in Fig. 1.6). Therefore *any* digital circuit can be constructed using only NAND or NOR gates. This fact is used in VLSI design where the IC manufacturers supply a 'sea' of universal gates which are then connected as necessary to implement the required digital circuit. This is called a *gate array*, and is discussed in more detail in Section 11.3.2.

**Example 1.29**

How can an AND gate be implemented from NOR gates?

*Solution*

This is shown in Fig. 1.20.

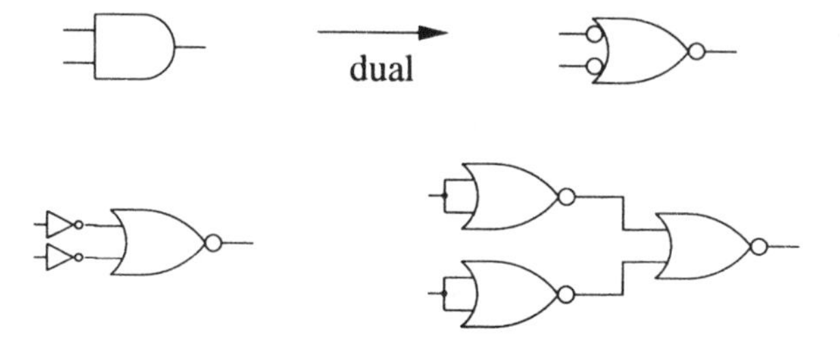


Fig. 1.20 Implementation of an AND gate using NOR gates (Example 1.29)

**Example 1.30**

How can $Y = (A + B) \cdot C$ be implemented only from NAND gates?

***Solution***

This is shown in Fig. 1.21.

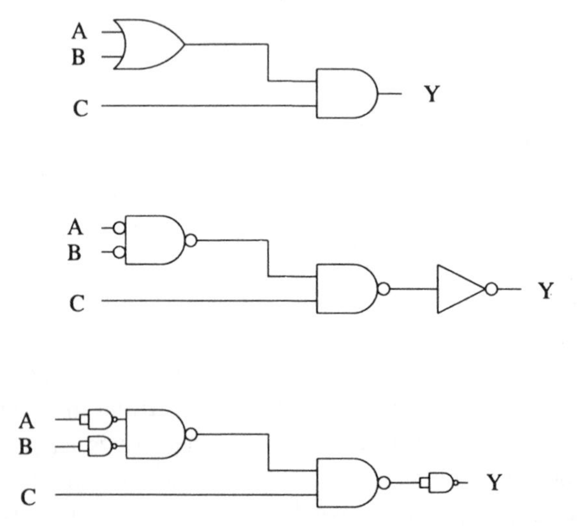


Fig. 1.21 Implementation of $Y = (A + B) \cdot C$ using NAND gates (see Example 1.30)

## 1.9 SELF-ASSESSMENT

1.1 What possible values can a Boolean variable take?

1.2 How many basic Boolean operations are there and what are they?

1.3 A logic circuit has four inputs and two outputs. How many rows and columns will the truth table have that describes its operation?

1.4 Why is a Boolean expression such as $(A \cdot \overline{B} \cdot C)$ referred to as a product term?

1.5 Draw out the truth tables and circuit symbols for two- and three-input AND, OR and XOR gates.

1.6 What is meant by a 'sum of products' expression?

1.7 Why can NAND and NOR gates act as universal gates?

1.8 Name the different ways that the operation of the same Boolean function can be described.

1.9 What is the result of (a) AND'ing a Boolean expression with 0?; (b) OR'ing a Boolean expression with 0?; (c) AND'ing a Boolean expression with 1?; (d) OR'ing a Boolean expression with 1?

## 1.10 PROBLEMS

1.1 Use Boolean algebra to rigorously prove the second absorption law:

$$A+(A \cdot B)=A$$

1.2 Use Boolean algebra and a truth table to rigorously prove the second 'other identity':

$$A+(\bar{A} \cdot B)=A+B$$

1.3 The XOR function is usually expressed as:

$$A \oplus B=A \cdot \bar{B}+\bar{A} \cdot B$$

Use Boolean algebra to show that this expression is also equivalent to:

(a) $\overline{A \cdot B+\bar{A} \cdot \bar{B}}$
(b) $(A+B) \cdot (\bar{A}+\bar{B})$
(c) $(A+B) \cdot \overline{(A \cdot B)}$

Draw the logic required to directly implement (a). Derive the dual of this circuit and state which of the two remaining expressions it directly represents. Finally, dual one of the gates in this second circuit to obtain the implementation of (c).

1.4 Using Boolean algebra, expand the Boolean function:

$$Y=A \oplus B \oplus C$$

using $P \oplus Q=P \cdot \bar{Q}+\bar{P} \cdot Q$ to show directly that the output from an XOR gate is high only if an odd number of inputs are high.

1.5 Use Boolean algebra to demonstrate that AND is distributive over XOR. That is:

$$A \cdot (B \oplus C)=A \cdot B \oplus A \cdot C$$

1.6 A combinational circuit responds as shown in the timing diagram in Fig. 1.22. Describe this circuit's function in the form of a truth table, a Boolean equation and a circuit diagram.

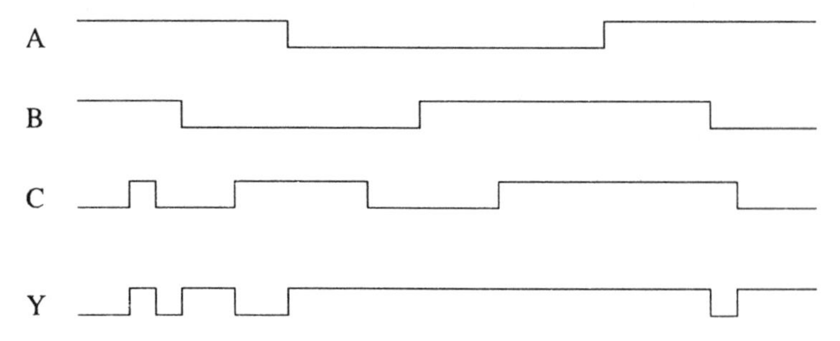


Fig. 1.22 Timing diagram for Problem 1.6

1.7 Draw the single gate implementation and its dual, of the function $Y=A+B$. Describe the operation of these gates in words, stating how the two descriptions differ, and relating this to the truth table of the function. How many NAND gates are required to implement this function?

1.8 A simple logic circuit is required to light an LED (via an active-LOW logic signal) if someone is sat in the driving seat of the car with their seatbelt undone. Sensors connected to the seat and belt give a logical HIGH if the seat is occupied and the belt fastened.

(a) Write out the appropriate truth table.
(b) Draw circuits implementing this truth table using NOT gates and (i) a single AND gate, and (ii) a single OR gate.
(c) Which implementation gives the greater understanding of the underlying operation of the circuit? (i.e. which uses assertion level logic)?
(d) How many (i) NAND and (ii) NOR gates would be needed to implement the complete circuit in b(i)?

1.9 Boolean algebra can be thought of as describing the operation of circuits containing switches. Draw the circuit containing two switches, a battery and a bulb which demonstrate the AND and OR operators. (Hint: if a switch is closed the logic variable associated with it is HIGH; if it is open the variable is LOW. So for the AND gate the bulb will be lit if switches $A$ and $B$ are closed.)

# 2 Arithmetic and digital electronics

## 2.1 INTRODUCTION

Many of the applications of digital electronic circuits involve representing and manipulating numbers as binary code (i.e. 0's and 1's). For instance in order to input any analogue value (e.g. a voltage or temperature) into a digital circuit it must be first encoded as a binary value, whilst subsequent arithmetic performed on such an input must be carried out by further digital circuits.

The way in which some arithmetic operations are implemented as digital electronic circuits is considered in the next chapter. Here, as a prelude to this, some of the many ways in which numbers can be represented as binary code are introduced, followed by a description of how to perform binary arithmetic; that is addition, subtraction, multiplication and division on numbers represented only by 0's and 1's.

## 2.2 BASES-2, 10 AND 16 (BINARY, DECIMAL AND HEXADECIMAL)

Numbers are most commonly represented using the 10 digits 0 to 9, that is in base-10 (or *decimal*). This widespread use is linked to our possession of 10 fingers and their value as a simple counting aid. However, from a purely mathematical viewpoint the base system used for counting is unimportant (indeed before metrification in Europe (use of base-10 in all weights and measures) many other bases were common). In digital electronics the only choice of base in which to perform arithmetic is base-2, that is *binary* arithmetic, using the only two digits available, 0 and 1.[1] Before continuing it is necessary to consider how to convert numbers from one base to another.

### 2.2.1 Conversion from base-*n* to decimal

In order to do this it is essential to realise what a number expressed in any base actually represents. For example the number $152_{10}$ in base-10 represents[2] the sum

[1]Digital systems using more than two logic levels (multilevel logic) have been proposed and built but are not considered here.

[2]A subscript is used to denote the base, so 152 in base 10 is written as $152_{10}$.

of 1 hundred, 5 tens and 2 units giving $152_{10}$ units. From this is can be seen that the least significant digit (the one furthest to the right) holds the number of units in the number, the digit next to that the number of 10's (i.e. $10^1$) and the next the number of 100's (i.e. $10^2$).

$$\begin{aligned}152_{10} &= (1\times 100) + (5\times 10)+(2\times 1)\\ &=(1\times 10^2)+(5\times 10^1)+(2\times 10^0)\end{aligned}$$

The same is true of any number in any base. So for a number in base-$n$, the least significant digit holds the units (i.e. $n^0$), the next the number of $n$'s (i.e. $n^1$) and the next the number of $n^2$'s. So in general the value of the three-digit number $abc$ in base $n$, i.e. $abc_n$, is given by:

$$abc_n=(a\times n^2)+(b\times n^1)+(c\times n^0)$$

In binary code this means that successive digits hold the number of 1's, 2's, 4's, 8's etc., that is quantities represented by 2 raised to successively higher powers.

In the above text and examples where it was necessary to use numeric representation (e.g. of $10^2=100$ and $2^3=8$) then base-10 was used. It should be appreciated that any base could have been chosen, with base-10 selected simply because it is the one we are most familiar with. Remember that any written number is basically a shorthand way of recording the total number of units with successive single digits representing larger and larger quantities (i.e. 1's, 10's, 100's etc., for base-10).

**Example 2.1**

What is $236_7$ in base-10?

***Solution***

$236_7=(2\times 49)+(3\times 7)+(6\times 1)=125_{10}$.

**Example 2.2**

What is $10011_2$ in base-10?

***Solution***

$10011_2=(1\times 16)+(1\times 2)+(1\times 1)=19_{10}$.

If a base larger than 10 is being used we have no numbers[3] (from base-10) to represent quantities larger that 9. Such a base commonly encountered in digital electronics is base-16, which is usually referred to as *hexadecimal*. The problem of

[3]Any base, $n$, will have $n$ units with the smallest equal to 0 and the largest equal to $(n-1)$.

representing numbers greater than 10 is overcome by using the letters A to F to replace the quantities 10 through to 15. Note that hexadecimal numbers are indicated by the use of the subscript H rather than 16.

**Example 2.3**

What is the hexadecimal number $AB1C_H$ in base-10?

***Solution***

$$\begin{aligned} AB1C_H &= (A\times 16^3)+(B\times 16^2)+(1\times 16^1)+(C\times 16^0) \\ &= (10\times 4096)+(11\times 256)+(1\times 16)+(12\times 1) \end{aligned}$$

Therefore $AB1C_H = 43804_{10}$.

### 2.2.2 Conversion from decimal to base-*n*

Conversion from decimal to base-*n* is performed by repeated subtraction of the closest approximate number given by a single digit (in base-*n*) from the remainder of the number to be converted. An example is the best way to illustrate this procedure and so the conversion of $125_{10}$ to base-7 is now given.

Firstly, since $125_{10}$ is less than $7^3 = 343$ it is the number of $7^2 = 49$s that are contained in $125_{10}$ that must first be found. There are 2 since $(2\times 49) = 98$ (which is less than 125). This leaves $27_{10}$.

$$\begin{array}{rr} & 125 \\ 2\times 49 \Rightarrow & \underline{-98} \longrightarrow 2XX_7 \\ & 27 \end{array}$$

The $27_{10}$ needs to be made up of $7^1$'s and $7^0$'s (i.e. units). Since $(3\times 7) = 21$ leaving 6 units, then $125_{10} = 236_7$.

$$\begin{array}{rr} & \overline{27} \\ 3\times 7 \Rightarrow & \underline{-21} \longrightarrow 23X_7 \\ & 6 \\ 6\times 1 \Rightarrow & \underline{-6} \longrightarrow 236_7 \\ & 0 \end{array}$$

**Example 2.4**

What is $82_{10}$ in base-5?

***Solution***

The first three digits in base-5 represent quantities (in base-10) of 1, 5, 25.
Since $82-(3\times 25) = 7$ and $7-(1\times 5) = 2$, then $82_{10} = 312_5$.

### Example 2.5

What is $153_{10}$ in binary?

***Solution***

$153_{10} = 128 + 16 + 8 + 1$. Therefore, $153_{10} = 10011001_2$.

### Example 2.6

What is $674_{10}$ in hexadecimal?

***Solution***

$674_{10} = (2 \times 256) + (10 \times 16) + (2 \times 1)$. Therefore, $674_{10} = 2A2_H$.

## 2.2.3 Binary and hexadecimal

The bases of most importance in digital electronics are binary and hexadecimal (base-2 and base-16). So, it is worth looking at conversion between these as a special case.[4] The reason hexadecimal is commonplace is because the wires in digitial circuits are often grouped into *buses* of 16 for addressing solid state memory and other devices.

The first four least significant digits of a binary number encode the numbers, in base-10, from 0 to 15. This is the range covered by the least significant digit in hexadecimal. The next four binary digits allow this range to be extended to cover up to $255_{10}$ (by using the numbers $16_{10}$, $32_{10}$, $64_{10}$ and $128_{10}$, i.e. the numbers represented by these binary digits, as appropriate). Correspondingly, the second hexadecimal digit enables numbers requiring up to $F_H = 15_{10}$ multiples of $16_{10}$ to be encoded. Hence, conversion from base-2 to hexadecimal can be performed by operating on blocks of four binary digits to produce the equivalent single hexadecimal digit.[5]

### Example 2.7

What is $A4E2_H$ in binary?

***Solution***

Since A = 1010, 4 = 0100, E = 1110 and 2 = 0010 then $A4E2_H = 1010010011100010_2$.

[4]Conversion between any other bases can always be performed via base-10. That is from base-*m* to base-10 and then from base-10 to base-*n*.

[5]This must be the case since four binary digits and one hexadecimal digit can both encode 16 different values.

**Example 2.8**

What is $100111110011_2$ in hexadecimal?

***Solution***

Since 1001 = 9, 1111 = F and 0011 = 3 then $100111110011_2 = 9F3_H$.

In concluding this section it is important to realise that to fully understand arithmetic operation in digital circuits, and the addressing of memory locations in computer systems, it is necessary to be able to readily convert numbers between bases-2, 10 and 16.

## 2.3 OTHER BINARY CODING SYSTEMS

We have just considered how quantities can be represented in binary (base-2) when only 0's and 1's are used. However, this is only one possible code (albeit the most logical from the arithmetic viewpoint) which can be used. In certain applications other forms of coding numbers, again using only 0's and 1's, are more appropriate. Two of the common alternatives, shown in Table 2.1 are now introduced.

Table 2.1 Decimal, hexadecimal, binary, Gray and binary coded decimal codes

| Decimal | Hexadecimal | binary | Gray | BCD 1st digit | BCD 2nd digit |
|---|---|---|---|---|---|
| 0 | 0 | 0000 | 0000 | 0000 | 0000 |
| 1 | 1 | 0001 | 0001 | 0000 | 0001 |
| 2 | 2 | 0010 | 0011 | 0000 | 0010 |
| 3 | 3 | 0011 | 0010 | 0000 | 0011 |
| 4 | 4 | 0100 | 0110 | 0000 | 0100 |
| 5 | 5 | 0101 | 0111 | 0000 | 0101 |
| 6 | 6 | 0110 | 0101 | 0000 | 0110 |
| 7 | 7 | 0111 | 0100 | 0000 | 0111 |
| 8 | 8 | 1000 | 1100 | 0000 | 1000 |
| 9 | 9 | 1001 | 1101 | 0000 | 1001 |
| 10 | A | 1010 | 1111 | 0001 | 0000 |
| 11 | B | 1011 | 1110 | 0001 | 0001 |
| 12 | C | 1100 | 1010 | 0001 | 0010 |
| 13 | D | 1101 | 1011 | 0001 | 0011 |
| 14 | E | 1110 | 1001 | 0001 | 0100 |
| 15 | F | 1111 | 1000 | 0001 | 0101 |

### Binary coded decimal

A problem of binary arithmetic is that direct conversion from binary to decimal (for numbers of many digits) requires a quite complex digital circuit. Therefore often when a number is being held in a digital circuit immediately before output to a display (in decimal form) binary coded decimal (BCD) rather than straight binary code is used.

BCD encodes each decimal digit with its binary equivalent using four bits. So decimal digits are simply represented in four bits by their direct binary values. A disadvantage of this is that only 10 of the possible 16 ($2^4$) codes that four bits can produce are used. Hence it is an inefficient code. Nevertheless, the advantages usually outweigh this disadvantage and so it is regularly used.

### Example 2.9

How would $916_{10}$ be represented in binary coded decimal?

#### *Solution*

Since the binary codes for 9, 1 and 6 are 1001, 0001 and 0110 respectively, then $916_{10} = 100100010110_{BCD}$. Note that the BCD code is 12 bits long since each of the decimal digits is coded by four bits.

### Gray code

As with binary code, Gray code uses $n$ digits to produce $2^n$ distinct codes all of which are used. The difference is that as successively higher numbers (in base-10) are represented in Gray code, only one bit is changed at a time. This is best seen by looking at the code which is given in Table 2.1.

The rule for generating Gray code is to start with all zeros, representing 0, and then change the lowest significant bit that will produce a code *not used before*. So, first the LSB is changed to give 0001, then the second LSB to give 0011 and then the LSB again to give 0010. The important thing is that only one bit changes between adjacent codes.

### Example 2.10

What is $8_{10}$ in Gray Code?

#### *Solution*

From Table 2.1, $8_{10} = 1100_{GRAY}$.

Gray code is of benefit when the $n$ digital signals from some device whose output consists of an $n$-bit binary code may not all attain their correct values at the same time. For instance consider the output of a four-bit code indicating the state of some device. If the output changed from 5 to 6 then using binary code this would mean a change in the bit pattern from 0101 to 0110, and so two bits changing their values. If the least significant bit changed more slowly than the second then this would lead to a transient indication of 0111, that is state 7.

If Gray code were used there would be no such problem of a transient state since there is only a one-bit difference between adjacent codes. (The only effect of a delayed change in one bit would be a correspondingly delayed indication of movement to that output.) Gray codes will be discussed again in the next chapter regarding their connection with the simplification of Boolean logic expressions.

## 2.4 OUTPUT FROM ANALOGUE-TO-DIGITAL CONVERTERS

Analogue-to-digital converters (ADC) take an analogue voltage and convert it to binary format (i.e. just 0's and 1's) for input into a digital circuit. The number of bits of the ADC gives the number of binary outputs and sets the resolution (i.e. how many discrete voltage levels can be represented). For example a four-bit ADC has four digital outputs and can represent $2^4 = 16$ distinct voltage levels.

**Example 2.11**

What is the resolution of an eight-bit ADC with an input voltage range of 10V?

***Solution***

An eight-bit ADC can encode $2^8 = 256$ analogue values in binary format. The resolution of the ADC is the smallest voltage that can be encoded digitally, in other words the voltage represented by one bit. This is given by $10 \div 255 \approx 0.04$ V.

The different voltage levels that the outputs from the ADC represent must be coded appropriately within the digital circuit. Several possible schemes exist of which three are considered here. The codes used are shown in Table 2.2. These are for the output from a four-bit ADC and so there are 16 voltage values to be coded. These values go from +7 to −8 (i.e. the ADC can accept positive and negative voltages).

**Sign magnitude**
In this scheme three bits are used to represent the magnitude of the signal (in binary code) with the fourth bit (the most significant) used to give the sign (with 0 indicating a positive voltage). With this code it is clear what the value represented is, but subtracting different stored inputs is not easy. Note also that there are two codes to represent zero (and therefore only 15 distinct voltage values).

**Offset binary**
Here the 16 possible voltages are simply represented by binary code with $0000_H$

Table 2.2 Possible coding schemes for the output from a four-bit analogue-to-digital converter

| Value | Sign magnitude | Offset binary | Two's complement |
|---|---|---|---|
| +7 | 0111 | 1111 | 0111 |
| +6 | 0110 | 1110 | 0110 |
| +5 | 0101 | 1101 | 0101 |
| +4 | 0100 | 1100 | 0100 |
| +3 | 0011 | 1011 | 0011 |
| +2 | 0010 | 1010 | 0010 |
| +1 | 0001 | 1001 | 0001 |
| 0 | 0000 | 1000 | 0000 |
| -1 | 1001 | 0111 | 1111 |
| -2 | 1010 | 0110 | 1110 |
| -3 | 1011 | 0101 | 1101 |
| -4 | 1100 | 0100 | 1100 |
| -5 | 1101 | 0011 | 1011 |
| -6 | 1110 | 0010 | 1010 |
| -7 | 1111 | 0001 | 1001 |
| -8 | | 0000 | 1000 |
| (-0) | 1000 | | |

representing the lowest. Advantages of offset binary coding are: it has only one code for zero; and it possesses a sign bit and the value represented can be obtained by simply subtracting the code for 0 (i.e. $8_{10}=1000_2$ in this case). This scheme is also obviously well suited to ADCs accepting only positive voltages, with all zeros representing ground and all ones the maximum voltage that can be input to the ADC.

**Two's complement**

Two's complement notation is the most commonly used for integer arithmetic (since it simplifies binary subtraction) and will be discussed in more detail in Section 2.5.2. The other benefit it offers is that it also only possesses one code for zero. Note that the most significant bit (MSB) acts as a sign bit and the positive values' codes are the same as for sign magnitude. The negative values are in two's complement notation (see Section 2.5.2).

## 2.5 BINARY ARITHMETIC

Binary arithmetic is in theory very simple since only 0's and 1's are used. However, subtraction can cause problems if performed conventionally, and so is usually carried out using two's complement arithmetic.

### 2.5.1 Addition

Binary addition is no different in principle than in base-10. The only potential problem is remembering that (1 + 1) gives (0 carry 1) that is $10_2$ (i.e. in decimal

$(1+1)=2$); and that $(1+1+1)$ gives (1 carry 1) that is $11_2$ (i.e. in decimal $(1+1+1)$ $=3$).

**Example 2.12**

What is 01100101100+01101101001?

***Solution***

$$\begin{array}{r} 01100101100 \\ +01101101001 \\ \hline =11010010101 \end{array}$$

### 2.5.2 Subtraction

Binary subtraction *can* be performed directly (as for base-10) but it is tricky due to the 'borrowing' process. The two's complement approach is easier, less error prone, and is therefore recommended. It is based upon the fact that for two numbers $A$ and $B$ then $A-B=A+(-B)$ where $-B$ is the complement of $B$. So rather than subtracting $B$ directly from $A$, the *complement* of $B$ is added to $A$. All that we need is a way of generating the complement of a number.

This can be achieved (in any base) by taking the number whose complement is required away from the next highest power of the base. So the complement of an $n$-digit number, $p$, in base-$m$ is given by:

$$m^n - p$$

**Example 2.13**

What is the complement (the ten's complement ) of $58_{10}$?

***Solution***

This is a two-digit number in base-10 and so must be subtracted from $10^2$ to give the complement of 42.

**Example 2.14**

What is the complement (the two's complement) of $110_2$?

***Solution***

This three-digit number must be subtracted from $2^3=8$ giving $2_{10}=10_2$.

Subtraction using the complement is then performed by adding the complement to the number you want it to be subtracted from. In practice it is not quite this simple as you need to remember how many digits are involved in the subtraction as the following examples demonstrate.

**Example 2.15**

Use ten's complement arithmetic to find $(68-35)$.

***Solution***

The complement of 35 is 65, so this gives $(68+65)=133$ which gives the correct answer 33 if the carry is ignored (i.e. only the 2 LSBs in the answer are used).

**Example 2.16**

Use ten's complement arithmetic to find $(42-75)$.

***Solution***

This gives $42+25=67$, clearly not the correct answer. However, in the previous answer one way of considering the carry produced is that it indicated a positive answer, so simply removing the 1 gave the correct solution. Here where there is no carry this means we have a negative answer which being negative is in ten's complement format. Taking the ten's complement of 67 gives 33 meaning that $42-75=-33$, the correct answer.

Quite why complement methods of arithmetic work is best seen by simply performing arithmetic graphically using number lines. That is lines are drawn whose length represents the numbers being added or subtracted. Using this method to perform complement arithmetic the actual processes involved become very clear.

**Obtaining the two's complement**

We have seen how the two's complement of an $n$-bit number is given by subtracting it from $2^n$.

**Example 2.17**

What is the two's complement of 1010?

***Solution***

$10000_2=2^4=16$ so the two's complement is $16_{10}-10_{10}=6_{10}=110_2$.

However, since we are trying to obtain the two's complement to perform a subtraction then using subtraction to obtain it is not ideal! Another, and indeed the most often quoted, way is to invert the number (i.e. rewrite it replacing all the 0's by 1's and all the 1's by 0's) and then add one. (The inverted number is called the one's complement.) This works because inverting an $n$-bit binary number is the same as subtracting it from the $n$-bit number composed totally of 1's, which is itself 1 less than $2^n$.

**Example 2.18**

What is the one's complement of 10110?

***Solution***

Inverting all bits gives $01001_2 = 9_{10}$. Note also that $(11111 - 10110) = (31 - 22) = 9_{10}$.

**Example 2.19**

What is the one's complement of 100101?

***Solution***

The answer is 11010. Note that $(63 - 37) = 26_{10}$.

Since adding 1 to an all 1 $n$-bit number gives $2^n$ then obviously forming the one's complement of an $n$-bit number and adding 1 is the same as subtracting it from $2^n$, which gives the two's complement. The advantage of using the one's complement is the ease with which it is obtained.

The most significant bit (MSB) in two's complement notation indicates the sign of the number, as in the above ten's complement subtractions. In two's complement code if it is 0 the number is positive and if it is 1 the number is negative and its two's complement must therefore be taken in order to find its *magnitude*. (Note that the sign bit was *not* included in Example 2.17.)

Because the MSB is used to give the sign of a two's complement number this means an $n$-digit number in two's complement form will code the numbers from $-(2^{n-1})$ to $+(2^{n-1} - 1)$ (e.g. for the four-digit numbers in Table 2.2 the range is from $-8$ to $+7$).

**Example 2.20**

What is the two's complement of 0101?

***Solution***

Inverting 0101 gives 1010 (the one's complement), adding 1 gives 1011, the two's complement of 0101. (The first bit is a sign bit which, being 1, indicates that this is a negative two's complement number.) Note that this shows the two's complement of (the four-bit number) 5 is 11, which is as expected since $16-5=11$.

**Example 2.21**

What number, given in two's complement notation, is represented by 10110?

***Solution***

Since the number is already in two's complement form and its most significant bit is 1 we know it is a negative number. Its two's complement will therefore give its magnitude.

Inverting 10110 gives 01001 which upon adding one becomes 01010. So the number represented is $-10_{10}$.

**Example 2.22**

What number, given in two's complement form, does 110011 represent?

***Solution***

The MSB is 1 indicating this is a negative number. It has six bits so subtracting its decimal equivalent (51) from $2^6=64$ gives its magnitude as 13. It therefore represents $-13$.

**Shorthand method of obtaining the two's complement**

A quicker method of obtaining the two's complement of a number is to invert all bits to the left of (i.e. bits more significant than) the first bit that is 1. (Hence, all bits less significant than and including the first bit that is 1 remain as they are).

**Example 2.23**

What is the two's complement of 10110100?

***Solution***

The first bit (starting with the LSB) that is 1 is the third LSB. Therefore, this and all bits to its right remain as they are giving $XXXXX100$. The bits more significant than this one are inverted giving $01001XXX$. Together this gives the two's complement of 10110100 as 01001100.

This can be confirmed by inverting all bits of 10110100 to give 01001011. Adding 1 then gives 01001100 confirming the above (also confirmed by $2^8=256$ with $256-(128+32+16+4)=76$).

**Example 2.24**

What is the two's complement of 010111?

***Solution***

Only the first bit in this case remains as it is: $XXXXX1$ with all others being inverted, $10100X$, giving 101001.

**Subtraction using two's complement**

Binary subtraction can be performed by simply adding the two's complement of a number to that from which it is to be subtracted, rather than subtracting it itself. The most likely source of error is in making sure the sign bit, which is necessary to indicate if a number is positive or negative, is produced and used properly. Confusion may arise because in the addition process digits can be 'carried' beyond the sign bit. They should be ignored. The following examples illustrate all possible cases.

**Example 2.25**

Subtract 5 from 8 using binary notation.

***Solution***

| | | | |
|---|---|---|---|
| 8 | 01000 | two's comp. | 01000 |
| −5 | −00101 | ⇒ | +11011 |
| +3 | | | 100011 |

The fifth bit is the sign bit (it is 0 for the binary codes of +8 and +5, but 1 (indicating a complemented number) for −5). Since it is 0 in the answer this shows that the result is positive, with a binary value of $0011 = 3_{10}$. The 'overflow' sixth bit in the result is discarded.

**Example 2.26**

Subtract 8 from 5 using binary notation.

***Solution***

| | | | |
|---|---|---|---|
| 5 | 00101 | two's comp. | 00101 |
| −8 | −01000 | ⇒ | +11000 |
| −3 | | | 11101 |

Again, the fifth bit is the sign bit which since it is 1 indicates the result is negative. In order to find the magnitude of the resulting negative number we must two's complement the result. This gives $0011 = 3_{10}$, so the answer is −3.

**Example 2.27**

Subtract –5 from –8 using binary notation.

***Solution***

$$\begin{array}{rrcr} -8 & 01000 & \text{two's comp.} & 11000 \\ \underline{-5} & -00101 & \Rightarrow & \underline{+11011} \\ -13 & & & 110011 \end{array}$$

The sign (fifth bit) indicates that the result is negative. The two's complement of 10011 is $01101 = 13_{10}$ giving the result as –13. (Note the sixth 'overflow' bit is discarded.)

## 2.5.3 Multiplication

Long multiplication in binary is performed in exactly the same way as in decimal arithmetic. However, in binary arithmetic digits can only be multiplied by 0 or 1. Multiplying by a power of two, e.g. $2^2 = 100_2$, simply results in the binary number being shifted $n$ digits to the left and $n$ zeros being added as the LSBs. Consequently long multiplication is simply performed by shifting (i.e. multiplying by powers of two) and adding.

Note that division by powers of two simply results in shifts to the right. Since logic circuits exist that perform these shift operations (see Section 6.3), multiplications and divisions by powers of two are very easy and quick to implement in a digital circuit.

**Example 2.28**

Calculate 6×5 using binary arithmetic.

***Solution***

$$\begin{array}{rr} & 110 \\ \times & \underline{101} \\ & 11000 \\ + & \underline{110} \\ = & 11110 \end{array}$$

**Example 2.29**

Calculate 6.5×2.75 using binary[6] arithmetic.

[6] In binary arithmetic $0.5 = 1/2^1$ which means the first digit after the decimal point is 1. Similarly $0.25 = 1/2^2$ so its binary equivalent is 0.01. 0.75 = 0.5 + 0.25; so in binary this is 0.11.

***Solution***

```
        110.10
×       010.11
      --------
      11010000
+       110100
+        11010
    ----------
=   10001.1110
```

So the product is 17.875. (Note that 0.875=0.5+0.25+0.125.)

### 2.5.4 Division

For completeness an example of binary long division is given.

**Example 2.30**

Perform 10.625 ÷ 2.5 using binary arithmetic.

***Solution***

To simplify the calculation it helps to turn the divisor into an integer by (in this case) multiplying both divisor and dividend by 2 (i.e. shifting to the left). This turns the calculation into 21.25 ÷ 5 which in binary is 10101.01 ÷ 101. The actual division process is shown in Table 2.3.

Table 2.3 Binary division of $21.25_{10} \div 5$ (Example 2.30)

```
          1 0 0 . 0 1
       ______________
1 0 1 |  1 0 1 0 1 . 0 1
         1 0 1
         -----
         0 0 0
               0 1  0 1
                 1  0 1
                 ------
                 0  0 0
```

## 2.6 SELF-ASSESSMENT

2.1 Why must digital circuits perform arithmetic in base-2?

2.2 How many units are used, and what are the smallest and largest, in base-8 arithmetic?

2.3 What numbers (in base-10) do the first three digits in a base-8 system represent?

2.4 How many units are used in the hexadecimal system and what are they?

2.5 How many binary digits are needed to represent a four-digit decimal number using BCD?

2.6 What is the significant feature of Gray code?

2.7 Write out the Gray code for a three-digit system?

2.8 How many distinct voltage levels does a 12-bit analogue-to-digital converter have?

2.9 Why can multiplication and division by powers of 2 be performed efficiently by a digital electronics circuit?

## 2.7 PROBLEMS

2.1 What is $346_8$ in base-10?

2.2 What is $632_7$ in base-10?

2.3 What is $235_{10}$ in base-5?

2.4 What is $824_{10}$ in base-6?

2.5 What is $300_{10}$ in binary?

2.6 What is $1246_{10}$ in hexadecimal?

2.7 What is $1010101_2$ in decimal?

2.8 What is $ABE_H$ in decimal?

2.9 What is $10100101001_2$ in hexadecimal?

2.10 What is $243_{10}$ in BCD?

2.11 A four-bit analogue-to-digital converter is used to sample a signal at 200 Hz (i.e. 200 samples per second are taken). How many bytes of data will be stored if the signal is sampled for 30 seconds? (1 byte = 8 bits).

2.12 A four-bit analogue-to-digital converter has an input voltage range of 5 V. What voltage is represented by 1 bit?

2.13 What binary pattern would be on a six-bit address bus when it is pointing to memory address $23_H$? How could a six-input NAND gate be used to decode the bus to produce an active-LOW signal to indicate when this memory location is accessed? (Inverters can also be used.)

2.14 Perform 011011101 + 101110110 using binary arithmetic.

2.15 Perform $76_{10} - 57_{10}$ using ten's complement arithmetic.

2.16 Perform $64_{10} - 83_{10}$ using ten's complement arithmetic.

2.17 Redo Examples 2.20, 2.21 and 2.22 using the alternative methods of obtaining the two's complements.

2.18 Perform the following using two's complement arithmetic (all numbers are given in *natural* binary notation):

(a) 10011 – 10101
(b) 10111 – 10110
(c) 1011 – 101101
(d) 10101 – 1110
(e) 1011 – 11011

2.19 Perform $3.5_{10} \times 7.25_{10}$ using binary arithmetic.

2.20 Perform $16.875_{10} \div 4.5_{10}$ using binary arithmetic.

# 3 Combinational logic basics

## 3.1 INTRODUCTION

At any time a combinational logic circuit's output(s) depends *only* upon the combination of its inputs at that time. The important point is that the output is *not* influenced by previous inputs, or in other words the circuit has *no memory*. The uses to which combinational logic circuits are put can be broadly classed as:

- data transfer circuits to control the flow of logic around a system;
- data processing circuits that process or transform data (i.e. perform useful computations).

Common examples are: multiplexers, encoders, adders, parity checkers and comparators, all of which we will look at in the next chapter.

**Summary of basic logic theory**

All of the Boolean expressions considered in Chapter 1 have been examples of combinational logic, with the output from a circuit described in terms of the Boolean operations (AND, OR and NOT) performed on its inputs.

For example, the Boolean expression

$$Y=(\overline{A}\cdot B)+(\overline{B}\cdot C)$$

tells us that the output, $Y$, will be 1 when either of the terms $(\overline{A}\cdot B)$ OR $(\overline{B}\cdot C)$ is 1. (That is either $((A=0)$ AND $(B=1))$ OR $((B=0)$ AND $(C=1)$.) Since $Y$ only depends upon the present inputs $A$, $B$ and $C$ this is a combinational logic expression.

As was described at the end of Section 1.3.2 this is in fact a *sum of products* combinational logic expression with $(\overline{A}\cdot B)$ and $(\overline{B}\cdot C)$ the *product* terms which are then *summed* (AND'd) together. It is called a sum of products expression because of the similarities between the AND and OR Boolean algebraic operations and multipication and addition in ordinary algebra (see Chapter 1).[1] This is the reason why the symbols $\cdot$ and $+$ are used to represent the AND and OR operations respectively.

[1]There is also an equivalence between the AND and OR Boolean operators and *the intersection*, $\cap$, and *union*, $\cup$, of sets in set theory.

**Shorthand notation for the AND operator**

Because of the similarity between the AND operation and multiplication, in the same way that the multiplication of two variables, $a$ and $b$, can be indicated by juxtaposition (placing the variables next to each other, e.g. $a \times b$ is represented by $ab$) then so too can the AND operator, $\cdot$, be omitted and the two variables to be AND'd simply placed together (e.g. $A \cdot B$ is represented by $AB$). *This convention will be adopted for the remainder of the book.*

Using this notation the above expression can be written as $Y=\bar{A}B+\bar{B}C$. The truth table, which describes the dependence of $Y$ on $A$, $B$ and $C$ is shown in Fig. 3.1. For this example the truth table gives the output, $Y$, for each of the $2^3=8$ possible combinations of the three inputs. Remember that in general the operation of any combinational circuit with $n$ inputs and $z$ outputs can be represented by a truth table with $(n+z)$ columns ($n$ for the inputs and $z$ for the outputs) and $2^n$ rows to show the outputs for each of the $2^n$ possible combinations of inputs.

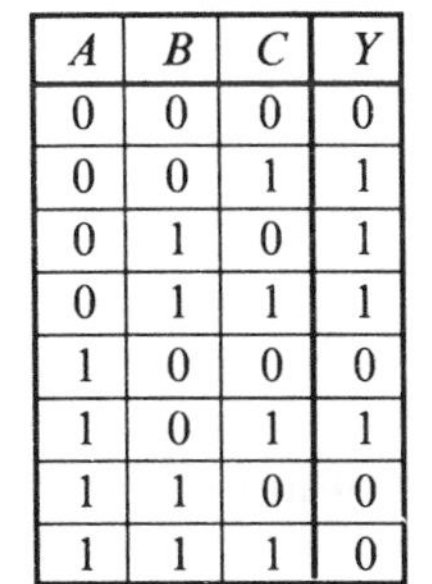

| $A$ | $B$ | $C$ | $Y$ |
|---|---|---|---|
| 0 | 0 | 0 | 0 |
| 0 | 0 | 1 | 1 |
| 0 | 1 | 0 | 1 |
| 0 | 1 | 1 | 1 |
| 1 | 0 | 0 | 0 |
| 1 | 0 | 1 | 1 |
| 1 | 1 | 0 | 0 |
| 1 | 1 | 1 | 0 |

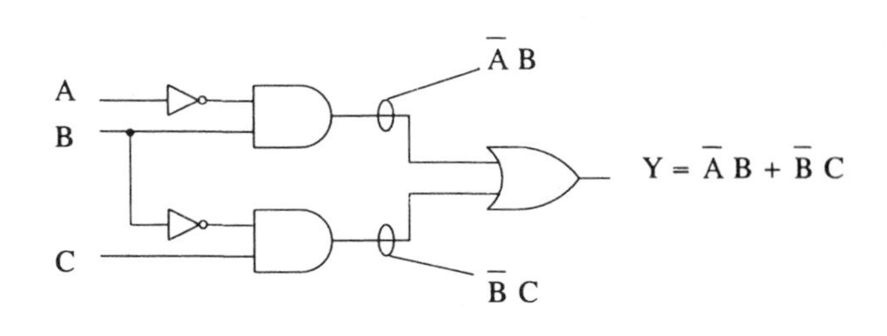


Fig. 3.1 Truth table and circuit for $Y=\bar{A}B+\bar{B}C$

Also shown in Fig. 3.1 is the circuit diagram that will implement $Y$. This consists of two AND gates, to produce the two product terms, and an OR gate to give the sum of these products.

**Example 3.1**

Write out the truth table and draw the circuit corresponding to the Boolean function $Y=\bar{A}C+\bar{A}\bar{B}$.

***Solution***

These are shown in Fig. 3.2. Note that the NOT gates to obtain $\bar{A}$, $\bar{B}$ and $\bar{C}$ have been omitted, and their availability for input into the AND gates simply assumed. This convention will also be used for other circuits in this book.

We will now look in more detail at the relationships between:

- the Boolean algebraic expression;
- the truth table;
- and the circuit diagram;

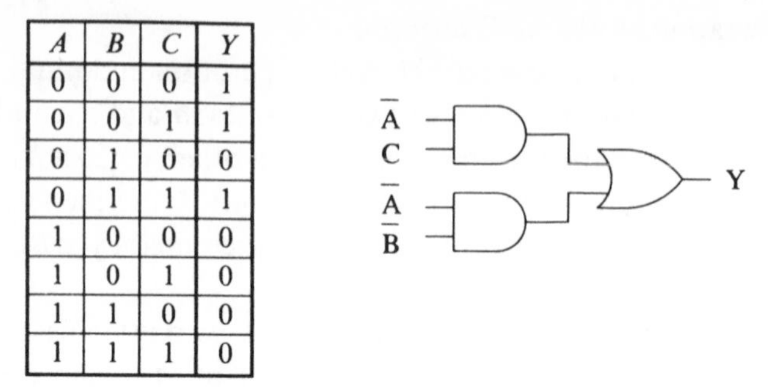

| $A$ | $B$ | $C$ | $Y$ |
|---|---|---|---|
| 0 | 0 | 0 | 1 |
| 0 | 0 | 1 | 1 |
| 0 | 1 | 0 | 0 |
| 0 | 1 | 1 | 1 |
| 1 | 0 | 0 | 0 |
| 1 | 0 | 1 | 0 |
| 1 | 1 | 0 | 0 |
| 1 | 1 | 1 | 0 |



Fig. 3.2 Truth table and circuit for $Y=\bar{A}C+\bar{A}\,\bar{B}$ (see Example 3.1)

which all describe the operation of sum of products combinational circuits. This will initially involve describing the circuit's operation in a more fundamental but expansive form.

## 3.2 COMBINATIONAL LOGIC THEORY

### 3.2.1 Fundamental sum of products expressions

We know from the Boolean algebraic expression, $Y=\bar{A}B+\bar{B}C$, introduced in Section 3.1, that $Y=1$ when either of the product terms $\bar{A}B$ or $\bar{B}C$ is 1. (This is due to the fact that any Boolean expression OR'd with 1 gives a result of 1, Equation 1.9.) Now if $(A=0)$ and $(B=1)$ are necessary for $Y=1$ (i.e. the product term $\bar{A}B$ being 1) then the value of $C$ does not matter.[2] Therefore there are two rows of the truth table ($\bar{A}B\bar{C}$ (010) and $\bar{A}BC$ (011)) which will give an output of 1 because $\bar{A}B=1$. Similarly the two rows corresponding to the product terms $A\bar{B}C$ and $\bar{A}\bar{B}C$ will also give an output of 1 since for both of these $\bar{B}C=1$ irrespective of $A$.

The two rows giving $Y=1$ for $\bar{A}B=1$ correspond to the two product terms, $\bar{A}B\bar{C}$ and $\bar{A}BC$, which contain all three input variables. Such product terms (which contain all of the input variables) are called *fundamental product terms.*[3]

Because each fundamental product term specifies the input conditions which produce an output of 1 in a single row of the truth table then there must be as many of these terms as there are 1's in the truth table. So, in this example, the product terms: $\bar{A}\bar{B}C$, $\bar{A}B\bar{C}$, $\bar{A}BC$ and $A\bar{B}C$ (reading them off from the truth table beginning at the top) are the fundamental product terms, one of which will be equal to 1 for each of the four input combinations that give an output of $Y=1$.

Now, since any Boolean expression OR'd with 1 gives a result of 1, if these fundamental product terms are OR'd together then if any one of them is 1 the result will be 1. The Boolean expression for this is:

$$Y=\bar{A}\bar{B}C+\bar{A}B\bar{C}+\bar{A}BC+A\bar{B}C$$

[2]The strict Boolean algebraic proof of this is as follows:

$$\bar{A}B=\bar{A}B\cdot 1=\bar{A}B(\bar{C}+C)=\bar{A}B\bar{C}+\bar{A}BC$$

which uses Equations 1.7, 1.4, 1.14.

[3]Sometimes the word 'fundamental' is replaced by 'canonical'.

and gives *exactly* the same information as the truth table with each fundamental product term corresponding to a row. Moreover, since each product term contains *all* of the three input variables these are fundamental product terms, and so this is a *fundamental sum of products* expression.

Since there is a one-to-one correspondence between each fundamental product term and a row in the truth table, then obviously the truth table of *any* combinational function can be used to obtain the fundamental Boolean logic expression for that function.

**Example 3.2**

Write out the Boolean expression for $Y=\bar{A}C+\bar{A}\bar{B}$ in fundamental sum of products form.

***Solution***

Using the truth table written out for this function in Fig. 3.2

$$Y=\bar{A}\bar{B}\bar{C}+\bar{A}\bar{B}C+\bar{A}BC$$

Note that, in this example, there are only three fundamental product terms. This is because the $\bar{A}\bar{B}C$ term is common to both $\bar{A}C$ (i.e. $\bar{A}(\bar{B})C$) and $\bar{A}\bar{B}$ (i.e. $\bar{A}\bar{B}(C)$).

**An alternative notation**

Since each fundamental product term corresponds to a row in the truth table an alternative way of describing a fundamental sum of products expression is simply to list the rows. To do this all that is required is to decide upon an appropriate code. The obvious choice is to use the decimal equivalent of the binary code held by the input variables. So for a truth table with three inputs we have row 0 ($\bar{A}\bar{B}\bar{C}$), row 1 ($\bar{A}\bar{B}C$) through to row 7 ($ABC$).

Using this notation, the expression $Y=\bar{A}B+\bar{B}C$ would be written as:

$$Y=\sum(1, 2, 3, 5)$$

**Example 3.3**

Express the function $Y=\bar{A}C+\bar{A}\bar{B}$ in this form.

***Solution***

$$Y=\sum(0, 1, 3)$$

### 3.2.2 Two-level circuits

The circuit to implement $Y=\bar{A}B+\bar{B}C$ was given in Fig. 3.1. It is shown again in Fig. 3.3 together with the circuit to implement the fundamental sum of products form. Since both of these circuits implement sum of products expressions they

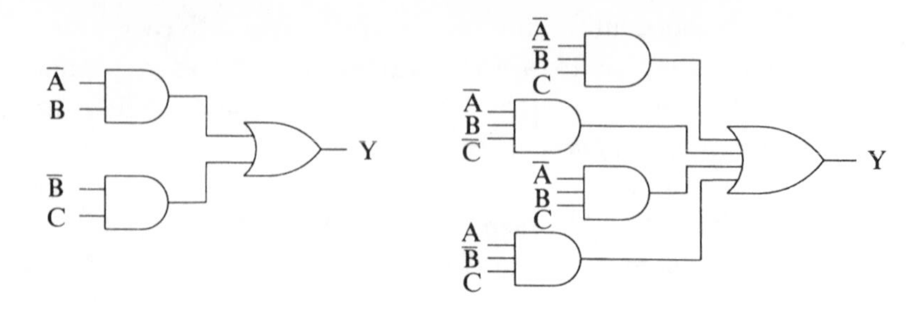


Fig. 3.3 $Y = \bar{A}B + \bar{B}C$ as a minimised and fundamental two-level circuit

have the same basic form of AND gates to produce the required product expressions, and a single OR gate to perform the summing.

Such implementations are called *two-level* circuits since they consist of a layer of AND gates feeding a single OR gate. Consequently each input signal only has to pass through two gates to reach the output. (Note that this assumes that all of the input variables and their complements are available for input to the AND gates.) Two-level implementations are important because since each signal only passes through two gates, and in real circuits it takes a finite time for a signal to propagate through a gate (see Section 9.7.4), this is (theoretically) the fastest possible implementation for a combinational logic circuit.

In practice, as we will discover via the example in Section 4.2, the practical implementation of two-level circuits often causes problems. The most obvious, and common, are that:

- an OR gate with too many inputs is required (to sum all of the required product terms);
- one input may have to feed into too many AND gates
- and that AND gates with too many inputs may be needed.

### 3.2.3 Minimisation of fundamental sum of products expressions

In this section we began with a Boolean expression containing three variables and have seen how it can be expanded, using either Boolean algebra or a truth table, to its fundamental sum of products form. When designing logic circuits it will usually be the truth table we need to implement that we have to begin with, from which we need to extract a Boolean expression.

We know we can always do this because the fundamental sum of products form can be produced directly from the truth table. However, this will in general give an unwieldy Boolean function with many fundamental product terms, with its two-level (AND-OR) implementation being correspondingly unsatisfactory. Clearly what is required is to be able to reverse the process we used earlier to generate the fundamental sum of products expression from its simpler form. This process is know as *minimisation.*

It is based upon the distributive law (Equation 1.14) and the property of inverse elements (Equation 1.4). Using the fundamental product terms $\bar{A}\bar{B}C$ and $A\bar{B}C$ as an example.

$$\begin{aligned} \bar{A}\bar{B}C+A\bar{B}C &= (\bar{A}+A)\cdot\bar{B}C && \text{Equation 1.14} \\ &= 1\cdot\bar{B}C && \text{Equation 1.4} \\ &= \bar{B}C && \text{Equation 1.7} \end{aligned}$$

An important point to note is that the fundamental terms which are combined differ only in that one of the variables in the two terms is complemented (e.g. in this example we have $A$ in one term and $\bar{A}$ in the other). This is essential for minimisation to be possible, and product terms which can be minimised because of this are said to be *logically adjacent*. Note that this is the reverse process of that described in an earlier footnote for rigorously producing the fundamental product terms (Section 3.2.1).

**Example 3.4**

Minimise $Y=\bar{A}B\bar{C}+\bar{A}BC$.

***Solution***

$$\begin{aligned} \bar{A}B\bar{C}+\bar{A}BC &= \bar{A}B\cdot(\bar{C}+C) \\ &= \bar{A}B\cdot 1 \\ &= \bar{A}B \end{aligned}$$

**Example 3.5**

Draw the circuit to implement the following fundamental sum of products expression and then minimise it to obtain the Boolean expression used in a previous example in this section.

$$Y=\bar{A}\bar{B}\bar{C}+\bar{A}\bar{B}C+\bar{A}BC$$

The circuit is shown in Fig. 3.4.

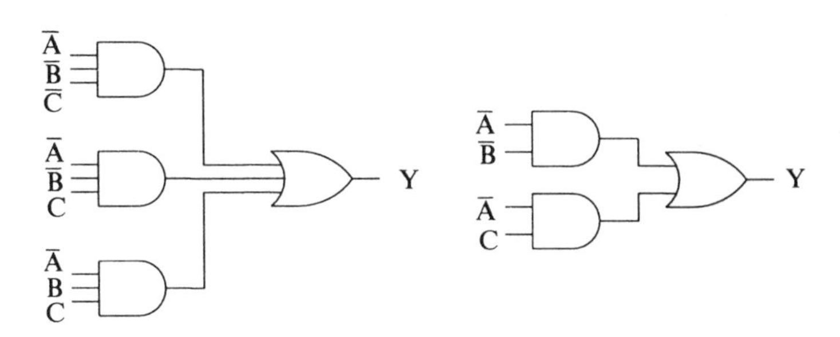


Fig. 3.4 Fundamental and minimised sum of product forms of the circuit discussed in Example 3.5

***Solution***

To minimise we must first use the fact the $X+X=X$ (Equation 1.2) to give

$$\bar{A}\bar{B}C=\bar{A}\bar{B}C+\bar{A}\bar{B}C$$

since this term is needed twice in the minimisation process. Then:

$$\begin{aligned}\bar{A}\bar{B}\bar{C}+\bar{A}\bar{B}C+\bar{A}BC &= \bar{A}\bar{B}\bar{C}+\bar{A}\bar{B}C+\bar{A}\bar{B}C+\bar{A}BC \\ &= \bar{A}\bar{B}\cdot(\bar{C}+C)+\bar{A}C\cdot(\bar{B}+B) \\ &= \bar{A}\bar{B}+\bar{A}C\end{aligned}$$

The circuit in its minimised form, which we have studied in earlier examples, is also shown.

Minimisation of combinational logic functions is a major part in the design of digital systems. For this reason, although Boolean algebra can always be used, alternative methods have been developed which are less cumbersome. However, they are still based upon the Boolean algebra described above, and essentially allow logically adjacent product terms to be easily and/or rigorously recognised and combined. Such methods are described in Section 3.3.

### 3.2.4 Summary

In summary, Boolean algebraic expressions, truth tables and circuit diagrams are all ways of describing the operation of a combinational logic circuit and all contain the same information. The Boolean algebraic expression may be in fundamental sum of products form, in which case it can be derived directly from the truth table, or in minimised (simplified) form. The fundamental sum of products form has a fundamental product term (containing all of the input variables) corresponding to each row of the truth table with an output of 1.

Minimisation of Boolean expressions is based upon recognising the occurrence of logically adjacent product terms and can be performed algebraically or using one of a number of other methods (see Section 3.3).

Sum of products expressions are implemented in two-level form with an AND gate used to produce each product term which are then summed by a single $n$-input OR gate where $n$ is the number of product terms. These are, in theory, the fastest possible general implementation of a combinational circuit although they often prove impractical.

## 3.3 MINIMISATION OF COMBINATIONAL LOGIC EXPRESSIONS

From the above we have seen how the operation of any combinational circuit can always be expressed in fundamental sum of products form. However, it is also clear (from the simple examples examined so far) that this is an extremely unwieldy form, both to write and implement. It is for this reason that the minimisation of combinational logic expressions is so important, allowing, as it does, circuits to be implemented in simpler forms.

Minimisation *can* be performed using Boolean algebra, or by inspection of truth tables, however the most commonly used manual method is the Karnaugh map. In this section we look at these methods of minimisation plus the Quine–McCluskey technique which is particularly suitable for algorithmic implemention in software.

### 3.3.1 Minimisation via Boolean algebra

This is best illustrated by an example.

**Example 3.6**

Use Boolean algebra to simplify the Boolean expression:

$$Y = AB + A\overline{B}$$

***Solution***

$$\begin{aligned} AB + A\overline{B} &= A \cdot (B + \overline{B}) && \text{distributive law} \\ &= A \cdot 1 && \text{Equation 1.4} \\ &= A && \text{Equation 1.7} \end{aligned}$$

Hence the expression originally containing two Boolean variables, but which significantly has two terms ($AB$) and ($A\overline{B}$) which only differ in that one contains the variable $B$ and the other its complement $\overline{B}$, now only has one.

What minimisation depends upon is the fact that if $A=1$ and also $B=1$, then $AB=1$ and so $Y=1$. If $B=0$ (with $A$ still 1) then yet again $Y=A\overline{B}=1$. So, $Y$ is independent of the value of $B$. Therefore the original function can be minimised to $Y=A$. All of the following examples of minimisation are based upon this principle.

**Example 3.7**

Minimise: $Y=\overline{A}B\overline{C}+\overline{A}BC+ABC$

***Solution***

$$\begin{aligned} Y &= \overline{A}B\overline{C}+\overline{A}BC+ABC \\ &= \overline{A}B\overline{C}+\overline{A}BC+\overline{A}BC+ABC && \text{Equation 1.2} \\ &= \overline{A}B(\overline{C}+C)+(\overline{A}+A)BC \\ &= \overline{A}B+BC \end{aligned}$$

Note that in this example the $\overline{A}BC$ term is used twice (Equation 1.2), and *all* of the AND operators have now been replaced by juxtaposition.

**Example 3.8**

Minimise: $Y = AB\bar{C} + ABC + A\bar{B}\bar{C} + A\bar{B}C$

***Solution***

$$\begin{aligned} Y &= AB(\bar{C}+C) + A\bar{B}(\bar{C}+C) \\ &= AB + A\bar{B} \\ &= A(B+\bar{B}) \\ &= A \end{aligned}$$

In this example four logically adjacent fundamental product terms have been combined into a single variable. What this tells us is that the expression for $Y$, which is a function of three variables, is independent of $B$ and $C$. Consequently, four rows of the (eight-row) truth table are 1 corresponding to $A=1$.

The above examples demonstrate the minimisation of fundamental sum of products expressions using Boolean algebra. It should be noted that it is also sometimes possible to spot how a function may be minimised directly from the truth table.

**Example 3.9**

Derive minimised expressions for $X$ and $Y$ from the truth table given in Table 3.1. Note that this truth table gives the outputs for two combinational logic functions, $X$ and $Y$.

Table 3.1 The truth table from which minimised functions for $X$ and $Y$ are found in Example 3.9

| $A$ | $B$ | $C$ | $X$ | $Y$ |
|---|---|---|---|---|
| 0 | 0 | 0 | 0 | 0 |
| 0 | 0 | 1 | 0 | 1 |
| 0 | 1 | 0 | 1 | 0 |
| 0 | 1 | 1 | 1 | 1 |
| 1 | 0 | 0 | 0 | 1 |
| 1 | 0 | 1 | 0 | 1 |
| 1 | 1 | 0 | 1 | 0 |
| 1 | 1 | 1 | 1 | 0 |

***Solution***

From the truth table it is clear that $X=B$ and $Y=\bar{A}C+A\bar{B}$. Note that the $\bar{A}C$ product term in $Y$ arises from rows 2 and 4, and the $A\bar{B}$ term from rows 5 and 6.

### 3.3.2 Minimisation via Karnaugh maps

Karnaugh maps contain *exactly* the same information as truth tables. The difference is that the Karnaugh map uses a 'matrix' format to hold the output values. The Karnaugh map is arranged so that, as far as possible, logically adjacent product terms are also adjacent in the 'matrix' and so can be logically combined and hence minimised. Essentially the minimisation process, described above using Boolean algebra, is performed visually.

**What does a Karnaugh map looks like?**

A truth table for two variables has two input columns, one output column and four rows (one for each possible combination of inputs). The corresponding Karnaugh map has four cells arranged as a square, with the two inputs, $A$ and $B$, used to label the columns and rows as shown in Table 3.2, which also shows the equivalence between the two for an AND gate. The information is transferred from a truth table to a Karnaugh map by simply entering the appropriate Boolean value for $Y$ into the corresponding cell of the map.

In the case of the AND gate the only cell with a 1 in is the one in the bottom right-hand corner since this holds the output when $A=1$ (the right-hand column) and $B=1$ (the bottom row).

Table 3.2 The truth table and Karnaugh map for a two-input AND gate

| $A$ | $B$ | $Y$ |
|---|---|---|
| 0 | 0 | 0 |
| 0 | 1 | 0 |
| 1 | 0 | 0 |
| 1 | 1 | 1 |

| Y | A=0 | A=1 |
|---|---|---|
| B=0 | 0 | 0 |
| B=1 | 0 | 1 |

## Example 3.10

Draw the Karnaugh maps for two-input OR and XOR gates.

### *Solution*

These are shown in Table 3.3. Note the distinctive 'chequerboard' pattern of alternate 0's and 1's that appears in the Karnaugh map of functions containing the XOR operator.

Table 3.3 Karnaugh maps for two-input OR and XOR gates (see Example 3.10)

| Y | A = 0 | A = 1 |
|---|---|---|
| B = 0 | 0 | 1 |
| B = 1 | 1 | 1 |

OR

| Y | A = 0 | A = 1 |
|---|---|---|
| B = 0 | 0 | 1 |
| B = 1 | 1 | 0 |

XOR

Obviously because each cell of the Karnaugh map corresponds to a row in the truth table it also corresponds to a fundamental product term, given by the column and row labels used to indicate its position. For example the top right-hand cell is indexed by $A=1$ and $B=0$, and indicates whether the fundamental product term $A\bar{B}$ gives 1 for the function $Y$ that the Karnaugh map is drawn for. In other words if the cell has a 1 in, then $Y$ contains this fundamental product term (i.e. $Y=1$ when $(A=1)$ AND $(B=0)$), but does not if the cell has a 0 in it.

This leads us on to an alternative way of indexing the Karnaugh map which is to replace $A=1$ by $A$, and $A=0$ by $\bar{A}$ for the columns, with $B$ and $\bar{B}$ used to index the rows. Table 3.4 shows the two notations with the fundamental product terms actually written in the apppropriate cells, together with the row numbers used in Section 3.2.1 to code fundamental sum of products expressions.

Table 3.4 Alternative notation for labelling a Karnaugh map

| Y | A = 0 | A = 1 |
|---|---|---|
| B = 0 | $_0\bar{A}\bar{B}$ | $_2A\bar{B}$ |
| B = 1 | $_1\bar{A}B$ | $_3AB$ |

| Y | $\bar{A}$ | A |
|---|---|---|
| $\bar{B}$ | $_0\bar{A}\bar{B}$ | $_2A\bar{B}$ |
| B | $_1\bar{A}B$ | $_3AB$ |

**Example 3.11**

Draw the truth table and Karnaugh map for $Y=\bar{A}\bar{B}+AB$. What function is this?

***Solution***

These are shown in Table 3.5 and are clearly for the XNOR function.

Table 3.5 Truth table and Karnaugh map for a two-input XNOR gate (see Example 3.11)

| $A$ | $B$ | $Y$ |
|---|---|---|
| 0 | 0 | 1 |
| 0 | 1 | 0 |
| 1 | 0 | 0 |
| 1 | 1 | 1 |

| Y | $\bar{A}$ | A |
|---|---|---|
| $\bar{B}$ | 1 | 0 |
| B | 0 | 1 |

**Minimisation using Karnaugh maps**

Although using a two-variable Karnaugh map for minimisation is a rather trivial process it nevertheless serves to illustrate the minimisation process, which as we have seen consists of combining logically adjacent fundamental product terms. We now look at examples of minimisation using Karnaugh maps.

**Example 3.12**

Draw the Karnaugh map for $Y=A\overline{B}+AB$, and then use it to obtain a minimised expression.

***Solution***

The Karnaugh map is shown in Table 3.6. From the minimisation examples using Boolean algebra given above it is clear that

$$Y=A\overline{B}+AB=A(\overline{B}+B)=A$$

Table 3.6 The Karnaugh map for $Y=A\overline{B}+AB$, discussed in Example 3.12

| Y | $\overline{A}$ | A |
|---|---|---|
| $\overline{B}$ | 0 | 1 |
| B | 0 | 1 |

Using the Karnaugh map for minimisation we see that the two fundamental product terms ($A\overline{B}$ and $AB$) are logically adjacent (they are in the same column and so only differ in one variable), and hence know they can be combined. The way we know that it is the variable $B$ that is eliminated is that the adjacent terms differ in this variable (one has $\overline{B}$, the other $B$) across the rows, whereas both contain $A$, since they are in the column labelled $A$. Hence, the expression minimises to $Y=A$.

It is important to fully appreciate that Karnaugh maps are simply a graphical aid to the minimisation process that could always be (although usually less simply) achieved using Boolean algebra.

**Example 3.13**

Use a Karnaugh map to minimise $Y=\overline{A}\overline{B}+A\overline{B}$.

***Solution***

From the Karnaugh map in Table 3.7 we see that the two fundamental product terms are logically adjacent and differ in $A$ across the columns, whilst $\overline{B}$ is the common term (i.e. they are in the top row). So $Y=\overline{B}$.

In these examples the expressions minimise to a single variable. We now return to the Karnaugh maps drawn earlier for the OR and XOR operators to demonstrate other possibilities.

Table 3.7 Karnaugh map for $Y=\bar{A}\bar{B}+A\bar{B}$, discussed in Example 3.13

| Y | $\bar{A}$ | A |
|---|---|---|
| $\bar{B}$ | 1 | 1 |
| B | 0 | 0 |

**Example 3.14**

Minimise the two-input OR function using the Karnaugh map drawn in Table 3.3.

***Solution***

Here we can combine the two terms in the $A$ column to obtain the 'product' term $A$ and the two terms in the $B$ row to obtain $B$. So, as expected, OR'ing these together gives $Y=A+B$ for the two-input OR gate.

The combination of logically adjacent fundamental product terms is indicated by a looping together of the combined terms as shown in the redrawn Karnaugh map in Table 3.8. Looking at this process using Boolean algebra:

$$
\begin{aligned}
Y &= A\bar{B}+\bar{A}B+AB \\
&= A\bar{B}+AB+\bar{A}B+AB \quad \text{Equation 1.2} \\
&= A(\bar{B}+B)+(\bar{A}+A)B \\
&= A+B
\end{aligned}
$$

Note that the $AB$ term is used twice in the minimisation processes using both Boolean algebra and the Karnaugh map (since it is included in both loops).

Table 3.8 Karnaugh map for a two-input OR gate demonstrating the grouping and looping of logically adjacent product terms (see Example 3.14)

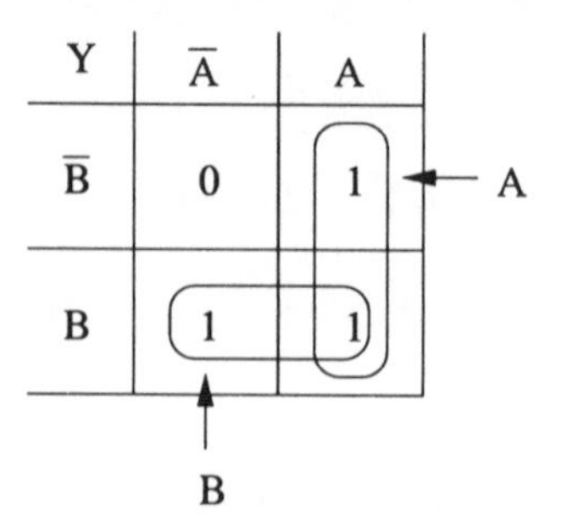


**Example 3.15**

Minimise the two-input XOR function using the Karnaugh map drawn in Table 3.3.

***Solution***

Here we see that the two fundamental product terms, $\bar{A}B$ and $A\bar{B}$, are *not* logically adjacent and so the expression cannot be minimised. Note that we can still read off the Boolean expression for the XOR gate directly from the Karnaugh map since the top right and bottom left cells correspond to the $A\bar{B}$ and $\bar{A}B$ fundamental product terms giving $Y=A\bar{B}+\bar{A}B$, the XOR function.

These simple examples serve to illustrate all there is to know about Karnaugh maps. This is basically:

- how to draw them;
- how to spot logically adjacent fundamental product terms that can be combined to minimise the fundamental sum of products expression corresponding to the map;
- and how to extract the product terms from the map to give a minimised sum of products expression for the Boolean function in question.

We now turn to examples of minimising expressions with more than two variables, and formalising some of the points that have been introduced above.

**Maps for more than two variables**

Since truth tables and Karnaugh maps contain exactly the same information, then obviously a Karnaugh map for $n$ variables must contain $2^n$ cells. In practice Karnaugh maps become unmanageable for more than five variables (i.e. 32 cells).

A three-variable map will have eight cells and its layout is shown in Table 3.9. We will now look at some examples of using three-variable Karnaugh maps for minimisation.

Table 3.9 Layout and labelling of a three-variable truth table and Karnaugh map illustrating the fundamental product terms occupying each cell

| $A$ | $B$ | $C$ | *row* |
|---|---|---|---|
| 0 | 0 | 0 | 0 |
| 0 | 0 | 1 | 1 |
| 0 | 1 | 0 | 2 |
| 0 | 1 | 1 | 3 |
| 1 | 0 | 0 | 4 |
| 1 | 0 | 1 | 5 |
| 1 | 1 | 0 | 6 |
| 1 | 1 | 1 | 7 |

| Y | $\bar{A}\bar{B}$ | $\bar{A}B$ | $AB$ | $A\bar{B}$ |
|---|---|---|---|---|
| $\bar{C}$ | 0 | 2 | 6 | 4 |
| C | 1 | 3 | 7 | 5 |

| Y | $\bar{A}\bar{B}$ | $\bar{A}B$ | $AB$ | $A\bar{B}$ |
|---|---|---|---|---|
| $\bar{C}$ | $\bar{A}\bar{B}\bar{C}$ | $\bar{A}B\bar{C}$ | $AB\bar{C}$ | $A\bar{B}\bar{C}$ |
| C | $\bar{A}\bar{B}C$ | $\bar{A}BC$ | $ABC$ | $A\bar{B}C$ |

**Example 3.16**

Draw the truth table and Karnaugh map for the function

$$Y=AB+\bar{A}C$$

***Solution***

This expression has three variables and so the Karnaugh map must contain eight cells. The variables $A$ and $B$ are used to label the four columns in the map (with the four possible combinations of these two variables), with the third variable, $C$, used to label the two rows (for $C=0$ and $C=1$) as shown in Table 3.9.

The outputs, $Y$, for the various inputs to the circuit represented by the truth table are entered into the corresponding cells in the Karnaugh map as shown in Table 3.10. To minimise we note that there are two logically adjacent fundamental product terms in column $AB$ ($AB\bar{C}$ and $ABC$), and a further two in row $C$ ($\bar{A}\bar{B}C$ and $\bar{A}BC$) which can be combined to give $\bar{A}C$ (as $\bar{B}+B=1$). Therefore $Y=AB+\bar{A}C$.

Table 3.10 Karnaugh map for $Y=AB+\bar{A}C$, discussed in Example 3.16

| $A$ | $B$ | $C$ | $Y$ |
|---|---|---|---|
| 0 | 0 | 0 | 0 |
| 0 | 0 | 1 | 1 |
| 0 | 1 | 0 | 0 |
| 0 | 1 | 1 | 1 |
| 1 | 0 | 0 | 0 |
| 1 | 0 | 1 | 0 |
| 1 | 1 | 0 | 1 |
| 1 | 1 | 1 | 1 |

| Y | $\bar{A}\bar{B}$ | $\bar{A}B$ | AB | $A\bar{B}$ |
|---|---|---|---|---|
| $\bar{C}$ | 0 | 0 | 1 | 0 |
| C | 1 | 1 | 1 | 0 |

The grouping of logically adjacent product terms for minimisation is indicated by looping them on the Karnaugh map, as illustrated in the above example. Obviously *all* fundamental product terms (i.e. 1's on the Karnaugh map) must be looped and so contribute to the final minimised sum of products expression. Two grouped and looped fundamental product terms are referred to as a *dual.*

To obtain the minimised product terms any variable existing in both uncomplemented and complemented forms within the looped product term is omitted. This is easy to see from the Karnaugh map because the labelling of adjacent columns and rows differs only by one variable (being in complemented form) which is therefore the one that is eliminated.

**Example 3.17**

Draw the Karnaugh map for the expression $Y=\bar{A}B\bar{C}+AC$.

***Solution***

This is shown in Table 3.11. Note that the $\bar{A}B\bar{C}$ term cannot be combined with any other product term and so remains in the minimised expression (as given).

Table 3.11 The Karnaugh map for $Y=\bar{A}B\bar{C}+AC$, discussed in Example 3.17

| Y | $\bar{A}\bar{B}$ | $\bar{A}B$ | $AB$ | $A\bar{B}$ |
|---|---|---|---|---|
| $\bar{C}$ | 0 | 1 | 0 | 0 |
| C | 0 | 0 | 1 | 1 |

### Example 3.18

Draw the Karnaugh map for $Y=\bar{A}\bar{B}\bar{C}+A\bar{B}\bar{C}$ and use it to minimise this expression.

### *Solution*

On first inspection it appears as if no fundamental product terms can be combined (grouped) in the map in Table 3.12. However, whereas in a two-variable map all logically adjacent product terms are actually next to each other, this cannot be the case for more than two variables.

For the three-variable map although $\bar{A}\bar{B}$ and $A\bar{B}$ are logically adjacent they are not physically next to each other in the Karnaugh map as they index the first and last columns. However, they are connected if the map is 'rolled' to effectively form a cylinder by connecting columns 1 and 4. In this case the terms can be looped as shown to give $Y=\bar{B}\bar{C}$.

Table 3.12 The Karnaugh map for $Y=\bar{A}\bar{B}\bar{C}+A\bar{B}\bar{C}$, discussed in Example 3.18

| Y | $\bar{A}\bar{B}$ | $\bar{A}B$ | $AB$ | $A\bar{B}$ |
|---|---|---|---|---|
| $\bar{C}$ | 1 | 0 | 0 | 1 |
| C | 0 | 0 | 0 | 0 |

### Example 3.19

Use the Karnaugh map in Table 3.13 to find a minimised expression for $Y$.

### *Solution*

Here we have four logically adjacent fundamental product terms ($AB\bar{C}$, $A\bar{B}\bar{C}$,

Table 3.13 Karnaugh map used to demonstrate minimisation in Example 3.19

| Y | $\bar{A}\bar{B}$ | $\bar{A}B$ | $AB$ | $A\bar{B}$ |
|---|---|---|---|---|
| $\bar{C}$ | 1 | 0 | 1 | 1 |
| C | 0 | 0 | 1 | 1 |

$ABC$ and $A\bar{B}C$) in which only $A$ does not appear in uncomplemented and complemented forms. This produces the term $A$ to be included in the sum of products expression. This grouping and looping of four fundamental product terms produces what is called a *quad*. We can also group and loop the $\bar{A}\bar{B}\bar{C}$ term with $A\bar{B}\bar{C}$ by rolling the map to give the dual $\bar{B}\bar{C}$. This gives the minimised form of $Y$ as $Y=A+\bar{B}\bar{C}$.

## Example 3.20

Draw the Karnaugh map for the function $Y=\bar{A}C+\bar{A}\bar{B}$, used in earlier examples in this chapter, and from it obtain the fundamental sum of product expression for $Y$.

### *Solution*

From the Karnaugh map in Table 3.14 we can see there are three fundamental product terms giving:

$$Y=\bar{A}\bar{B}\bar{C}+\bar{A}\bar{B}C+\bar{A}BC$$

Table 3.14 Karnaugh map for function $Y=\bar{A}C+\bar{A}\bar{B}$, discussed in Example 3.20

| Y | $\bar{A}\bar{B}$ | $\bar{A}B$ | $AB$ | $A\bar{B}$ |
|---|---|---|---|---|
| $\bar{C}$ | 1 | 0 | 0 | 0 |
| C | 1 | 1 | 0 | 0 |

### Four-variable Karnaugh maps

For a four-variable Karnaugh map, two pairs of variables are used to label the four columns and four rows respectively as shown in Table 3.15. Continuing the map 'rolling' idea introduced above, note that the top and bottom rows are logically connected.

Table 3.15 Layout and labelling of a four-variable Karnaugh map

| *A* | *B* | *C* | *D* | *row* |
|---|---|---|---|---|
| 0 | 0 | 0 | 0 | 0 |
| 0 | 0 | 0 | 1 | 1 |
| 0 | 0 | 1 | 0 | 2 |
| 0 | 0 | 1 | 1 | 3 |
| 0 | 1 | 0 | 0 | 4 |
| 0 | 1 | 0 | 1 | 5 |
| 0 | 1 | 1 | 0 | 6 |
| 0 | 1 | 1 | 1 | 7 |

| *A* | *B* | *C* | *D* | *row* |
|---|---|---|---|---|
| 1 | 0 | 0 | 0 | 8 |
| 1 | 0 | 0 | 1 | 9 |
| 1 | 0 | 1 | 0 | 10 |
| 1 | 0 | 1 | 1 | 11 |
| 1 | 1 | 0 | 0 | 12 |
| 1 | 1 | 0 | 1 | 13 |
| 1 | 1 | 1 | 0 | 14 |
| 1 | 1 | 1 | 1 | 15 |

| Y | $\bar{A}\bar{B}$ | $\bar{A}B$ | $AB$ | $A\bar{B}$ |
|---|---|---|---|---|
| $\bar{C}\bar{D}$ | 0 | 4 | 12 | 8 |
| $\bar{C}D$ | 1 | 5 | 13 | 9 |
| $CD$ | 3 | 7 | 15 | 11 |
| $C\bar{D}$ | 2 | 6 | 14 | 10 |

| Y | $\bar{A}\bar{B}$ | $\bar{A}B$ | $AB$ | $A\bar{B}$ |
|---|---|---|---|---|
| $\bar{C}\bar{D}$ | $\bar{A}\bar{B}\bar{C}\bar{D}$ | $\bar{A}B\bar{C}\bar{D}$ | $AB\bar{C}\bar{D}$ | $A\bar{B}\bar{C}\bar{D}$ |
| $\bar{C}D$ | $\bar{A}\bar{B}\bar{C}D$ | $\bar{A}B\bar{C}D$ | $AB\bar{C}D$ | $A\bar{B}\bar{C}D$ |
| $CD$ | $\bar{A}\bar{B}CD$ | $\bar{A}BCD$ | $ABCD$ | $A\bar{B}CD$ |
| $C\bar{D}$ | $\bar{A}\bar{B}C\bar{D}$ | $\bar{A}BC\bar{D}$ | $ABC\bar{D}$ | $A\bar{B}C\bar{D}$ |

## Example 3.21

What Boolean expression is represented by the Karnaugh map in Table 3.16?

Table 3.16 Karnaugh map of the function minimised in Example 3.21

| Y | $\bar{A}\bar{B}$ | $\bar{A}B$ | $AB$ | $A\bar{B}$ |
|---|---|---|---|---|
| $\bar{C}\bar{D}$ | 1 | 0 | 1 | 0 |
| $\bar{C}D$ | 0 | 0 | 1 | 0 |
| $CD$ | 1 | 1 | 0 | 0 |
| $C\bar{D}$ | 0 | 0 | 0 | 0 |

### *Solution*

This map contains two duals (two grouped and looped fundamental product terms) and a fundamental product term that has no logically adjacent product terms giving:

$$Y = AB\bar{C} + \bar{A}CD + \bar{A}\bar{B}\bar{C}\bar{D}$$

### Example 3.22

What Boolean expression is represented by the Karnaugh map in Table 3.17?

Table 3.17 Karnaugh map of the function minimised in Example 3.22

| Y | $\bar{A}\bar{B}$ | $\bar{A}B$ | $AB$ | $A\bar{B}$ |
|---|---|---|---|---|
| $\bar{C}\bar{D}$ | 0 | 0 | 1 | 1 |
| $\bar{C}D$ | 1 | 1 | 1 | 1 |
| $CD$ | 0 | 0 | 1 | 1 |
| $C\bar{D}$ | 0 | 0 | 1 | 1 |

***Solution***

This map contains a quad, $\bar{C}D$ (four grouped and looped fundamental product terms) and an octet, $A$, (eight grouped and looped fundamental product terms) giving:

$$Y=A+\bar{C}D$$

### Example 3.23

What Boolean expression is represented by the Karnaugh map in Table 3.18?

Table 3.18 Karnaugh map of the function minimised in Example 3.23

| Y | $\bar{A}\bar{B}$ | $\bar{A}B$ | $AB$ | $A\bar{B}$ |
|---|---|---|---|---|
| $\bar{C}\bar{D}$ | 0 | 1 | 1 | 0 |
| $\bar{C}D$ | 1 | 0 | 0 | 1 |
| $CD$ | 1 | 0 | 0 | 1 |
| $C\bar{D}$ | 0 | 1 | 1 | 0 |

***Solution***

This Karnaugh map contains two quads both of which are obtained by 'rolling' the map; firstly around the first and last columns, secondly between the top and bottom rows. This gives:

$$Y=\bar{B}D+B\bar{D}$$

Note that this is an XOR function, in $B$ and $D$, which is to be expected because of the 'chequerboard' pattern in the map.

**Five-variable Karnaugh maps**

For a five-variable map, two four-variable maps must be drawn with the fifth variable $E$ used to index the two maps. The layout is shown in Tables 3.19 and 3.20.

Table 3.19 Row numbering for a five-variable truth table. See Table 3.20 for the corresponding Karnaugh map

| *A* | *B* | *C* | *D* | *E* | *row* |
|---|---|---|---|---|---|
| 0 | 0 | 0 | 0 | 0 | 0 |
| 0 | 0 | 0 | 0 | 1 | 1 |
| 0 | 0 | 0 | 1 | 0 | 2 |
| 0 | 0 | 0 | 1 | 1 | 3 |
| 0 | 0 | 1 | 0 | 0 | 4 |
| 0 | 0 | 1 | 0 | 1 | 5 |
| 0 | 0 | 1 | 1 | 0 | 6 |
| 0 | 0 | 1 | 1 | 1 | 7 |
| 0 | 1 | 0 | 0 | 0 | 8 |
| 0 | 1 | 0 | 0 | 1 | 9 |
| 0 | 1 | 0 | 1 | 0 | 10 |
| 0 | 1 | 0 | 1 | 1 | 11 |
| 0 | 1 | 1 | 0 | 0 | 12 |
| 0 | 1 | 1 | 0 | 1 | 13 |
| 0 | 1 | 1 | 1 | 0 | 14 |
| 0 | 1 | 1 | 1 | 1 | 15 |

| *A* | *B* | *C* | *D* | *E* | *row* |
|---|---|---|---|---|---|
| 1 | 0 | 0 | 0 | 0 | 16 |
| 1 | 0 | 0 | 0 | 1 | 17 |
| 1 | 0 | 0 | 1 | 0 | 18 |
| 1 | 0 | 0 | 1 | 1 | 19 |
| 1 | 0 | 1 | 0 | 0 | 20 |
| 1 | 0 | 1 | 0 | 1 | 21 |
| 1 | 0 | 1 | 1 | 0 | 22 |
| 1 | 0 | 1 | 1 | 1 | 23 |
| 1 | 1 | 0 | 0 | 0 | 24 |
| 1 | 1 | 0 | 0 | 1 | 25 |
| 1 | 1 | 0 | 1 | 0 | 26 |
| 1 | 1 | 0 | 1 | 1 | 27 |
| 1 | 1 | 1 | 0 | 0 | 28 |
| 1 | 1 | 1 | 0 | 1 | 29 |
| 1 | 1 | 1 | 1 | 0 | 30 |
| 1 | 1 | 1 | 1 | 1 | 31 |

Table 3.20 Layout and labelling of a five-variable Karnaugh map. See Table 3.19 for the corresponding row numbering of the truth table

| Y | $\bar{A}\bar{B}$ | $\bar{A}B$ | $AB$ | $A\bar{B}$ |
|---|---|---|---|---|
| $\bar{C}\bar{D}$ | 0 | 4 | 12 | 8 |
| $\bar{C}D$ | 1 | 5 | 13 | 9 |
| $CD$ | 3 | 7 | 15 | 11 |
| $C\bar{D}$ | 2 | 6 | 14 | 10 |

$\bar{E}$

| Y | $\bar{A}\bar{B}$ | $\bar{A}B$ | $AB$ | $A\bar{B}$ |
|---|---|---|---|---|
| $\bar{C}\bar{D}$ | 16 | 20 | 28 | 24 |
| $\bar{C}D$ | 17 | 21 | 29 | 25 |
| $CD$ | 19 | 23 | 31 | 27 |
| $C\bar{D}$ | 18 | 22 | 30 | 26 |

$E$

## Example 3.24

The five-variable truth table in Table 3.21 gives the outputs, $X$ and $Y$, from two combinational logic circuits. Use Karnaugh maps to minimise these functions.

Table 3.21 Truth tables of the functions minimised in Example 3.24

| $A$ | $B$ | $C$ | $D$ | $E$ | $X$ | $Y$ |
|---|---|---|---|---|---|---|
| 0 | 0 | 0 | 0 | 0 | 0 | 1 |
| 0 | 0 | 0 | 0 | 1 | 1 | 1 |
| 0 | 0 | 0 | 1 | 0 | 1 | 1 |
| 0 | 0 | 0 | 1 | 1 | 0 | 1 |
| 0 | 0 | 1 | 0 | 0 | 0 | 1 |
| 0 | 0 | 1 | 0 | 1 | 1 | 0 |
| 0 | 0 | 1 | 1 | 0 | 0 | 0 |
| 0 | 0 | 1 | 1 | 1 | 0 | 0 |
| 0 | 1 | 0 | 0 | 0 | 1 | 1 |
| 0 | 1 | 0 | 0 | 1 | 1 | 0 |
| 0 | 1 | 0 | 1 | 0 | 1 | 1 |
| 0 | 1 | 0 | 1 | 1 | 0 | 1 |
| 0 | 1 | 1 | 0 | 0 | 1 | 1 |
| 0 | 1 | 1 | 0 | 1 | 1 | 1 |
| 0 | 1 | 1 | 1 | 0 | 0 | 0 |
| 0 | 1 | 1 | 1 | 1 | 0 | 0 |

| $A$ | $B$ | $C$ | $D$ | $E$ | $X$ | $Y$ |
|---|---|---|---|---|---|---|
| 1 | 0 | 0 | 0 | 0 | 1 | 1 |
| 1 | 0 | 0 | 0 | 1 | 0 | 1 |
| 1 | 0 | 0 | 1 | 0 | 1 | 1 |
| 1 | 0 | 0 | 1 | 1 | 0 | 1 |
| 1 | 0 | 1 | 0 | 0 | 0 | 0 |
| 1 | 0 | 1 | 0 | 1 | 0 | 0 |
| 1 | 0 | 1 | 1 | 0 | 0 | 0 |
| 1 | 0 | 1 | 1 | 1 | 0 | 1 |
| 1 | 1 | 0 | 0 | 0 | 1 | 0 |
| 1 | 1 | 0 | 0 | 1 | 1 | 0 |
| 1 | 1 | 0 | 1 | 0 | 0 | 1 |
| 1 | 1 | 0 | 1 | 1 | 1 | 1 |
| 1 | 1 | 1 | 0 | 0 | 1 | 0 |
| 1 | 1 | 1 | 0 | 1 | 1 | 1 |
| 1 | 1 | 1 | 1 | 0 | 1 | 0 |
| 1 | 1 | 1 | 1 | 1 | 0 | 1 |

### *Solution*

From the Karnaugh map in Table 3.22:

$$X = A\bar{C} + \bar{C}D\bar{E} + \bar{A}\bar{B}C\bar{D} + \bar{B}\bar{D}E + ABDE$$

Table 3.22 Karnaugh map for the function $X$, minimised in Example 3.24, whose truth table is given in Table 3.21

$\bar{E}$

| X | $\bar{A}\bar{B}$ | $\bar{A}B$ | $AB$ | $A\bar{B}$ |
|---|---|---|---|---|
| $\bar{C}\bar{D}$ | 0 | 0 | 1 | 1 |
| $\bar{C}D$ | 1 | 1 | 1 | 1 |
| $CD$ | 0 | 0 | 0 | 0 |
| $C\bar{D}$ | 1 | 0 | 0 | 1 |

$E$

| X | $\bar{A}\bar{B}$ | $\bar{A}B$ | $AB$ | $A\bar{B}$ |
|---|---|---|---|---|
| $\bar{C}\bar{D}$ | 1 | 0 | 1 | 1 |
| $\bar{C}D$ | 0 | 0 | 1 | 1 |
| $CD$ | 0 | 0 | 1 | 0 |
| $C\bar{D}$ | 1 | 0 | 0 | 1 |

The product terms containing $\bar{E}$ and $E$ are obtained solely from the Karnaugh maps for $\bar{E}$ and $E$ respectively. Those terms not containing either $\bar{E}$ or $E$ are for product terms obtained by grouping and looping the same fundamental product terms on *both* maps. For example the term $A\bar{C}$ occurs because of the quads in the top right-hand corner of the maps for both $\bar{E}$ and $E$. From the Karnaugh map in Table 3.23:

$$Y=\bar{A}\bar{B}+\bar{C}\bar{D}\bar{E}+\bar{B}C+AB\bar{C}D+CDE$$

Table 3.23 Karnaugh map for the function Y, minimised in Example 3.24, whose truth table is given in Table 3.21

| Y | $\bar{A}\bar{B}$ | $\bar{A}B$ | $AB$ | $A\bar{B}$ |
|---|---|---|---|---|
| $\bar{C}\bar{D}$ | 1 | 1 | 1 | 1 |
| $\bar{C}D$ | 1 | 0 | 1 | 0 |
| $CD$ | 1 | 0 | 0 | 1 |
| $C\bar{D}$ | 1 | 0 | 0 | 1 |

$\bar{E}$

| Y | $\bar{A}\bar{B}$ | $\bar{A}B$ | $AB$ | $A\bar{B}$ |
|---|---|---|---|---|
| $\bar{C}\bar{D}$ | 1 | 0 | 0 | 0 |
| $\bar{C}D$ | 1 | 0 | 1 | 0 |
| $CD$ | 1 | 1 | 1 | 1 |
| $C\bar{D}$ | 1 | 0 | 0 | 1 |

$E$

**The layout of Karnaugh maps**

It may seem as if the labelling of the columns and rows of the above Karnaugh maps was chosen at random, but this is certainly not so. In all the examples in this book the same format for Karnaugh maps will be used which is that truth tables will be drawn with $A$ as the most significant bit. Then for three-variable maps, as used above, $A$ and $B$ will be used to index the columns and $C$ the rows, whilst for four-variable maps $C$ and $D$ will be used to index the rows.

A significant advantage of keeping to the same convention for labelling Karnaugh maps is that it simplifies greatly the process of writing them out from truth tables since the filling pattern remains the same. This pattern can be seen by observing Tables 3.9, 3.15, 3.19 and 3.20, which show which rows of the truth tables (and hence fundamental products) correspond to which cells in the Karnaugh maps.

This is an appropriate point to bring attention to the link between Karnaugh maps and Gray code, introduced in Section 2.3. This coding scheme is distinguished by the fact that adjacent codes only differ in one variable, and it is therefore of no surprise to discover a connection with Karnaugh maps since these are also based upon this principle. The link can be seen by filling in the cells of a four-variable Karnaugh map *with the decimal codes used to index the Gray code* given in Table 2.1 as shown in Table 3.24. Note that the codes representing all adjacent

Table 3.24 The link between Gray code and Karnaugh maps

| Y | $\bar{A}\bar{B}$ | $\bar{A}B$ | $AB$ | $A\bar{B}$ |
|---|---|---|---|---|
| $\bar{C}\bar{D}$ | 0 | 7 | 8 | 15 |
| $\bar{C}D$ | 1 | 6 | 9 | 14 |
| $CD$ | 2 | 5 | 10 | 13 |
| $C\bar{D}$ | 3 | 4 | 11 | 12 |

cells differ by only one variable (e.g. 0 and 7 have Gray codes 0000 and 0100 (corresponding to fundamental product terms $\bar{A}\bar{B}\bar{C}\bar{D}$ and $\bar{A}B\bar{C}\bar{D}$), and 9 and 14 have codes 1101 and 1001).

**Minterms, prime implicants and essential prime implicants**

Since each fundamental product term occupies a single cell in the Karnaugh map it is called a *minterm* (as it specifies the minimum area of 1's, i.e. a single cell, in the Karnaugh map). Once the minterms have been looped and grouped in the minimisation process (to duals, quads and octets) then the resulting minimised (simplified) products are known as *prime implicants*.[4]

In order to minimise a function all of the minterms in the Karnaugh map must be covered (i.e. grouped and looped), since they must be contained in the minimised Boolean expression. However, if *all* of the possible groups, that is the prime implicants, are used in the final minimised sum of product expression there may be more of them than are strictly necessary to cover the whole map (in the previous examples we have used just sufficient prime implicants to cover all of the minterms). Those prime implicants which describe product terms which *must* be used for all minterms to be covered are called *essential prime implicants*. This is best illustrated by example.

## Example 3.25

Obtain minimised expressions for $X$ and $Y$ from the Karnaugh map in Table 3.25.

### *Solution*

The map for $X$ contains four prime implicants: the quad, $BD$; and duals $AB\bar{C}$, $\bar{A}BC$ and $\bar{A}C\bar{D}$. However, only three of these are essential prime implicants since

[4]An implicant refers to any product term consisting of looped and grouped minterms. This includes a product term that may be able to be combined with another implicant to produce a prime implicant (e.g. two duals combined to give a quad which is a prime implicant).

Table 3.25 Karnaugh maps of the functions which are minimised in Example 3.25

| X | $\bar{A}\bar{B}$ | $\bar{A}B$ | $AB$ | $A\bar{B}$ |
|---|---|---|---|---|
| $\bar{C}\bar{D}$ | 0 | 0 | 1 | 0 |
| $\bar{C}D$ | 0 | 1 | 1 | 0 |
| $CD$ | 0 | 1 | 1 | 0 |
| $C\bar{D}$ | 1 | 1 | 0 | 0 |

| Y | $\bar{A}\bar{B}$ | $\bar{A}B$ | $AB$ | $A\bar{B}$ |
|---|---|---|---|---|
| $\bar{C}\bar{D}$ | 0 | 0 | 1 | 0 |
| $\bar{C}D$ | 1 | 1 | 1 | 1 |
| $CD$ | 1 | 0 | 1 | 0 |
| $C\bar{D}$ | 1 | 0 | 1 | 1 |

$\bar{A}BC$ is also covered by $BD$ and $\bar{A}C\bar{D}$. The minimised expression is therefore:

$$X = BD + AB\bar{C} + \bar{A}C\bar{D}$$

The map for $Y$ contains six prime implicants: the quads $\bar{C}D$ and $AB$; and duals $\bar{A}\bar{B}D$, $\bar{A}\bar{B}C$, $AC\bar{D}$ and $\bar{B}C\bar{D}$. Only two of these are essential prime implicants, namely: $AB$ and $\bar{C}D$. The remaining three minterms $\bar{A}\bar{B}CD$, $\bar{A}\bar{B}C\bar{D}$ and $A\bar{B}C\bar{D}$ must also be covered by choosing appropriate non-essential prime implicants from the four remaining. This can be achieved a number of ways, any of which provide a complete and therefore adequate minimised expression. These are:

$$Y = \bar{C}D + AB + \bar{A}\bar{B}D + \bar{B}C\bar{D}$$
$$Y = \bar{C}D + AB + \bar{A}\bar{B}C + \bar{B}C\bar{D}$$
$$Y = \bar{C}D + AB + \bar{A}\bar{B}C + AC\bar{D}$$

It is instructive to compare minimisation of an expression with more prime implicants than are required using both a Karnaugh map and Boolean algebra.

**Example 3.26**

List all of the prime implicants from the Karnaugh map in Table 3.26, and then give a minimised expression for $Y$. Then beginning with the expression containing all of the prime implicants minimise this to the form produced from the Karnaugh map.

***Solution***

The map contains three prime implicants $AB$, $\bar{A}C$ and $BC$. Of these $AB$ and $\bar{A}C$ are essential prime implicants with $BC$ non-essential since it is also covered by these two. Therefore, the minimised form is:

$$Y = AB + \bar{A}C$$

Table 3.26 Karnaugh map of the function minimised in Example 3.26

| Y | $\bar{A}\bar{B}$ | $\bar{A}B$ | $AB$ | $A\bar{B}$ |
|---|---|---|---|---|
| $\bar{C}\bar{D}$ | 0 | 0 | 1 | 0 |
| $\bar{C}D$ | 0 | 0 | 1 | 0 |
| $CD$ | 1 | 1 | 1 | 0 |
| $C\bar{D}$ | 1 | 1 | 1 | 0 |

Using Boolean algebra:

$$\begin{aligned} Y &= AB + BC + \bar{A}C \\ &= AB + \bar{A}C + BC(\bar{A} + A) \\ &= AB + \bar{A}C + \bar{A}BC + ABC \\ &= AB(1 + C) + \bar{A}C(1 + B) \\ &= AB + \bar{A}C \end{aligned}$$

This demonstrates how the layout of the Karnaugh map allows this process to be performed 'visually'.

**'Don't Care' product terms**

Sometimes, when defining a combinational logic function, the output for certain inputs will not matter (e.g. certain input combinations may never occur). In this case these outputs are known as 'don't care' terms, and can be made either 0 or 1.

It is usual to choose the output which will aid the minimisation process.[5]

**Example 3.27**

Obtain a minimised expression for $Y$ from the Karnaugh map in Table 3.27.

Table 3.27 Karnaugh map of a function containing 'don't care' terms which is minimised in Example 3.27

| Y | $\bar{A}\bar{B}$ | $\bar{A}B$ | $AB$ | $A\bar{B}$ |
|---|---|---|---|---|
| $\bar{C}$ | 1 | 1 | 0 | X |
| $C$ | 1 | X | 0 | 0 |

[5]For some circuits a more important factor than aiding the minimisation process is to ensure the circuit *always* functions correctly. This is particularly true in the design of sequential circuits.

***Solution***

It is clear that if we set $\bar{A}BC$ to 1 then we can group the quad, $A$, which, if the other don't care term, $A\bar{B}\bar{C}$, is set to 0 will cover all minterms in the map giving a minimised form of $Y=\bar{A}$.

**Karnaugh maps: summary of rules for minimisation**

Karnaugh maps 'work' because adjacent cells (and therefore rows and columns) are the same except that one contains one of the variables in its complemented form. Tables 3.9, 3.15, 3.19 and 3.20 show the notation used for drawing Karnaugh maps with three, four and five variables, respectively.

When used for five variables, two maps of four variables are employed with looping and grouping across the two maps. For more than five variables the benefit of Karnaugh maps (which is their ease of use for looping adjacent minterms) disappears. The following section introduces Quine–McCluskey minimisation which is an alternative that is theoretically not limited in the number of variables that can be used.

In summary, the rules for using Karnaugh maps are as follows:

- Draw the map, remembering the 'pattern' when filling from a truth table. (For this to work the same layout of Karnaugh map *must* be adhered to.)
- Loop all octets, quads and duals (groups of eight, four and two adjacent minterms). These are the prime implicants. It does not matter if some minterms are covered more than once, although duals should not be totally enclosed in quads, and quads in octets as this simply means full minimisation has not been performed.
- Remember the map can be 'rolled' across its edges, or across the two maps in the case of a five-variable map.
- Remember that you can set 'don't care' or 'can't happen' terms to either 0 or 1 to aid minimisation.
- Determine which prime implicants are essential and so must be used in the minimised sum of products expression.
- Pick enough prime implicants which together with the essential prime implicants already selected will cover the whole map. Other non-essential prime implicants need not be included in the minimised expression.[6]
- Also bear in mind:
    - Although the minimised expression is obtained it may not be the best expression for your particular problem (e.g. you may already have generated other product terms, for some other circuit, which could be used).
    - Look out for the characteristic XOR pattern that cannot be minimised but which is implemented easily using XOR gates.
    - Do not forget that it is sometimes easier to minimise by inspection from the

[6]We will see in the next chapter how the inclusion of non-essential prime implicants can sometimes ensure the correct operation of a real circuit.

truth table (i.e. a Karnaugh map may not offer the best route to minimisation).

### 3.3.3 Minimisation via the Quine–McCluskey technique

The Quine–McCluskey method of minimisation is introduced here for two reasons. Firstly it provides a method by which Boolean expressions with more than five variables can be minimised. Secondly it relies upon the systematic combination of adjacent minterms, then duals, then quads, then octets, etc., which is an algorithmic method which can be readily programmed, thus allowing minimisation to be performed automatically by computer. (Programs offering this facility are readily available.)

The basic procedure is:

- Find all of the prime implicants by systematically combining logically adjacent product terms (this produces firstly duals, then quads, then octets, etc.).
- Select a minimal set of prime implicants to cover all minterms (using all the essential prime implicants, and sufficient non-essential prime implicants for sufficient coverage).

The stages are as follows:

1. Firstly express the function to be minimised in terms of its minterms, $Y=\Sigma(\ )$ (as described in Section 3.2.1).
2. The minterms must then be grouped in tabular form according to how many 1's the codes for these minterms contain (e.g. minterm 13 (fundamental product $AB\bar{C}D$) which has four variables is coded by 1101 and so has three 1's).[7]
3. Then *each* term in each group (of terms with the same number of 1's) is compared with *all* terms in the next group (containing terms with an additional 1).

   *If* they differ by one term only (and so can be minimised like a dual on a Karnaugh map) then they are combined and retabulated as duals. The common terms should be marked (here a dash is used), and combined minterms in the first table marked (here with a cross) to indicate that they have been combined into a dual.

   This is the systematic combination of logically adjacent terms, and the process (i.e. this stage) is repeated until only one combined group is left.
4. Once this phase has been completed then all uncrossed (unmarked) terms from all of the earlier tables and (non-duplicated) terms from the final table should be collected as these are the prime implicants. (Duplicated terms may appear because of terms being combined in more than one way.)
5. All that now remains is to choose sufficient prime implicants to effect total coverage of the Karnaugh map. To aid in this process it is helpful to draw up a

[7]The significance of this is that fundamental product terms differing only by one variable being in complemented form in one of the terms must be logically adjacent, and so can be minimised.

table indicating which minterms (used to label the table's columns) are included in which minimised (duals, quads, octets, etc.) terms, which are used to label the table's rows.

A cross can then be entered into this table to indicate coverage of a minterm by a prime implicant. For total coverage *all* columns must contain at least one cross.

Those columns containing only one cross indicate minterms covered by only one, and therefore an essential, prime implicant. So these essential prime implicants *must* be included in the final minimised expression. If these essential prime implicants cover all of the minterms then no other prime implicants are needed, otherwise additional prime implicants must be used to effect total coverage.

The implementation of this procedure is best illustrated by an example.

**Quine–McCluskey minimisation: Case 1**

Minimise $Y$, given in fundamental sum of products form, using the Quine–McCluskey method.

$$Y=\bar{A}\bar{B}C\bar{D}+\bar{A}\bar{B}CD+\bar{A}B\bar{C}D+\bar{A}BC\bar{D}+\bar{A}BCD \\ +A\bar{B}\bar{C}\bar{D}+A\bar{B}\bar{C}D+A\bar{B}C\bar{D}+A\bar{B}CD$$

This Boolean logic expression can be written as:

$$Y=\sum(2, 3, 5, 6, 7, 8, 9, 10, 11)$$

Of the codes for these minterms:

- two have a single 1 (2 and 8)
- five have two 1's (3, 5, 6, 9 and 10)
- two have three 1's (7 and 11).

This information can then be tabulated as shown in Table 3.28. The next stage is to combine logically adjacent product terms, which is achieved by comparing terms with a single 1 with those with two, then those with two with those with three. So, for example, minterms 2 and 3 differ only in $D$ and so can be combined to give $2\cdot3$ which is the dual $\bar{A}\bar{B}C$. This dual is therefore entered into the next table (of duals), a cross used in the first table to indicate the inclusion of minterms 2 and 3 in a dual, and the fact that $D$ has been eliminated by a dash in the table of duals for $2\cdot3$.

This procedure is carried out for all minterms, thus producing a table of all duals as shown in Table 3.29. Note that this table is still split into those terms with a single 1 and those with two 1's. This is essential since it is the members of these two groups that must now be compared to see if any duals can be combined to produce a quad.

Repeating the process for the table of duals, to produce a table of quads shown in Table 3.30, we see, for example, that duals $2\cdot3$ and $6\cdot7$ only differ in $B$ (the dashes must match) and so can be combined to give $2\cdot3\cdot6\cdot7$ which is quad $\bar{A}C$.

Table 3.28 Collation and tabulation of minterms for Case 1

| | A | B | C | D |
|---|---|---|---|---|
| 2 | 0 | 0 | 1 | 0 |
| 8 | 1 | 0 | 0 | 0 |
| 3 | 0 | 0 | 1 | 1 |
| 5 | 0 | 1 | 0 | 1 |
| 6 | 0 | 1 | 1 | 0 |
| 9 | 1 | 0 | 0 | 1 |
| 10 | 1 | 0 | 1 | 0 |
| 7 | 0 | 1 | 1 | 1 |
| 11 | 1 | 0 | 1 | 1 |

Table 3.29 Logical combination of minterms, for Case 1, producing all dual implicants for the function

minterms

| | | A | B | C | D |
|---|---|---|---|---|---|
| 2 | x | 0 | 0 | 1 | 0 |
| 8 | x | 1 | 0 | 0 | 0 |
| 3 | x | 0 | 0 | 1 | 1 |
| 5 | x | 0 | 1 | 0 | 1 |
| 6 | x | 0 | 1 | 1 | 0 |
| 9 | x | 1 | 0 | 0 | 1 |
| 10 | x | 1 | 0 | 1 | 0 |
| 7 | x | 0 | 1 | 1 | 1 |
| 11 | x | 1 | 0 | 1 | 1 |

duals

| | A | B | C | D |
|---|---|---|---|---|
| 2.3 | 0 | 0 | 1 | - |
| 2.6 | 0 | - | 1 | 0 |
| 2.10 | - | 0 | 1 | 0 |
| 8.9 | 1 | 0 | 0 | - |
| 8.10 | 1 | 0 | - | 0 |
| 3.7 | 0 | - | 1 | 1 |
| 3.11 | - | 0 | 1 | 1 |
| 5.7 | 0 | 1 | - | 1 |
| 6.7 | 0 | 1 | 1 | - |
| 9.11 | 1 | 0 | - | 1 |
| 10.11 | 1 | 0 | 1 | - |

Table 3.30 Minimisation process for Case 1 resulting in tabulation of all dual and quad implicants

minterms

| | | A | B | C | D |
|---|---|---|---|---|---|
| 2 | x | 0 | 0 | 1 | 0 |
| 8 | x | 1 | 0 | 0 | 0 |
| | | | | | |
| 3 | x | 0 | 0 | 1 | 1 |
| 5 | x | 0 | 1 | 0 | 1 |
| 6 | x | 0 | 1 | 1 | 0 |
| 9 | x | 1 | 0 | 0 | 1 |
| 10 | x | 1 | 0 | 1 | 0 |
| | | | | | |
| 7 | x | 0 | 1 | 1 | 1 |
| 11 | x | 1 | 0 | 1 | 1 |

duals

| | | A | B | C | D | |
|---|---|---|---|---|---|---|
| 2.3 | x | 0 | 0 | 1 | - | |
| 2.6 | x | 0 | - | 1 | 0 | |
| 2.10 | x | - | 0 | 1 | 0 | |
| 8.9 | x | 1 | 0 | 0 | - | |
| 8.10 | x | 1 | 0 | - | 0 | |
| | | | | | | |
| 3.7 | x | 0 | - | 1 | 1 | |
| 3.11 | x | - | 0 | 1 | 1 | |
| 5.7 | | 0 | 1 | - | 1 | ← Prime Implicant |
| 6.7 | x | 0 | 1 | 1 | - | |
| 9.11 | x | 1 | 0 | - | 1 | |
| 10.11 | x | 1 | 0 | 1 | - | |

quads

| | A | B | C | D | |
|---|---|---|---|---|---|
| 2.3.6.7 | 0 | - | 1 | - | ← Prime Implicants |
| 2.3.10.11 | - | 0 | 1 | - | ← Prime Implicants |
| 2.6.3.7 | 0 | - | 1 | - | duplicate |
| 2.10.3.11 | - | 0 | 1 | - | duplicate |
| 8.9.10.11 | 1 | 0 | - | - | ← Prime Implicants |
| 8.10.9.11 | 1 | 0 | - | - | duplicate |

This is entered into the table of quads and the inclusion of the two duals in this quad indicated again by crosses in the dual table, with the elimination of $B$ shown by the dash.

From the table of quads all terms now have a single 1 which means none can be combined to give an octet (since this would require a quad differing in only one variable, which must therefore have either two 0's or two 1's). The combination of logically adjacent product terms is now complete and we are left with a list of prime implicants.

Note:

- The quad table contains duplicates (e.g. $2 \cdot 3 \cdot 6 \cdot 7$ and $2 \cdot 6 \cdot 3 \cdot 7$). This will happen for all quads because there are always two ways in which the four minterms can be combined into pairs of duals.
- There is a dual, $5 \cdot 7$, which cannot be combined into a quad and so is itself a prime implicant. Such duals and minterms which cannot be reduced must always be looked for.

We can now produce the table of prime implicants (the rows) and minterms (the columns) in Table 3.31. A cross indicates that a minterm is included in a prime implicant. A circle around a cross indicates that the row containing it is an essential prime implicant (epi). This is so if there is only a single cross in the column and therefore this is the only prime implicant containing this minterm.

Table 3.31 Prime implicant table for Case 1

| | 2 | 3 | 5 | 6 | 7 | 8 | 9 | 10 | 11 |
|---|---|---|---|---|---|---|---|---|---|
| 5.7 | | | (X) | | X | | | | |
| 2.3.6.7 | X | X | | (X) | X | | | | |
| 2.3.10.11 | X | X | | | | | | X | X |
| 8.9.10.11 | | | | | | (X) | (X) | X | X |

All that remains is to pick out the essentiai prime implicants from this table, plus sufficient prime implicants (if required) to ensure all minterms are included in the final expression.

Table 3.32 Karnaugh map for the function minimised in Case 1 which illustrates the process employed in Quine–McCluskey minimisation

| Y | $\bar{A}\bar{B}$ | $\bar{A}B$ | $AB$ | $A\bar{B}$ |
|---|---|---|---|---|
| $\bar{C}\bar{D}$ | 0 | 0 | 0 | 1 |
| $\bar{C}D$ | 0 | 1 | 0 | 1 |
| $CD$ | 1 | 1 | 0 | 1 |
| $C\bar{D}$ | 1 | 1 | 0 | 1 |

| Y | $\bar{A}\bar{B}$ | $\bar{A}B$ | $AB$ | $A\bar{B}$ |
|---|---|---|---|---|
| $\bar{C}\bar{D}$ | 0 | 4 | 12 | 8 |
| $\bar{C}D$ | 1 | 5 | 13 | 9 |
| $CD$ | 3 | 7 | 15 | 11 |
| $C\bar{D}$ | 2 | 6 | 14 | 10 |

For this example we get three essential prime implicants, 5.7, 2.3.6.7 and 8.9.10.11, which cover all minterms which means the remaining prime implicant, 2.3.10.11, is non-essential. This gives:

$$Y = 5.7 + 2.3.6.7 + 8.9.10.11$$
$$Y = \bar{A}BD + \bar{A}C + A\bar{B}$$

It is instructive to look at the Karnaugh map that would be used to minimise this function which is shown in Table 3.32 together with the codes for the minterms in their corresponding cells in the map. From these the process used in Quine–McCluskey minimisation can clearly be seen.

**Quine-McCluskey minimisation: Case 2**
Minimise $Y = \sum(0, 4, 5, 7, 10, 12, 13, 14, 15)$.

- one has no 1's (0)
- one has a single 1 (4)
- three have two 1's (5,10 and 12)
- three have three 1's (7, 13 and 14)
- one has four 1's (15).

The minimisation process is shown in Table 3.33. The prime implicants are:

0.4, 10.14, 4.5.12.13, 5.7.13.15 and 12.13.14.15

The table of prime implicants, Table 3.34, is now used to produce a minimised expression with coverage of all minterms. From Table 3.34 we see there are three essential prime implicants, 0.4, 10.14 and 5.7.13.15, with either of the non-essential prime implicants, 4.5.12.13 or 12.13.14.15, also required to give full coverage. Therefore:

$$Y = \bar{A}\bar{C}\bar{D} + AC\bar{D} + BD + AB$$

or

$$Y = \bar{A}\bar{C}\bar{D} + AC\bar{D} + BD + B\bar{C}$$

Drawing the Karnaugh map, Table 3.35 (although not necessary), serves again to illustrate the stages of minimisation used in the Quine–McCluskey method.

## 3.4 PRODUCT OF SUMS: THE NEGATIVE LOGIC APPROACH

So far in this chapter we have approached all topics using positive level logic. In other words we have always considered the output and minimisation of combinational logic expressions in terms of the input combinations giving an output of 1. (For example we have considered the position of minterms in the Karnaugh maps.)

Given the principle of duality it should come as no surprise to learn that we could instead have used a negative level logic appproach. As we will now see this

Table 3.33 Quine–McCluskey minimisation process for Case 2

minterms

| | | A | B | C | D |
|---|---|---|---|---|---|
| 0 | x | 0 | 0 | 0 | 0 |
| 4 | x | 0 | 1 | 0 | 0 |
| 5 | x | 0 | 1 | 0 | 1 |
| 10 | x | 1 | 0 | 1 | 0 |
| 12 | x | 1 | 1 | 0 | 0 |
| 7 | x | 0 | 1 | 1 | 1 |
| 13 | x | 1 | 1 | 0 | 1 |
| 14 | x | 1 | 1 | 1 | 0 |
| 15 | x | 1 | 1 | 1 | 1 |

duals

| | | A | B | C | D | |
|---|---|---|---|---|---|---|
| 0.4 | | 0 | - | 0 | 0 | ← Prime Implicant |
| 4.5 | x | 0 | 1 | 0 | - | |
| 4.12 | x | - | 1 | 0 | 0 | |
| 5.7 | x | 0 | 1 | - | 1 | |
| 5.13 | x | - | 1 | 0 | 1 | |
| 10.14 | | 1 | - | 1 | 0 | ← Prime Implicant |
| 12.13 | x | 1 | 1 | 0 | - | |
| 12.14 | x | 1 | 1 | - | 0 | |
| 7.15 | x | - | 1 | 1 | 1 | |
| 13.15 | x | 1 | 1 | - | 1 | |
| 14.15 | x | 1 | 1 | 1 | - | |

quads

| | A | B | C | D | |
|---|---|---|---|---|---|
| 4.5.12.13 | - | 1 | 0 | - | ← Prime Implicants |
| 4.12.5.13 | - | 1 | 0 | - | duplicate |
| 5.7.13.15 | - | 1 | - | 1 | ← Prime Implicants |
| 5.13.7.15 | - | 1 | - | 1 | duplicate |
| 12.13.14.15 | 1 | 1 | - | - | ← Prime Implicants |
| 12.14.13.15 | 1 | 1 | - | - | duplicate |

approach is linked to the use of *fundamental product of sums* expressions, and the position of *maxterms* (the maximum area of 1's, and hence the position of a single zero) on a Karnaugh map.

Table 3.34 Prime implicant table for Case 2

| | 0 | 4 | 5 | 7 | 10 | 12 | 13 | 14 | 15 |
|---|---|---|---|---|---|---|---|---|---|
| 0.4 | (X) | X | | | | | | | |
| 10.14 | | | | | (X) | | | X | |
| 4.5.12.13 | | X | X | | | X | X | | |
| 5.7.13.15 | | | X | (X) | | | X | | X |
| 12.13.14.15 | | | | | | X | X | X | X |

Table 3.35 Karnaugh map of the function minimised in Case 2

| Y | $\bar{A}\bar{B}$ | $\bar{A}B$ | AB | $A\bar{B}$ |
|---|---|---|---|---|
| $\bar{C}\bar{D}$ | 1 | 1 | 1 | 0 |
| $\bar{C}D$ | 0 | 1 | 1 | 0 |
| CD | 0 | 1 | 1 | 0 |
| $C\bar{D}$ | 0 | 0 | 1 | 1 |

| Y | $\bar{A}\bar{B}$ | $\bar{A}B$ | AB | $A\bar{B}$ |
|---|---|---|---|---|
| $\bar{C}\bar{D}$ | 0 | 4 | 12 | 8 |
| $\bar{C}D$ | 1 | 5 | 13 | 9 |
| CD | 3 | 7 | 15 | 11 |
| $C\bar{D}$ | 2 | 6 | 14 | 10 |

**Fundamental product of sums form**

Using the function considered earlier:

$$Y = \bar{A}B + \bar{B}C$$

from its truth table, shown again in Table 3.36, we can see (in a similar way to that used when obtaining the fundamental sum of products form) that $Y=0$ if:

$$((A=0) \text{ AND } (B=0) \text{ AND } (C=0)) \text{ OR}$$
$$((A=1) \text{ AND } (B=0) \text{ AND } (C=0)) \text{ OR}$$
$$((A=1) \text{ AND } (B=1) \text{ AND } (C=0)) \text{ OR}$$
$$((A=1) \text{ AND } (B=1) \text{ AND } (C=1)$$

which in Boolean form is:

$$\bar{Y} = \bar{A}\bar{B}\bar{C} + A\bar{B}\bar{C} + AB\bar{C} + ABC$$

This means that $\bar{Y}=1$ (i.e. $Y=0$) if any of these fundamental product terms are 1.

Table 3.36 Truth table of $Y=\overline{A}B+\overline{B}C$

| $A$ | $B$ | $C$ | $Y$ |
|---|---|---|---|
| 0 | 0 | 0 | 0 |
| 0 | 0 | 1 | 1 |
| 0 | 1 | 0 | 1 |
| 0 | 1 | 1 | 1 |
| 1 | 0 | 0 | 0 |
| 1 | 0 | 1 | 1 |
| 1 | 1 | 0 | 0 |
| 1 | 1 | 1 | 0 |

(This is a fundamental sum of products expression for $\overline{Y}$.) This expression can be dualled to give:

$$Y=(A+B+C)\cdot(\overline{A}+B+C)\cdot(\overline{A}+\overline{B}+C)\cdot(\overline{A}+\overline{B}+\overline{C})$$

which expresses $Y$ in *fundamental product of sums* form (i.e. fundamental sum terms (as they contain all three variables) are AND'd together to give their product). (Note that the · has been used to emphasise that the sum terms are being AND'd together.) This, and the fundamental sum of products expression

$$Y=\overline{A}\overline{B}C+\overline{A}B\overline{C}+\overline{A}BC+A\overline{B}C$$

are *identical* expressions for $Y$ in different forms.

Although the above dualling process can be performed directly, by swapping all operators and inverting all variables, it is instructive to consider it in more detail. Firstly, De Morgan's theorem, $\overline{P}\cdot\overline{Q}=\overline{P+Q}$, is applied to the individual fundamental product terms to give:

$$\overline{Y}=\overline{(A+B+C)}+\overline{(\overline{A}+B+C)}+\overline{(\overline{A}+\overline{B}+C)}+\overline{(\overline{A}+\overline{B}+\overline{C})}$$

Then the second of De Morgan's theorems, $\overline{P}+\overline{Q}=\overline{P\cdot Q}$, is used to give:

$$\overline{Y}=\overline{(A+B+C)\cdot(\overline{A}+B+C)\cdot(\overline{A}+\overline{B}+C)\cdot(\overline{A}+\overline{B}+\overline{C})}$$

Finally, inverting both sides gives:

$$Y=(A+B+C)\cdot(\overline{A}+B+C)\cdot(\overline{A}+\overline{B}+C)\cdot(\overline{A}+\overline{B}+\overline{C})$$

**Product of sums and maxterms**

With the sum of products form, if any one of the product terms is 1 then the output will be 1 because any Boolean expression OR'd with 1 gives a result of 1 (Equation 1.9). Regarding the product of sums form, the significant point is that anything AND'd with 0 gives 0 (Equation 1.6). Consequently, in the fundamental product of sums form if *any* of the sum terms is 0 then $Y=0$.

These processes are illustrated in Table 3.37. This shows how OR'ing (on the left-hand side) the Karnaugh maps for the *individual* product terms in the fundamental sum of products expression for $Y$ leads to the overall map for $Y$, together with how AND'ing (forming the product) of the individual sum terms in the fundamental product of sums expression for $Y$ also leads to the same map.

Table 3.37 The production of the overall Karnaugh map for $Y$ from the maps of the individual fundamental product and sum terms. The sum of products form is shown on the left, with the product of sums form on the right

| Y | $\overline{A}\,\overline{B}$ | $\overline{A}B$ | $AB$ | $A\overline{B}$ |
|---|---|---|---|---|
| $\overline{C}$ | 0 | 0 | 0 | 0 |
| C | 1 | 0 | 0 | 0 |

OR

| Y | $\overline{A}\,\overline{B}$ | $\overline{A}B$ | $AB$ | $A\overline{B}$ |
|---|---|---|---|---|
| $\overline{C}$ | 0 | 1 | 0 | 0 |
| C | 0 | 0 | 0 | 0 |

OR

| Y | $\overline{A}\,\overline{B}$ | $\overline{A}B$ | $AB$ | $A\overline{B}$ |
|---|---|---|---|---|
| $\overline{C}$ | 0 | 0 | 0 | 0 |
| C | 0 | 1 | 0 | 0 |

OR

| Y | $\overline{A}\,\overline{B}$ | $\overline{A}B$ | $AB$ | $A\overline{B}$ |
|---|---|---|---|---|
| $\overline{C}$ | 0 | 0 | 0 | 0 |
| C | 0 | 0 | 0 | 1 |

=

| Y | $\overline{A}\,\overline{B}$ | $\overline{A}B$ | $AB$ | $A\overline{B}$ |
|---|---|---|---|---|
| $\overline{C}$ | 0 | 1 | 0 | 0 |
| C | 1 | 1 | 0 | 1 |

| Y | $\overline{A}\,\overline{B}$ | $\overline{A}B$ | $AB$ | $A\overline{B}$ |
|---|---|---|---|---|
| $\overline{C}$ | 0 | 1 | 1 | 1 |
| C | 1 | 1 | 1 | 1 |

AND

| Y | $\overline{A}\,\overline{B}$ | $\overline{A}B$ | $AB$ | $A\overline{B}$ |
|---|---|---|---|---|
| $\overline{C}$ | 1 | 1 | 1 | 0 |
| C | 1 | 1 | 1 | 1 |

AND

| Y | $\overline{A}\,\overline{B}$ | $\overline{A}B$ | $AB$ | $A\overline{B}$ |
|---|---|---|---|---|
| $\overline{C}$ | 1 | 1 | 0 | 1 |
| C | 1 | 1 | 1 | 1 |

AND

| Y | $\overline{A}\,\overline{B}$ | $\overline{A}B$ | $AB$ | $A\overline{B}$ |
|---|---|---|---|---|
| $\overline{C}$ | 1 | 1 | 1 | 1 |
| C | 1 | 1 | 0 | 1 |

=

| Y | $\overline{A}\,\overline{B}$ | $\overline{A}B$ | $AB$ | $A\overline{B}$ |
|---|---|---|---|---|
| $\overline{C}$ | 0 | 1 | 0 | 0 |
| C | 1 | 1 | 0 | 1 |

The important points to note are that the fundamental product terms specify where the minterms are in the final map, whereas the fundamental sum terms specify where a zero appears in the final map. However, an alternative way of viewing this is that the fundamental sum terms rather specify that all cells except one have a 1 in them. It is for this reason that the fundamental sum terms are known as *maxterms*, since they specify the maximum area of 1's (i.e. all cells except one) in the Karnaugh map.

In the same way that we could describe any combinational logic expression as a list of minterms, we can also describe it as a list of sum terms (maxterms). These will be those which were not minterms. So $Y=\bar{A}B+\bar{B}C$ can be written as either $Y=\Sigma(1, 2, 3, 5)$ or $Y=\Pi(0, 4, 6, 7)$.

**Example 3.28**

In Sections 3.1 and 3.2.1 the Boolean expression $Y=\bar{A}C+\bar{A}\bar{B}$ was used as an example. Write out the fundamental sum of products expression for $\bar{Y}$ and then dual it to give the fundamental product of sums expression for $Y$. Also give an expression for $Y$ in terms of the codes for the maxterms.

***Solution***

From the earlier truth table in Fig. 3.2:

$$\bar{Y}=\bar{A}B\bar{C}+A\bar{B}\bar{C}+A\bar{B}C+AB\bar{C}+ABC$$

Dualling this gives:

$$Y=(A+\bar{B}+C)\cdot(\bar{A}+B+C)\cdot(\bar{A}+B+\bar{C})\cdot(\bar{A}+\bar{B}+C)\cdot(\bar{A}+\bar{B}+\bar{C})$$

Finally, $Y=\Pi(2, 4, 5, 6, 7)$.

**Minimisation of product of sums expressions using Boolean algebra**

Minimisation of the fundamental product of sums expression can of course be performed algebraically. To do this it is easiest to simplify the expression for $\bar{Y}$:

$$\begin{aligned}\bar{Y}&=\bar{A}B\bar{C}+A\bar{B}\bar{C}+AB\bar{C}+ABC\\&=(\bar{A}+A)B\bar{C}+(\bar{B}+B)A\bar{C}+AB(\bar{C}+C)\\&=B\bar{C}+A\bar{C}+AB\end{aligned}$$

Dualling gives:

$$Y=(B+C)\cdot(\bar{A}+C)\cdot(\bar{A}+\bar{B})$$

The final expression for $\bar{Y}$ gives the product terms which correspond to the three 'prime implicants' for the 0's in the Karnaugh map. This is because this is a sum of products expression for $\bar{Y}$. The product of sums expression for $Y$ is composed of the *prime implicates* which are the corresponding sum expressions to prime implicants. It is important to note that whereas the 'prime implicants' for $\bar{Y}$ specify where the 0's are in the Karnaugh map (for each product term), see Table 3.37, the prime implicates rather specify where the 1's are.

It is clear from the Karnaugh map that this is not a minimised product of sums expression for $Y$ because $A\bar{C}$ is a non-essential prime implicate. This is confirmed by Boolean algebra as follows:

$$\begin{aligned}\bar{Y} &= \bar{B}\bar{C}+A\bar{C}+AB \\ &= \bar{B}\bar{C}+A\bar{B}\bar{C}+AB\bar{C}+AB \\ &= \bar{B}\bar{C}(1+A)+AB(1+\bar{C}) \\ &= \bar{B}\bar{C}+AB\end{aligned}$$

Dualling gives:

$$Y=(B+C)\cdot(\bar{A}+\bar{B})$$

The sum of products and product of sums form are complementary and both produce the same Boolean function with two-level circuits. The circuit for the minimised product of sums form of $Y$ is shown in Fig. 3.5.

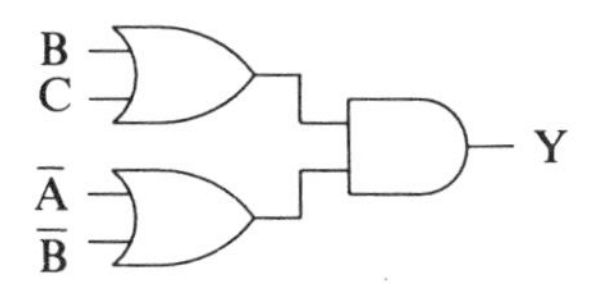


Fig. 3.5 The two-level circuit for $Y=(B+C)\cdot(\bar{A}+\bar{B})$

**Minimisation using Karnaugh maps**

The above illustrates how we can use a Karnaugh map to produce a minimised product of sums expression. Basically the process is exactly the same as usual except the 0's are grouped and looped rather than the 1's. This gives a sum of products expression for $\bar{Y}$, which is then simply dualled to give the minimised product of sums form for $Y$.

**Example 3.29**

Minimise the fundamental product of sums expression from Example 3.28:

$$Y=(A+\bar{B}+C)\cdot(\bar{A}+B+C)\cdot(\bar{A}+B+\bar{C})\cdot(\bar{A}+\bar{B}+C)\cdot(\bar{A}+\bar{B}+\bar{C})$$

firstly using Boolean algebra and then a Karnaugh map. Then draw the circuit which implements the minimised form of $Y$.

***Solution***

Using the dual of $Y$ (from the previous example)

$$\begin{aligned}\bar{Y} &= \bar{A}B\bar{C}+A\bar{B}\bar{C}+A\bar{B}C+AB\bar{C}+ABC \\ &= (\bar{A}+A)B\bar{C}+(\bar{B}+B)A\bar{C}+(\bar{B}+B)AC \\ &= B\bar{C}+A\bar{C}+AC \\ &= A+B\bar{C}\end{aligned}$$

Dualling this expression gives:

$$Y=\bar{A}\cdot(\bar{B}+C)$$

From the Karnaugh map for $Y$ in Table 3.38, looping and grouping the 0's gives

$$\bar{Y}=A+B\bar{C}$$

Table 3.38 Karnaugh map for the function discussed in Example 3.29

| Y | $\bar{A}\bar{B}$ | $\bar{A}B$ | $AB$ | $A\bar{B}$ |
|---|---|---|---|---|
| $\bar{C}$ | 1 | 0 | 0 | 0 |
| $C$ | 1 | 1 | 0 | 0 |

which can be dualled to give the same result as above. The circuit to implement this minimised product of sums form is shown in Fig. 3.6.

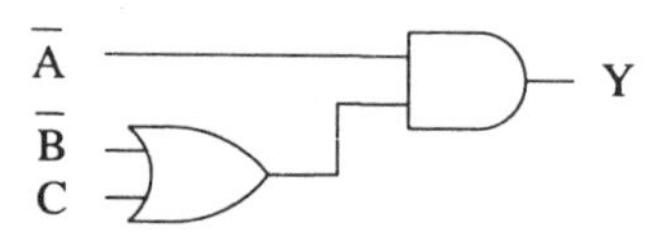


Fig. 3.6 Minimised product of sums form of Boolean expression considered in Example 3.29

## 3.5 SELF-ASSESSMENT

3.1 What characteristic defines a combinational logic circuit?

3.2 What are product terms and sum terms?

3.3 What are the two (functionally identical) forms that can be used to describe the operation of a combinational logic circuit?

3.4 What different ways are there of describing the operation of a combinational logic circuit?

3.5 How does the number of columns and rows in a truth table relate to the number of input and output variables?

3.6 What is meant by a fundamental product term, and how are they related to the outputs from a truth table?

3.7 What form of circuit does a two-level circuit implement and what are the advantages of this type of circuit?

3.8 What is the effect of minimising a fundamental sum of products expression?

3.9 Which single-variable theorem is the minimisation of fundamental sum of products expressions dependent upon?

3.10 Why can Karnaugh maps be used for 'visually' minimising Boolean expressions.

3.11 Draw out the basic form of Karnaugh maps for three and four variables, together with the corresponding truth tables, indicating the link between the cells of the Karnaugh maps and rows of the truth table.

3.12 What are octets, quads and duals?

3.13 What are minterms, prime implicants and essential prime implicants?

3.14 What is meant by 'don't care' product terms and what is their significance regarding minimising Boolean expressions?

3.15 How does the description and implementation of a combinational logic circuit differ when a product of sums rather than sum of products form is used?

3.16 The minimisation of fundamental forms via a Karnaugh map relies upon grouping and looping either minterms or maxterms. If in general $n$ terms are looped how many variables are eliminated?

## 3.6 PROBLEMS

The first eight problems refer to the truth tables in Table 3.39.

Table 3.39 Truth tables of functions referred to in the following problems

| $A$ | $B$ | $C$ | $J$ | $K$ | $L$ | $M$ | $N$ | $P$ | $Q$ |
|---|---|---|---|---|---|---|---|---|---|
| 0 | 0 | 0 | 1 | 0 | 1 | 0 | 0 | 1 | 1 |
| 0 | 0 | 1 | 1 | 0 | 0 | 1 | 1 | 0 | 1 |
| 0 | 1 | 0 | 0 | 1 | 0 | 0 | 1 | 1 | 0 |
| 0 | 1 | 1 | 0 | 1 | 1 | 1 | 1 | 0 | 0 |
| 1 | 0 | 0 | 1 | 1 | 1 | 1 | 0 | 0 | 1 |
| 1 | 0 | 1 | 0 | 0 | 0 | 1 | 1 | 0 | 0 |
| 1 | 1 | 0 | 0 | 1 | 1 | 1 | 1 | 1 | 1 |
| 1 | 1 | 1 | 1 | 1 | 1 | 0 | 0 | 1 | 0 |

| $A$ | $B$ | $C$ | $D$ | $R$ | $S$ | $T$ | $U$ | $V$ | $W$ | $X$ | $Y$ | $Z$ |
|---|---|---|---|---|---|---|---|---|---|---|---|---|
| 0 | 0 | 0 | 0 | 0 | 0 | 1 | 0 | 1 | 1 | 1 | 1 | 1 |
| 0 | 0 | 0 | 1 | 1 | 0 | 0 | 0 | 0 | 1 | 1 | 1 | 0 |
| 0 | 0 | 1 | 0 | 0 | 0 | 0 | 0 | 1 | 0 | 0 | 1 | 0 |
| 0 | 0 | 1 | 1 | 1 | 0 | 0 | 0 | 0 | 0 | 0 | 0 | 0 |
| 0 | 1 | 0 | 0 | 1 | 0 | 1 | 1 | 1 | 1 | 1 | 1 | 1 |
| 0 | 1 | 0 | 1 | 1 | 1 | 0 | 1 | 1 | 1 | 1 | 1 | 1 |
| 0 | 1 | 1 | 0 | 1 | 0 | 0 | 1 | 1 | 0 | 0 | 1 | 0 |
| 0 | 1 | 1 | 1 | 0 | 0 | 1 | 1 | 1 | 0 | 0 | 1 | 1 |
| 1 | 0 | 0 | 0 | 0 | 1 | 1 | 1 | 1 | 1 | 1 | 0 | 0 |
| 1 | 0 | 0 | 1 | 1 | 0 | 1 | 1 | 0 | 1 | 1 | 0 | 0 |
| 1 | 0 | 1 | 0 | 0 | 0 | 1 | 1 | 1 | 0 | 0 | 0 | 1 |
| 1 | 0 | 1 | 1 | 1 | 0 | 1 | 1 | 1 | 0 | 0 | 1 | 0 |
| 1 | 1 | 0 | 0 | 1 | 1 | 0 | 0 | 1 | 0 | 1 | 0 | 1 |
| 1 | 1 | 0 | 1 | 1 | 1 | 0 | 0 | 0 | 1 | 0 | 1 | 1 |
| 1 | 1 | 1 | 0 | 1 | 1 | 0 | 0 | 1 | 0 | 0 | 0 | 1 |
| 1 | 1 | 1 | 1 | 1 | 0 | 0 | 0 | 0 | 1 | 0 | 1 | 1 |

3.1 For all of the outputs ($J$ to $Q$) from the three input variable truth table above:

- Give the output from the the truth table as a fundamental sum of products expression.
- Use Boolean algebra to simplify this expression.
- Confirm that this is correct by minimising using a Karnaugh map.
- Using OR and AND gates draw the two-level circuits required to implement the functions described by these truth tables.
- Use the Karnaugh maps to derive minimised product of sums expressions for the outputs, and then use Boolean algebra to confirm their equivalence to the minimised sum of products form.

3.2 Use the Quine–McCluskey method to minimise $M$.

3.3 Draw circuits to implement $J$ using firstly only NAND gates and then only NOR gates.

3.4 Draw the Karnaugh map for $R$ and use it to determine all of the prime implicants. Which ones are essential prime implicants? Give a minimised sum of products expression for $R$.

Verify that all of the prime implicants have been found using the Quine–McCluskey technique.

Use the Karnaugh map to derive a minimised product of sums expression for $R$, and then demonstrate its equivalence to the sum of products form using Boolean algebra.

How could $R$ be implemented using one each of: a two-input OR gate, a two-input AND gate, a four-input NAND gate and an inverter?

3.5 Write out expressions for $S$ and $T$ in sum of fundamental products form and then minimise them using Boolean algebra. Check the results using Karnaugh maps.

3.6 Minimise functions $U$, $V$ and $W$. Draw the circuit required to implement $U$ using firstly AND and OR gates and then only NAND gates. What single gate could be used instead? Express $U$ in minimised product of sums form.

3.7 Minimise $X$ into both sum of products and products of sums form. Show, via Boolean algebra, the relation between the two forms and how many gates are necessary to implement them. Then consider the product of sums implementation using a maxterm and an octet of 0's (i.e. a non-minimised implementation) which leads to a form for $X$ consisting of an octet of 1's minus a single minterm (produced by AND'ing the octet with the complement of the missing minterm).

3.8 Using Karnaugh maps find the prime implicants of $Y$ and $Z$. Which are essential prime implicants? Give minimised expressions for these functions.

Confirm these results using the Quine–McCluskey method.

3.9 Boolean expressions can be factored via the distributive law with a basic example being:

$$AC+AD+BC+BD=(A+B)\cdot(C+D)$$

Draw Karnaugh maps of the left-hand side of this equation and the two sum terms (i.e. factors) on the right-hand side to prove that this equation is correct.

3.10 Use Karnaugh maps to show that $(AC+\bar{A}\bar{C})$ and $(C\oplus D)$ are factors of $(\bar{A}B\bar{C}D+A\bar{B}C\bar{D})$ and find a suitable third and final factor.

3.11 Minimise the following function using firstly a Karnaugh map and then the Quine–McCluskey method.

$$Y=\Sigma(1, 2, 5, 8, 9, 10, 12, 13, 16, 18, 24, 25, 26, 28, 29, 31)$$

# 4 Combinational logic circuits

## 4.1 COMMON COMBINATIONAL LOGIC CIRCUITS

There are a number of combinational logic circuits which are used so frequently that they are often considered to exist as circuit elements (like logic gates) in their own right. Note that the forms of these circuits given here are those that implement the basic functions. When provided as 'building blocks' for digital design some of these may have additional combinational logic circuits attached to their inputs that allow extra control of their action.

### 4.1.1 Multiplexers

Multiplexers provide a way of selecting one out of many digital signals. A multiplexer will in general have $n$ inputs, and obviously one output, with $m$ control lines which are used to select one of the $n$ inputs. The block diagram of a multiplexer (mux) is shown in Fig. 4. 1.

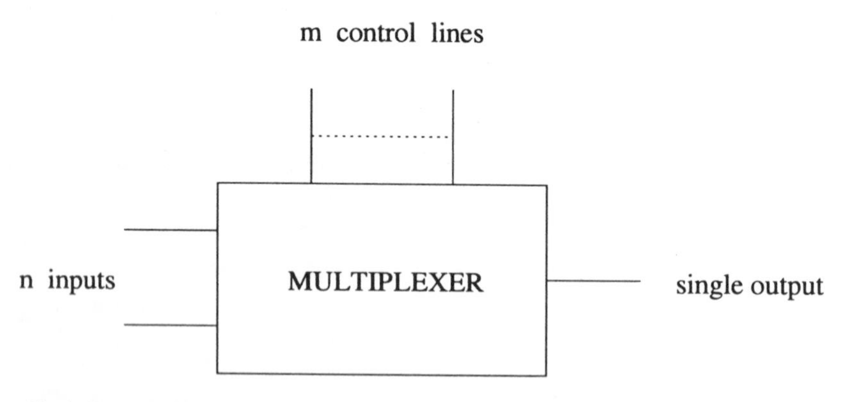


Fig. 4.1 Block diagram of an $n$-to-1 multiplexer

Which of the $n$-input channels is routed through to the output is determined by the bit pattern on the $m$ control lines. Hence, $n$, the number of input lines that can be multiplexed is $2^m$. The basic structure of an $n$-input multiplexer is $n$ $(m+1)$-input AND gates (that is one AND gate to decode each of the $n=2^m$ possible combinations of the $m$ control inputs), all feeding into a single OR gate. The extra (to the $m$ control lines) input to each gate is connected to one of the $n$ inputs.

Multiplexers are usually referred to as $n$-to-1 or 1-of-$n$ multiplexers or data selectors.

The operation is based upon the fact that only one of the $2^m$ possible input combinations can ever be applied to the control inputs at any one time, and therefore only the corresponding AND gate will be capable of giving an output other than 0. This is the gate whose input will be routed through to the output.

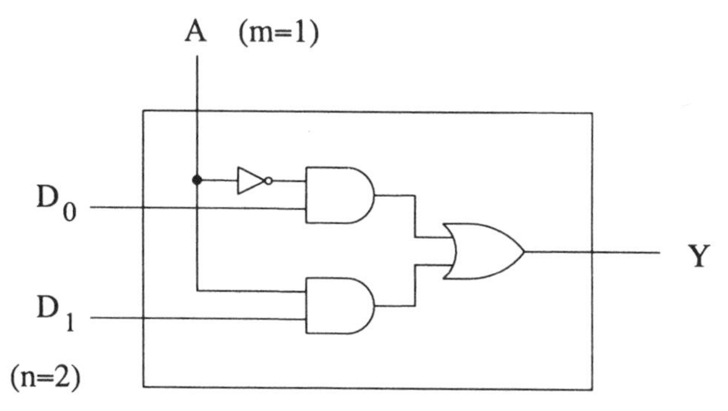


Fig. 4.2 A 2-to-1 multiplexer

**2-to-1 multiplexer**

Figure 4.2 is the circuit diagram of a 2-to-1 multiplexer. Note that it has two inputs ($n=2$), with a single control line ($m=1$). If $A=0$ then the output from the AND gate with $D_1$ as an input must be 0 (since anything AND'd with 0 is 0, Equation 1.6) whilst the output from the other AND gate will be $\bar{A} \cdot D_0 = 1 \cdot D_0 = D_0$. So, the output from the multiplexer is $Y = D_0 + 0 = D_0$ (Equation 1.8). By similar reasoning if $A=1$ then $Y=D_1$. In Boolean algebraic terms:

$$Y = \bar{A} \cdot D_0 + A \cdot D_1$$

One way of thinking of the action of a multiplexer is that only one of the AND gates is ever activated and so allows the input signal fed to it through to the OR gate. This is illustrated in Fig. 4.3 which shows one of the AND gates from an 8-to-1 multiplexer which therefore has three control signals $A$, $B$ and $C$. The gate shown controls the passage of input $D_3$ which will be selected for an input of $\bar{A}BC$. The output from this gate is $((\bar{A}BC) \cdot D_3)$,which will be $D_3$ when the product term $\bar{A}BC=1$, and 0 otherwise. Hence the presence of this product term effectively 'activates' the gate, meaning the output is then $D_3$. Any other input combination

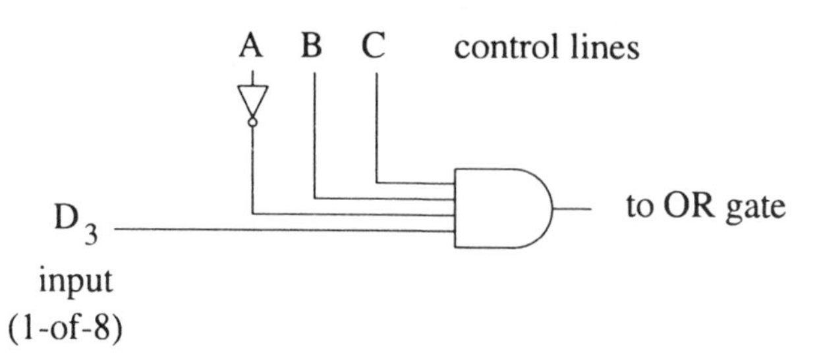


Fig. 4.3 The 'activation' of an AND gate in an 8-to-1 multiplexer

means the output from the gate is always 0. Only one AND gate in a multiplexer is activated at a time and it is therefore its output that appears as $Y$, the output from the OR gate and hence the multiplexer.

### Example 4.1

Draw the circuit diagram and truth table, and give the Boolean equation describing the output, of a 4-to-1 multiplexer.

#### *Solution*

These are shown in Fig. 4.4

$$Y=\bar{A}\bar{B}D_0+\bar{A}BD_1+A\bar{B}D_2+ABD_3$$

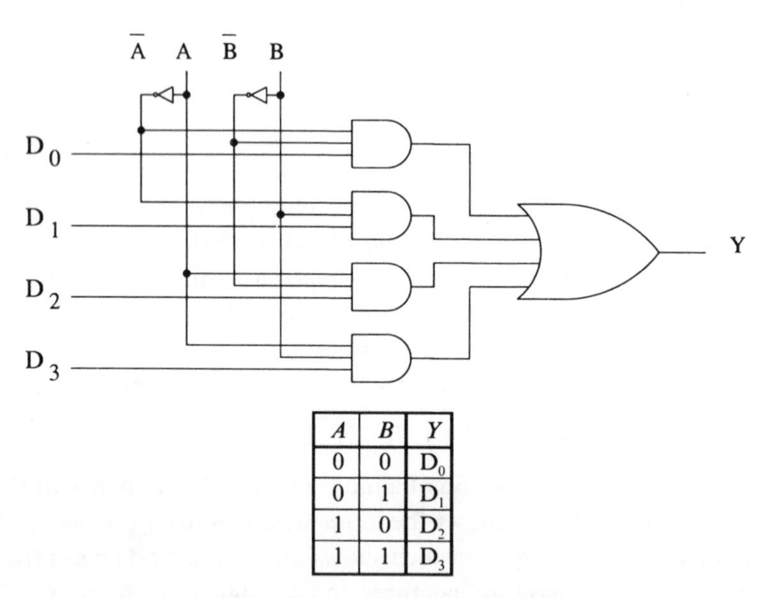


| $A$ | $B$ | $Y$ |
|---|---|---|
| 0 | 0 | $D_0$ |
| 0 | 1 | $D_1$ |
| 1 | 0 | $D_2$ |
| 1 | 1 | $D_3$ |

Fig. 4.4 A 4-to-1 multiplexer and its truth table, as discussed in Example 4.1

**The multiplexer as a universal logic solution**

Because all $2^m$ possible combinations of the $m$ control lines of a multiplexer are fed to $2^m$ AND gates then there is an AND gate for all of the fundamental product terms of the $m$ variables. A multiplexer therefore provides a way of synthesising the logic function of any $m$-input truth table in fundamental sum of products form. All that has to be done is to connect the input lines of the multiplexer to either 0 or 1 depending upon the desired output for the particular fundamental product. So any $m$-input ($n$-row) truth table can be implemented by a $n$-input multiplexer. The advantage of this type of implementation of a combinational logic circuit is that it requires only a single circuit element and that no minimisation is required since the circuit is in fundamental sum of products form.

**Example 4.2**

Implement the truth table in Table 4.1 using a multiplexer.

Table 4.1 Truth table implemented via a multiplexer in Example 4.2

| $A$ | $B$ | $Y$ |
|---|---|---|
| 0 | 0 | 1 |
| 0 | 1 | 0 |
| 1 | 0 | 1 |
| 1 | 1 | 1 |

***Solution***

This will require a 4-to-1 multiplexer (i. e. two control inputs) with inputs $D_0$ through to $D_3$ tied to 1, 0, 1 and 1, respectively (i.e. the output from the truth table) as shown in Fig. 4.5.

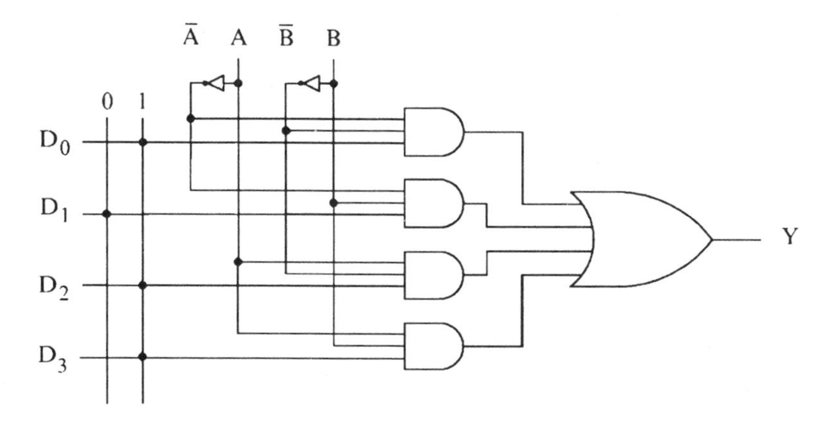


Fig. 4.5 Multiplexer used as a universal logic solution, as described in Example 4.2

Furthermore, an $n$-input multiplexer *and an inverter* can be used to implement any $2n$ row truth table. To achieve this all inputs to the truth table *except one* are connected to the multiplexer's control lines. This means that each AND gate is now activated for two rows of the truth table, i.e. two input patterns differing in the variable *not* connected to a control line.

These two rows have four possible output combinations: both 0; both 1; one 0 and the other 1; or vice versa. For the same value in both rows the activated AND gate can be tied to either 0 or 1 as required, whilst for different values it can be connected to the least significant input *or* its inverse (this is why the inverter is needed).

**Example 4.3**

How can the truth table in Table 4.2 be implemented using a four-input multiplexer and an inverter?

Table 4.2 The eight-row truth table implemented using a four-input multiplexer as described in Example 4.3, and shown in Figure 4.6

| $A$ | $B$ | $C$ | $Y$ |
|---|---|---|---|
| 0 | 0 | 0 | 1 |
| 0 | 0 | 1 | 1 |
| 0 | 1 | 0 | 0 |
| 0 | 1 | 1 | 0 |
| 1 | 0 | 0 | 1 |
| 1 | 0 | 1 | 0 |
| 1 | 1 | 0 | 0 |
| 1 | 1 | 1 | 1 |

***Solution***

This is a three input, eight row ($2n=8$) truth table which we are going to implement using a four-input ($n=4$ and $m=2$) multiplexer. The two most significant inputs to the truth table, $A$ and $B$, are connected to the two control lines of the multiplexer.

This relates each of the four AND gates in the multiplexer to a *pair* of rows in the truth table (for fixed values of $A$ and $B$). For $\bar{A}\bar{B}$ we see that $Y=1$ and for $\bar{A}B$ that $Y=0$. For $A\bar{B}$, $Y=\bar{C}$ and for $AB$, $Y=C$. Consequently, the multiplexer must be wired as in Fig. 4.6.

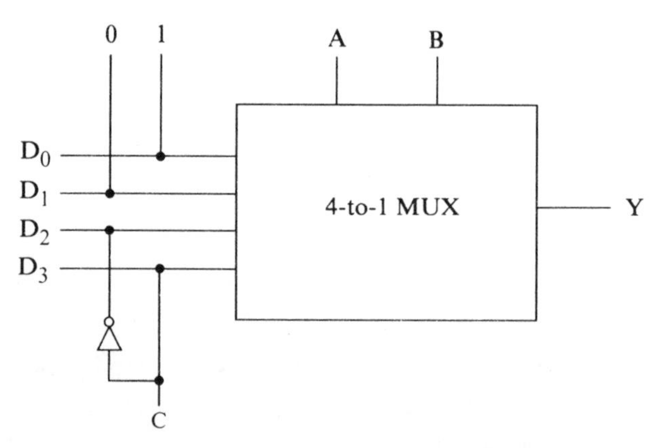


Fig. 4.6 Use of a four-input multiplexer to implement the truth table shown in Table 4.2, as described in Example 4.3

## 4.1.2 Demultiplexers

Demultiplexers provide the reverse operation of multiplexers since they allow a single input to be routed to one of $n$ outputs, selected via $m$ control lines ($n=2^m$). This circuit element is usually referred to as a 1-of-$n$ demultiplexer. The circuit basically consists of $n$ AND gates, one for each of the $2^m$ possible combinations of the $m$ control inputs, with the single line input fed to all of these gates. Since only one AND gate will ever be active this determines which output the input is fed to. The block, and circuit, diagram of a 1-of-4 demultiplexer is shown in Fig. 4.7.

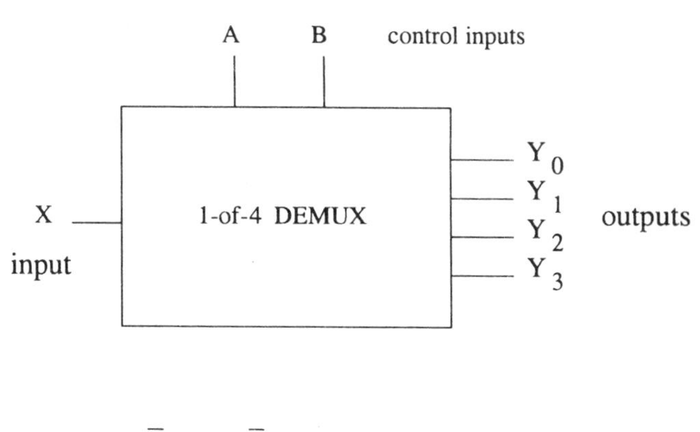


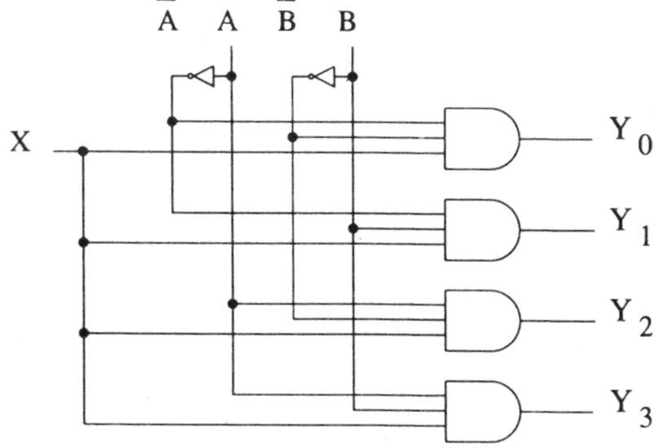


Fig. 4.7 The block and circuit diagrams of a 1-of-4 demultiplexer

**Decoders**

A decoder is essentially a demultiplexer with no input line. So instead of feeding an input line through to the selected output, rather the selected output will simply become active (this may be either active-HIGH or LOW, with a NAND rather than an AND used for the latter). Obviously a demultiplexer can be used as a decoder by tying the input line to the appropriate value. A 2-to-4 line decoder implemented using a 1-of-4 demultiplexer is shown in Fig. 4.8.

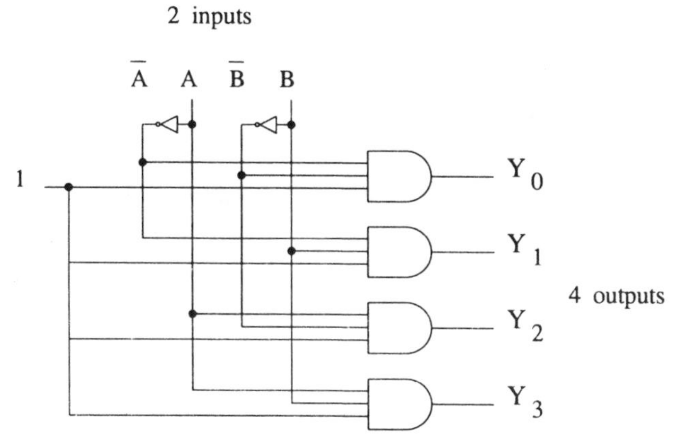


Fig. 4.8 An active-HIGH 2-to-4 line decoder implemented using a 1-of-4 demultiplexer. (Note that an active-LOW device would have the AND gates replaced with NAND gates)

## Example 4.4

Draw the truth table of a BCD-to-decimal decoder, and show how the generation of the first four codes could be achieved using a demultiplexer.

### *Solution*

Table 4.3 shows the truth table. To produce a decoder for the first four codes (0 to 3) requires a 2-to-4 decoder (i. e. a 1-of-4 demultiplexer). Note that the first four codes have $A=B=0$ so these two inputs are not needed. The circuit to implement this using a demultiplexer is shown in Fig. 4.9.

Table 4.3 Truth table for a BCD-to-decimal decoder as discussed in Example 4.4

| | BCD | | | | Decimal | | | | | | | | | |
|---|---|---|---|---|---|---|---|---|---|---|---|---|---|---|
| | $A$ | $B$ | $C$ | $D$ | 0 | 1 | 2 | 3 | 4 | 5 | 6 | 7 | 8 | 9 |
| 0 | 0 | 0 | 0 | 0 | 1 | 0 | 0 | 0 | 0 | 0 | 0 | 0 | 0 | 0 |
| 1 | 0 | 0 | 0 | 1 | 0 | 1 | 0 | 0 | 0 | 0 | 0 | 0 | 0 | 0 |
| 2 | 0 | 0 | 1 | 0 | 0 | 0 | 1 | 0 | 0 | 0 | 0 | 0 | 0 | 0 |
| 3 | 0 | 0 | 1 | 1 | 0 | 0 | 0 | 1 | 0 | 0 | 0 | 0 | 0 | 0 |
| 4 | 0 | 1 | 0 | 0 | 0 | 0 | 0 | 0 | 1 | 0 | 0 | 0 | 0 | 0 |
| 5 | 0 | 1 | 0 | 1 | 0 | 0 | 0 | 0 | 0 | 1 | 0 | 0 | 0 | 0 |
| 6 | 0 | 1 | 1 | 0 | 0 | 0 | 0 | 0 | 0 | 0 | 1 | 0 | 0 | 0 |
| 7 | 0 | 1 | 1 | 1 | 0 | 0 | 0 | 0 | 0 | 0 | 0 | 1 | 0 | 0 |
| 8 | 1 | 0 | 0 | 0 | 0 | 0 | 0 | 0 | 0 | 0 | 0 | 0 | 1 | 0 |
| 9 | 1 | 0 | 0 | 1 | 0 | 0 | 0 | 0 | 0 | 0 | 0 | 0 | 0 | 1 |

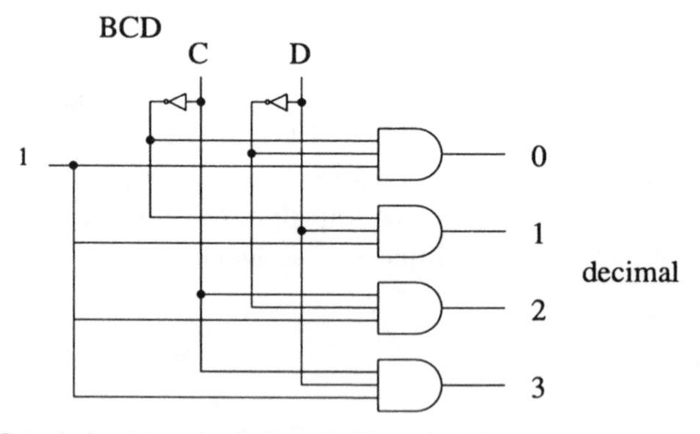


Fig. 4.9 The BCD-to-decimal decoder designed in Example 4.4

Some decoders have more than one output active at a time, an example being a BCD to 7-segment decimal display decoder. Here, rather than a single output layer of AND gates (one for each possible input pattern), a two-level output is required to allow the OR'ing together of the outputs that may become active for many different input patterns. (See Problem 4.3.)

### 4.1.3 Encoders

These are the opposite of decoders in that they convert a single active signal (out of $r$ inputs) into a coded binary, $s$-bit, output (this would be referred to as an $r$-line-to-$s$-line encoder). Often encoders are of a type called *priority encoders* which means that more than one of the $r$ inputs may be active, in which case the output pattern produced is that for the highest priority input.

Encoders have a less general form than multiplexers and demultiplexers, being specifically designed for the required task. Their usual form is of $s$ combinational circuits (e. g. AND-OR design),with $r$ inputs.

## Example 4.5

Write out the truth table for the 4-line-to-2-line encoder that takes a four-line decimal signal and converts it to binary code. Design, and draw, the circuit to implement this encoder.

### *Solution*

The required truth table is shown in Table 4.4. This truth table is incomplete since it has four input columns but only four rows. However, we know that for all of the input combinations not given we need $A=0$ and $B=0$. So we can pick out the fundamental sum of product terms for $A$ and $B$ directly from the truth table to give:

$$A=\overline{0}\,\overline{1}\,2\,\overline{3}+\overline{0}\,\overline{1}\,\overline{2}\,3$$
$$B=\overline{0}\,1\,\overline{2}\,\overline{3}+\overline{0}\,\overline{1}\,\overline{2}\,3$$

Table 4.4 Truth table for a 4-line decimal-to-binary encoder as discussed in Example 4.5

| | Decimal | | | | Binary | |
|---|---|---|---|---|---|---|
| | 0 | 1 | 2 | 3 | *A* | *B* |
| 0 | 1 | 0 | 0 | 0 | 0 | 0 |
| 1 | 0 | 1 | 0 | 0 | 0 | 1 |
| 2 | 0 | 0 | 1 | 0 | 1 | 0 |
| 3 | 0 | 0 | 0 | 1 | 1 | 1 |

The circuit is shown in Fig. 4.10.

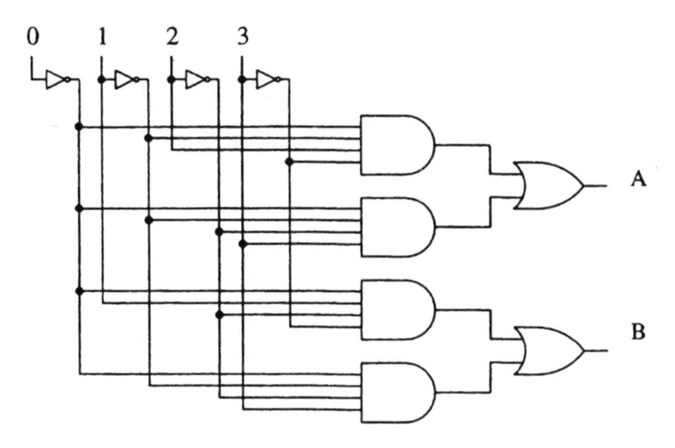


Fig. 4.10 Circuit that implements the truth table shown in Table 4.4, discussed in Example 4.5

As with decoders, encoders are often used in an active-LOW input and/or output form.

### 4.1.4 XOR gate based circuits

The truth table for the XOR operation, $Y = A\bar{B} + \bar{A}B$, is given in Table 4.5.

Table 4.5 The truth table for a two-input XOR gate

| $A$ | $B$ | $Y$ |
|---|---|---|
| 0 | 0 | 0 |
| 0 | 1 | 1 |
| 1 | 0 | 1 |
| 1 | 1 | 0 |

**Controlled inverter**

From either the truth table or the Boolean logic expression for the XOR gate it is clear that if $A=0$ then $Y=B$, whereas if $A=1$ then $Y=\bar{B}$. Consequently a two-input XOR gate can be used as a *controlled inverter*, with, in this example, the value of $A$ used to determine whether the output, $Y$, is equal to $B$ or its complement.

**Comparator**

The output of a two-input XOR gate is 0 if the inputs are the same and 1 if they differ. This means that XOR gates can be used as the basis of comparators, which are circuits used to check whether two digital words (a sequence of binary digits, i.e. bits) are the same.

**Example 4.6**

Design, using two-input XOR gates, a comparator which will give an active-LOW output if two four-bit words, $A$ and $B$, are the same.

***Solution***

The necessary circuit is shown in Fig. 4.11. $Y$ will be 0 only if the outputs from all of the XOR gates are 0, that is if all corresponding bits (e. g. $A_0$ and $B_0$) in the two words are the same.

**Example 4.7**

A two-bit comparator gives an active-HIGH output, $Y$, if two two-bit words, $A$ and $B$, are the same. Give $Y$ in fundamental sum of products form and then use Boolean algebra to show that

$$Y = \overline{(A_0 \oplus B_0) + (A_1 \oplus B_1)}$$

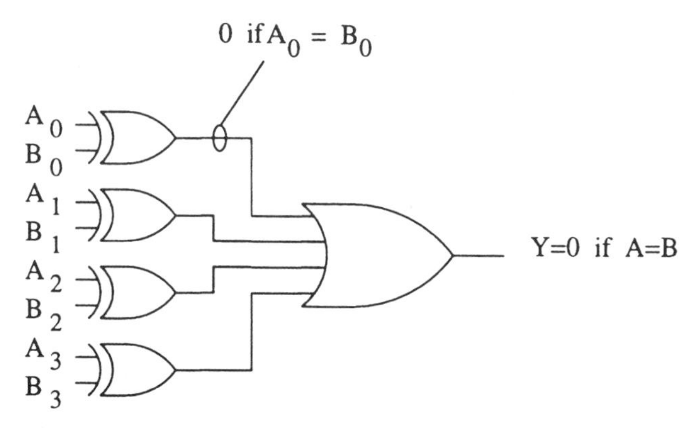


Fig. 4.11 A four-bit comparator constructed using XOR gates, discussed in Example 4.6

***Solution***

$$
\begin{aligned}
Y &= \overline{A_0}\,\overline{A_1}\,\overline{B_0}\,\overline{B_1} + \overline{A_0}A_1\overline{B_0}\,B_1 + A_0\overline{A_1}\,B_0\overline{B_1} + A_0A_1B_0B_1 \\
&= \overline{A_0}\,\overline{B_0}(\overline{A_1}\,\overline{B_1} + A_1B_1) + A_0B_0(\overline{A_1}\,\overline{B_1} + A_1B_1) \\
&= (\overline{A_0}\,\overline{B_0} + A_0B_0)\cdot(\overline{A_1}\,\overline{B_1} + A_1B_1) \\
&= (\overline{(A_0 \oplus B_0)}\cdot\overline{(A_1 \oplus B_1)}) \\
&= \overline{(A_0 \oplus B_0) + (A_1 \oplus B_1)} && \text{De Morgan's theorem}
\end{aligned}
$$

**Parity generators and checkers**

When sending *n* bits of data along a serial line (i.e. one bit after another) a simple way of checking whether an error (in a single bit) has occurred is to use a parity bit. This is an extra bit which is added on to the *n*-bit word, and whose value is set at either 0 or 1, to ensure that the total number of bits that are 1 in the complete $(n+1)$-bit word sent is either odd or even (i.e. odd or even parity).

The XOR gate can be used for this purpose since it only gives an output of 1 if an odd number of its inputs are 1. Parity generation refers to the process of determining the value of the parity bit; parity checking is performed on the received $(n+1)$-bit word to see if an error has occurred.

**Example 4.8**

Use two-input XOR gates to produce the parity generation and checking systems to ensure the four-bit data sent over a serial link is transmitted and received as an odd parity word.

***Solution***

The circuit is shown in Fig. 4.12. The output from gates 1 and 2 will be 1 if their inputs have odd parity (i.e. a 0 and a 1), as will the output from gate 3. (Note that if gates 1 and 2 both output 1 to indicate odd parity of the two bits they have compared, then when used as the inputs to gate 3, its output will indicate even parity, which is correct.) If the four-bit word has odd parity then the parity bit must

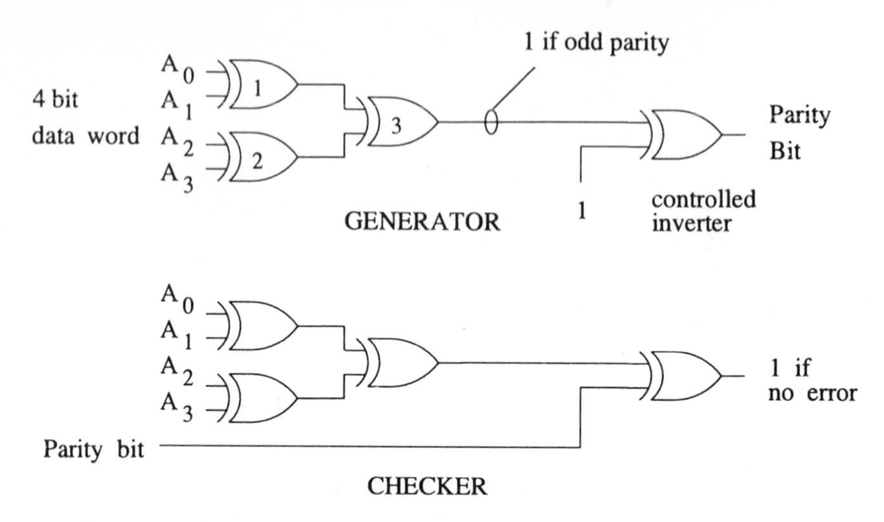


Fig. 4.12 Parity generator and checker constructed using XOR gates as discussed in Example 4.8

be zero and so the output from gate 3 is passed through an inverter to generate the parity bit. (The use of a controlled inverter means that by simply changing the control bit to 0 the parity bit can be generated for an even parity system.)

An output of 1 from the parity checker indicates that the data has been correctly received with odd parity.

### 4.1.5 Full adders

A full adder circuit is central to most digital circuits that perform addition or subtraction. It is so called because it adds together two binary digits, *plus* a carry-in digit to produce a sum and carry-out digit.[1] It therefore has three inputs and two outputs. The truth table and corresponding Karnaugh maps for it are shown in Table 4.6.

**Example 4.9**

Two 1's with no carry-in are added using a full adder. What are the outputs?

***Solution***

Adding two 1's in binary gives a result of 0 with a carry-out of 1. So $S=0$ and $C_{out}=1$. (In decimal this is saying $1+1=2$, in binary $01+01=10$.)

**Example 4.10**

Two 1's with a carry-in of 1 are added using a full adder. What are the outputs?

[1]A *half adder* only adds two bits together with no carry-in.

Table 4.6 The truth table and Karnaugh maps for a full adder. $X$ and $Y$ are the two bits to be added, $C_{in}$ and $C_{out}$ the carry-in and carry-out bits, and $S$ the sum

| $X$ | $Y$ | $C_{in}$ | $S$ | $C_{out}$ |
|---|---|---|---|---|
| 0 | 0 | 0 | 0 | 0 |
| 0 | 0 | 1 | 1 | 0 |
| 0 | 1 | 0 | 1 | 0 |
| 0 | 1 | 1 | 0 | 1 |
| 1 | 0 | 0 | 1 | 0 |
| 1 | 0 | 1 | 0 | 1 |
| 1 | 1 | 0 | 0 | 1 |
| 1 | 1 | 1 | 1 | 1 |

| S | $\overline{X}\overline{Y}$ | $\overline{X}Y$ | $XY$ | $X\overline{Y}$ |
|---|---|---|---|---|
| $\overline{C}_{in}$ | 0 | 1 | 0 | 1 |
| $C_{in}$ | 1 | 0 | 1 | 0 |

| $C_{out}$ | $\overline{X}\overline{Y}$ | $\overline{X}Y$ | $XY$ | $X\overline{Y}$ |
|---|---|---|---|---|
| $\overline{C}_{in}$ | 0 | 0 | 1 | 0 |
| $C_{in}$ | 0 | 1 | 1 | 1 |

### *Solution*

Here the result is 1 carry 1, that is $S=1$ and $C_{out}=1$. (In decimal $1+1+1$ (carry-in) $=3$; in binary $01+01+1$ (carry-in)$=11$.)

Using the Karnaugh maps to obtain minimised expressions for $S$ and $C_{out}$, we notice the chequerboard pattern of an XOR gate in the sum term to give:

$$S=X\oplus Y\oplus C_{in}$$

whilst

$$C_{out}=XY+XC_{in}+YC_{in}$$

The circuit to implement the full adder is shown in Fig. 4.13.

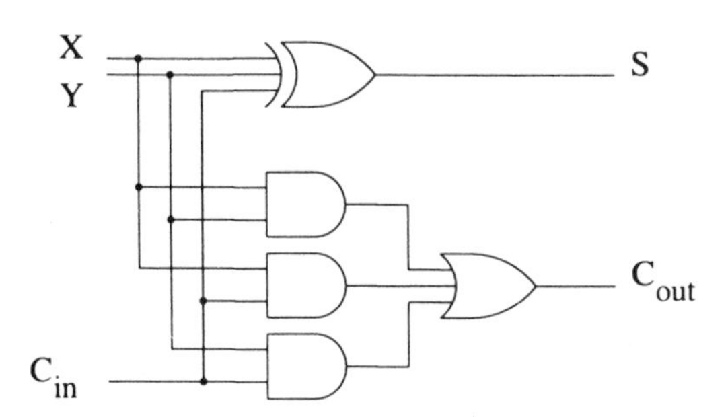


Fig. 4.13 Circuit diagram of a full adder

# 4.2 COMBINATIONAL LOGIC DESIGN EXAMPLE: A FOUR-BIT ADDER

In this section we consider the design of a four-bit adder; i.e. a circuit that adds together two four-bit binary numbers. This needs to be a combinational logic circuit and therefore serves as a useful exercise to apply what we have learnt.

To recap, we know that any truth table can be implemented using a product of sums or sum of products expression in either a fundamental or minimised (via Boolean algebra or Karnaugh maps for example) form. Using this approach we end up with a two-level circuit implementation of AND-OR, OR-AND, NAND-NAND or NOR-NOR. We have not yet considered the practicalities of any circuits we have designed or analysed, which is one of the purposes of this section.

We begin by looking again at both the benefits and problems of two-level circuits, before considering this means of implementation for the four-bit adder. We then move on to two other methods of implementation which rely upon a more thorough look at what we want the circuit to do, rather than simply treating it as a combinational logic problem to be approached using fixed 'rules'.

## 4.2.1 Two-level circuits

Two-level circuits are direct implementations of sum of products and product of sums forms, either in fundamental form (straight from the truth table) or after minimisation. We now consider the advantages and disadvantages of this type of circuit:

- Advantages:
  - Any combinational logic function can be realised as a two-level circuit.
  - This is theoretically the fastest implementation since signals have only to propagate through two gates.[2]
  - They can be implemented in a variety of ways, e. g. AND-OR, OR-AND, etc.
- Disadvantages:
  - A very large number of gates may be required.
  - Gates with a prohibitively large number of inputs may be needed.
  - Signals may be required to feed to more gates than is possible (because of the electrical characteristics of the circuit).
  - The task of minimisation increases exponentially with the number of input variables (although computer programs can obviously help reduce this problem).

The effect of minimising a fundamental two-level circuit is to reduce the first three disadvantages although it cannot be guaranteed to remove them. Note that the

[2]Note however that a single large multi-input gate may be *slower* than the equivalent constructed from several gates with fewer inputs (see Section 9. 3. 5).

second disadvantage can always be overcome by using more gates (e. g. by using three two-input AND gates to implement a four-input AND gate) but that this means a single-level gate has itself become a two-level circuit.

### 4.2.2 Specification for the four-bit binary adder

A four-bit binary adder is required to add together two four-bit binary numbers plus one carry-in bit, and produce a four-bit sum plus a carry-out bit. This is shown diagramatically in Fig. 4.14.

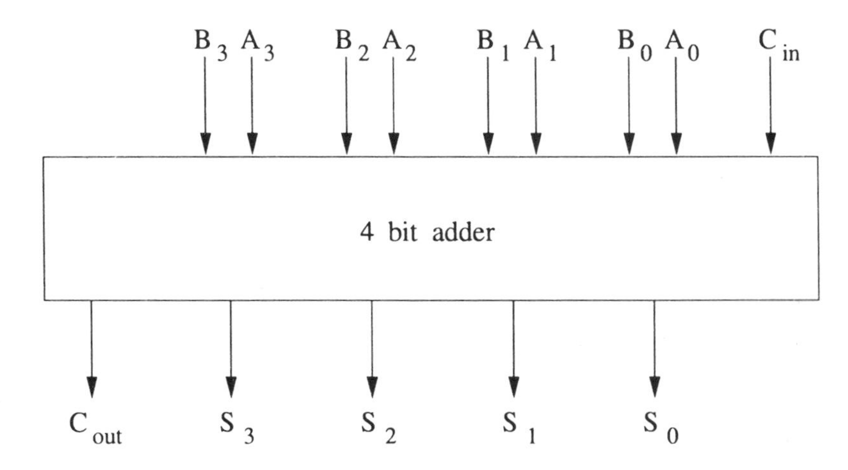


Fig. 4.14 Block diagram of a four-bit adder

By definition this is a combinational logic problem as no memory is involved, and the outputs (the sum and carry-out) depend solely upon the inputs. The truth table for this circuit will have nine input columns, and hence $2^9=512$ rows, and five output columns. We will now look at four different ways this four-bit adder could be constructed. The first two consider the use of fundamental and then minimised two-level circuits; the second two are developed by taking a closer look at the mechanics of the addition process.

### 4.2.3 Two-level circuit implementation

**Fundamental form**

We begin to consider the feasibility of constructing the four-bit adder in fundamental two-level form by looking at the output from the first sum bit, $S_0$. Since the result of the addition will be equally odd and even then the output column for $S_0$ will contain 256 1's and 256 0's. Since the truth table has nine inputs, and we need to use all of these as we are considering a fundamental sum of products implementation, then our two-level circuit will need 256 nine-input AND gates plus a 256-input OR gate to perform the summing. This is clearly impractical so we immediately rule out this method.

**Minimised form**

The most complex Boolean function in the circuit is the one for $C_{out}$ since it depends on all of the nine inputs. The minimised expression for $C_{out}$ contains over 30 essential prime implicants, which means that this many AND gates plus an OR gate with this number of inputs would be needed for a minimised two-level implementation. Furthermore, some of the input variables (or their complements) must be fed to up to 15 of the 31 essential prime implicants.

Clearly the large number of gates required, the large number of inputs they must possess, and the fact that some signals must feed into many gates, means that this implementation is also impractical, although it is an improvement on the fundamental two-level form. So although the two-level implementation is theoretically the fastest (assuming ideal gates) we see that for this application it is not really practical.

## 4.2.4 Heuristic implementation

Heuristic implementations can broadly be considered as those that are not produced by rigorous application of Boolean logic theory. (Remember that *any* truth table can, in theory, be implemented in fundamental or minimised two-level form.) Heuristic implementations are found by looking at the overall problem, often considering the way in which the task would be tackled manually.

**Ripple carry adder**

The ripple carry, or parallel adder, arises out of considering how we perform addition, and is therefore a heuristic solution. Two numbers can be added by beginning with the two least significant digits to produce their sum, plus a carry-out bit (if necessary). Then the next two digits are added (together with any carry-in bit from the addition of the first two digits) to produce the next sum digit and any carry-out bit produced at this stage. This process is then repeated until the most significant digits are reached (see Section 2. 5. 1).

To implement this procedure, for binary arithmetic, what is required is a logic block which can take two input bits, and add them together with a carry-in bit to produce their sum and a carry-out bit. This is exactly what the full adder, described earlier in Section 4.1.5, does. Consequently by joining four full adders together, with the carry-out from one adder connected to the carry-in of the next, a four-bit adder can be produced. This is shown in Fig. 4.15.

This implementation is called a parallel adder because all of the inputs are entered into the circuit at the same time (i.e. in parallel, as opposed to serially which means one bit entered after another). The name *ripple carry adder* arises because of the way the carry signal is passed along, or ripples, from one full adder to the next. This is in fact the main disadvantage of the circuit because the output, $S_3$, may depend upon a carry out which has rippled along, through the second and third adders, after being generated from the addition of the first two bits. Consequently the outputs from the ripple carry adder cannot be guaranteed

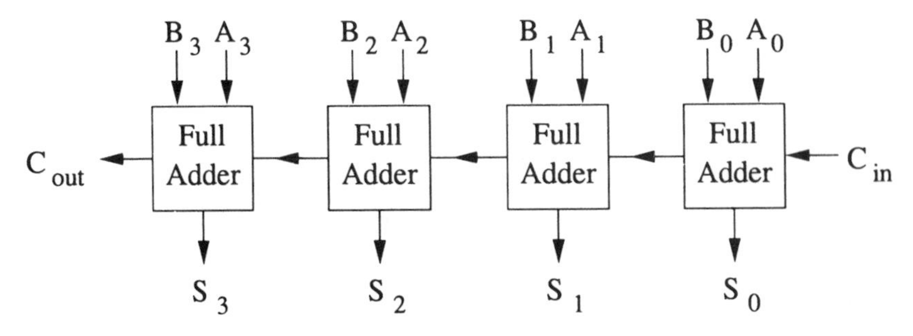


Fig. 4.15 A four-bit ripple carry adder constructed from four full adders

stable until enough time has elapsed[3] to ensure that a carry, if generated, has propagated right through the circuit.

This rippling limits the operational speed of the circuit which is dependent upon the number of gates the carry signal has to pass through. Since each full adder is a two-level circuit, the full four-bit ripple carry adder is an eight-level implementation. So after applying the inputs to the adder, the correct output cannot be guaranteed to appear until a time equal to eight propagation delays of the gates being used has elapsed.

The advantage of the circuit is that as each full adder is composed of five gates[4] then only 20 gates are needed. The ripple carry, or parallel adder, is therefore a practical solution to the production of a four-bit adder. This circuit is an example of an *iterative* or *systolic* array, which is the name given to a combinational circuit that uses relatively simple blocks (the full adders) connected together to perform a more complex function.

**Look-ahead carry adder**

The fourth possible implementation of a four-bit binary adder bears some resemblance to the ripple carry adder, but overcomes the problem of the 'rippling' carry by using extra circuitry to predict this rippling in advance. This gives a speed advantage at the expense of a more complex circuit, which is a demonstration of a general rule that any gain in performance in some aspect of a circuit is usually matched by a loss in performance of another.

Reconsidering the ripple carry adder and denoting the carry-out from each stage by $C_n$, where $n$ is the stage number, and the initial carry-in bit as $C_i$, we begin with the first stage and derive the Boolean expression for $C_0$. (We know what it is from the Karnaugh map in Section 4.1.5). So:

$$\begin{aligned} C_0 &= A_0B_0 + A_0C_i + B_0C_i \\ &= A_0B_0 + (A_0 + B_0) \cdot C_i \end{aligned}$$

[3]Remember that it takes a finite time for a logic signal to propagate through a real logic gate.

[4]Note however that the XOR is not strictly a single gate, with a two-input XOR gate requiring two two-input ANDs and a two-input OR gate for its implementaton.

Similarly for the second stage:

$$\begin{aligned} C_1 &= A_1B_1 + A_1C_0 + B_1C_0 \\ &= A_1B_1 + (A_1 + B_1) \cdot C_0 \end{aligned}$$

This expression demonstrates the problem of the ripple carry adder, because $C_1$ depends upon $C_0$ which must be produced first. However, we already have an expression for $C_0$ in terms of the actual inputs to the adder, so we can substitute this into $C_1$ so removing the rippling problem. This gives:

$$C_1 = A_1B_1 + (A_1 + B_1) \cdot (A_0B_0 + (A_0 + B_0) \cdot C_i)$$

This is a rather unwieldy expression, but we can simplify it by letting, for a general stage, $j$:

$$G_j = A_jB_j \quad \text{and} \quad P_j = A_j + B_j$$

This gives

$$C_0 = G_0 + P_0 \cdot C_i$$

$$\begin{aligned} C_1 &= G_1 + P_1\,C_0 \\ &= G_1 + P_1 \cdot (G_0 + P_0\,C_i) \\ &= G_1 + P_1G_0 + P_1P_0C_i \end{aligned}$$

Continuing this process also gives:

$$\begin{aligned} C_2 &= G_2 + P_2\,C_1 \\ &= G_2 + P_2(G_1 + P_1G_0 + P_1P_0C_i) \\ &= G_2 + P_2G_1 + P_2P_1G_0 + P_2P_1P_0C_i \end{aligned}$$

$$\begin{aligned} C_3 &= G_3 + P_3C_2 \\ &= G_3 + P_3(G_2 + P_2G_1 + P_2P_1G_0 + P_2P_1P_0C_i) \\ &= G_3 + P_3G_2 + P_3P_2G_1 + P_3P_2P_1G_0 + P_3P_2P_1P_0C_i \end{aligned}$$

This gives all four carry-outs in terms of the inputs, which means they can be produced *as soon as the inputs are applied to the circuit*. Hence, there is no 'rippling' delay, although there will still be a delay given by the number of levels required to implement these expressions. (Two-level circuits could be used but, as shown in the following circuit diagram, other implementations are usually employed.) From the above it is clear that there is a distinct pattern for the carry-outs which can be used to continue this process further if required.

The use of $P$ and $G$ to simplify the above expressions was not an arbitrary choice and by looking again at the truth table for the carry-out, shown in Table 4.7, we can see their origin. By looping the minterms to produce the $AB$ product we note that a carry-out is *generated* (hence the use of $G$) if:

$$G = AB$$

that is if both inputs are 1. The remaining two minterms for $C_0$ ($\bar{A}BC$ and $A\bar{B}C$) show that a carry-in to the full adder (i.e. $C_i = 1$) is *propagated* (hence the use of $P$)

Table 4.7 Truth table and Karnaugh map for the carry-out bit, $C_{out}$, and its use in the look-ahead carry adder

| A | B | $C_i$ | $C_o$ | |
|---|---|---|---|---|
| 0 | 0 | 0 | 0 | |
| 0 | 0 | 1 | 0 | |
| 0 | 1 | 0 | 0 | |
| 0 | 1 | 1 | 1 | ← Propagate |
| 1 | 0 | 0 | 0 | |
| 1 | 0 | 1 | 1 | ← Propagate |
| 1 | 1 | 0 | 1 | ← Generate |
| 1 | 1 | 1 | 1 | ← Generate, Propagate |

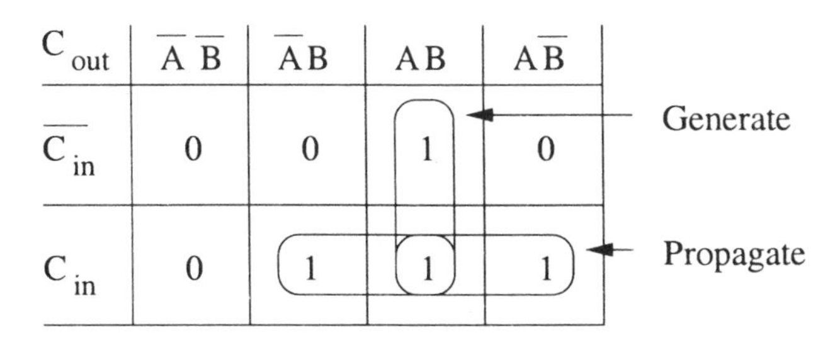

| $C_{out}$ | $\bar{A}\,\bar{B}$ | $\bar{A}B$ | $AB$ | $A\bar{B}$ |
|---|---|---|---|---|
| $\overline{C}_{in}$ | 0 | 0 | 1 | 0 |
| $C_{in}$ | 0 | 1 | 1 | 1 |



if either $A$ or $B$ are 1. So, these two minterms are covered by the Boolean expression:

$$AC_i + BC_i = (A+B) \cdot C_i = PC_i$$

where $P=(A+B)$. Note that this expression for $P$ means that the $ABC_i$ minterm is covered by both $G$ and $P$.[5]

**Implementation of the look-ahead carry adder**

The implementation of the four-bit look-ahead carry adder using the forms of the carry-outs derived above is shown in Fig. 4.16. As shown the circuit requires 19 gates with a maximum delay of four levels. Note: this does not include production of the final carry-out ($C_3$); that some gates have four inputs; and that this implementation requires four three-input XOR gates which we know is not a basic Boolean operator.

**A more practical implementation**

The four-bit look-ahead carry adder is available (as are many of the circuits we have already discussed) as a single integrated circuit (IC).[6] It is instructive to consider this circuit since it employs a different implementation that eliminates

[5]This fact is mentioned because some texts use $P=A \oplus B$ to explicitly exclude (from $PC_i$) the $ABC_i$ minterm, since it is included in the generate, $G$, term (as $AB=1$).

[6]This IC is the '283' four-bit adder which belongs to a family of logic devices. Such 'families' are discussed in more detail in Chapter 9.

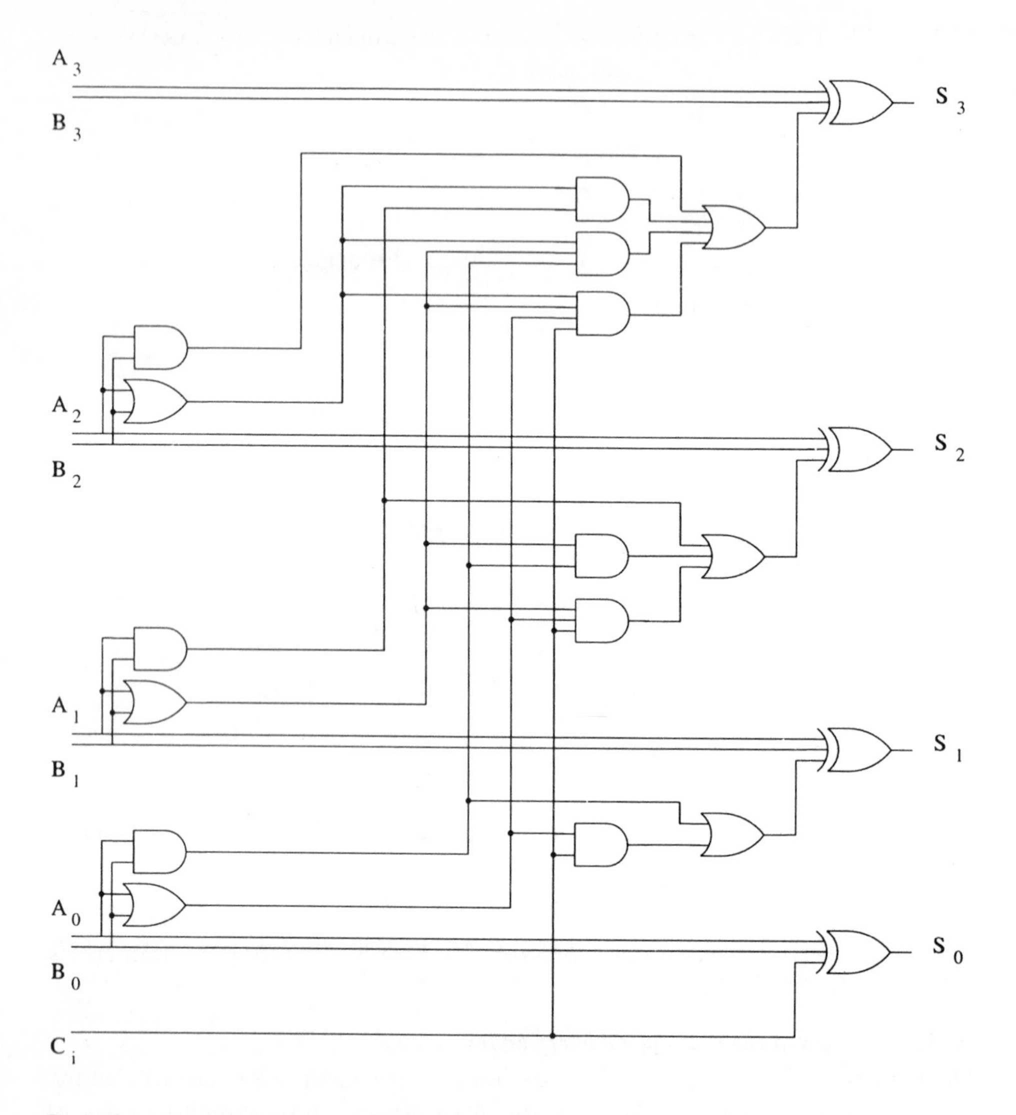


Fig. 4.16 A circuit for a four-bit look-ahead carry adder

the need for three-input XOR gates. Looking at how this is achieved serves as a useful exercise in Boolean algebra.

The sum of the two bits $A$ and $B$ can be written as:

$$\begin{aligned} A \oplus B &= A\bar{B} + \bar{A}B \\ &= (A+B)\cdot(\bar{A}+\bar{B}) && \text{distributive law} \\ &= (A+B)\cdot(\overline{AB}) && \text{De Morgan's theorem} \\ &= P\bar{G} \end{aligned}$$

This means that to produce the sum term, $S$, rather than use a three-input XOR gate, a two-input one fed with the above result (generated from the $P$ and

$G$ terms which are needed anyway for the look-ahead carry) and the carry-in can be used.

In addition:[7]

$$
\begin{aligned}
C_0 &= G_0 + P_0 \cdot C_i \\
&= P_0 G_0 + P_0 C_i \qquad \text{see footnote}^{7} \\
&= P_0 \cdot (G_0 + C_i) \\
&= \overline{\overline{P_0} + \overline{(G_0 + C_i)}} \qquad \text{De Morgan's theorem}
\end{aligned}
$$

Hence:

$$\overline{C_0} = \overline{P_0} + \overline{G_0}\,\overline{C_i}$$

Similarly it can be shown that

$$
\begin{aligned}
\overline{C_1} &= \overline{P_1} + \overline{G_1}\,\overline{C_0} \\
&= \overline{P_1} + \overline{G_1}(\overline{P_0} + \overline{G_0}\,\overline{C_i}) \\
&= \overline{P_1} + \overline{G_1}\,\overline{P_0} + \overline{G_1}\,\overline{G_0}\,\overline{C_i}
\end{aligned}
$$

From this it can be seen that a pattern is emerging, as before, but in this case for $\overline{C_j}$. To implement the circuit in this form requires 25 gates (excluding inverters) and has a longest delay, for $S_3$, of four levels. Although this is more gates than the previous implementation, only two-input XOR gates are needed. (Remember that three-input XOR gates actually implement quite a complex Boolean function.[8])

### 4.2.5 Summary

In this section we have considered four ways of implementing a combinational digital circuit, namely a four-bit binary adder. This serves to illustrate that there are always several ways any circuit can be designed. We firstly considered a two-level implementation in both fundamental and minimised sum of products form, which both proved impractical. (A product of sums approach would have had the same outcome.)

A heuristic approach was then tried via consideration of the mechanics of the addition process. This led directly to the ripple carry, or parallel, adder which produces a practical implementation with a reasonable number of gates, but suffers from the problem of a rippling carry which reduces the speed of the circuit.

Finally in order to overcome the problem of the rippling carry, we developed the look-ahead carry adder which calculates the carry-out at each stage using the initial inputs, by 'looking ahead'. This produces a faster design but does require more gates.

[7]Note that:

$$PG = (A+B)(AB) = AAB + ABB = AB + AB = AB = G$$

[8]This demonstrates that when comparing a circuit's relative complexity a simple gate count is not sufficient since some gates have a more complex construction than others. For example a NAND gate may actually be simpler than an AND gate.

## 4.3 HAZARDS

### 4.3.1 Introduction

So far we have usually considered the logic gates we have discussed to be ideal. One of the characteristics of such gates is that their outputs change instantaneously in response to changes in their inputs. In practice this is not the case because the actual circuit which performs the required operation of the gate cannot respond immediately, and so there will be an inevitable delay between a change in the inputs affecting the output.

These delays can lead to *race conditions*, so called because two signals, originally from the same source, may take different paths through the circuit and so end up 'racing' each other. *If* one wins then this may lead to a *hazard* which basically means the output may have the wrong value for a short period of time (until the other signal catches up).

The basic idea is introduced in Fig. 4.17 which shows how, in practice, the response to a change in the input, $A$, of the output, $Y$, from a NOT gate will be delayed by a small amount. This delay, labelled $\tau_{PHL}$, is the time it takes for the circuit's output to go from a HIGH to LOW state (1 to 0), whilst $\tau_{PLH}$ is the time taken for the output to go from a LOW to HIGH state. In general these will not be the same, and will depend upon how the analogue circuit implementing the gate actually operates.

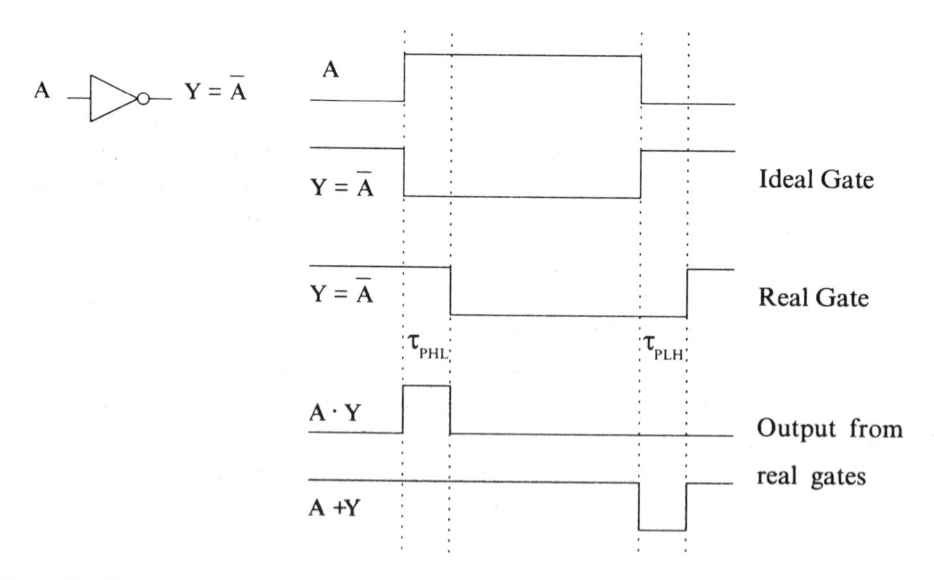


Fig. 4.17 The output from a real NOT gate

An important point to note, since it is the fundamental cause of some hazards, is that because of this delay $A \cdot \bar{A} \neq 0$, since the delay in the output falling to 0 means that for $\tau_{PHL}$ both $A$ and $Y = \bar{A}$ (from the real gate) are 1, and therefore

$A \cdot Y = 1$. Also $A + \bar{A} \neq 1$ since the delay in the output rising to 1 means that for $\tau_{PLH}$ both $A$ and $Y = \bar{A}$ are 0 and so $A + Y = 0$.

The approach adopted in this section is to firstly investigate examples of hazards in a simple and intuitive manner, and then to look into their cause in more detail.

### 4.3.2 Static hazards

Consider the circuit shown in Fig. 4.18 where the NOT gate is considered real (and so has a delay) but the XNOR gate is ideal. (Although an artificial situation this serves to illustrate all of the important problems of real circuits.) We know that the NOT gate will delay the signal $\bar{A}$ into the XOR gate, and so for a short period of time after a rising edge input $A \cdot \bar{A} \neq 0$, and similarly after a falling edge, $A + \bar{A} \neq 1$.

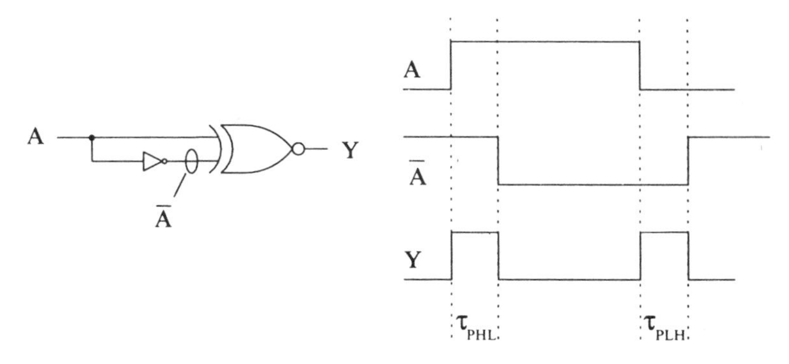


Fig. 4.18 An example of a static-0 hazard

The effects of these anomalies are shown in the timing diagram in Fig. 4.18 and are that every time $A$ changes state a short pulse (equal to $\tau_{PHL}$ or $\tau_{PLH}$) is produced at the output of the XNOR gate. This is because $A$ and $\bar{A}$ are *not* always the complement of each other, which the XNOR gate responds to with an output of 1. These short pulses are known as *spikes* or *glitches*, which is a general name for any unwanted short pulses that appear in a digital electronics circuit.

This non-ideal operation of the circuit is an example of a *static hazard* which refers to two changes in the logic level of a digital signal when none is expected. (There are two changes because the signal should ideally stay at 0 but rather goes first to 1 and then back to 0.) This is in fact a static-0 hazard because the signal was expected to stay at 0. Its occurrence is due to the fact that the input $A$ traces two different paths through the circuit (one straight into the XNOR gate and the other via the NOT gate) and so there is the opportunity for race conditions. It is important to realise that when $A$ changes state, eventually this will be reflected in all paths traced by $A$ changing state, and so the output goes to its expected value. The hazard occurs because the different paths traced by $A$ attain their final values at different times because of the race.

## Example 4.11

What type of hazard is obtained if the XNOR gate in Fig. 4.18 is replaced by an XOR gate.

### *Solution*

The circuit and associated timing diagrams are shown in Fig. 4.19. These show that every change in the input produces a static-1 hazard in the output.

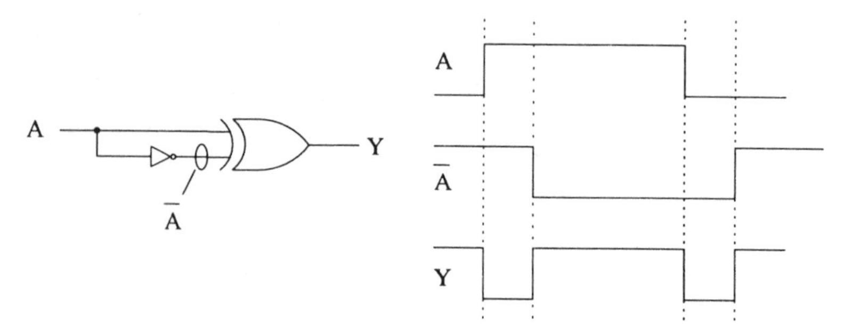


Fig. 4.19 Static-1 hazard produced by a delay into an XOR gate (see Example 4.11)

We can look at the cause of the static-0 hazard from the XNOR based circuit (Fig. 4.18) using Boolean algebra. The output from the circuit is:

$$\begin{aligned} Y &= \overline{A \oplus \overline{A}} \\ &= \overline{A\overline{\overline{A}} + \overline{A}\,\overline{A}} \\ &= \overline{AA + \overline{A}\,\overline{A}} \\ &= \overline{A + \overline{A}} \\ &= \overline{1} \qquad \text{note the use of } A + \overline{A} = 1 \\ &= 0 \end{aligned}$$

This shows that for an ideal gate the output $Y$ will always be 0 because $A + \overline{A} = 1$. However, with a real NOT gate this cannot be guarateed (because of the propagation delay of the gate) and so the hazard occurs when this condition is not met.

### A practical use of hazards

Although not particularly recommended *unless* the propagation delay of the NOT gates being used can be guaranteed, and even then not an ideal method of logic design, the hazards that occur in circuits can be used to generate pulses from an edge signal.

We have already seen how this is achieved in Fig. 4.17 where it can be seen that if the input and output from the NOT gate are fed through an AND gate a positive pulse will be produced on the input's leading edge. If an OR gate is used a negative pulse will be produced on the input's falling edge. Using a NAND or NOR gate will give the opposite polarity pulses from the same edges.

**Example 4.12**

What simple circuit will produce a short negative pulse on the input of a rising edge. (Note that this is *not* a recommended design procedure.)

### *Solution*

It can be seen from Fig. 4.17 that a rising edge is delayed upon passing through an inverter, and so if AND'd with the circuit's input gives a glitch since $A \cdot \overline{A} \neq 0$. Using a NAND gate will produce the required negative pulse. (Note that to produce similar pulses from a falling edge an (N)OR gate must be used since then $A + \overline{A} \neq 1$.

Although hazards are used in this manner by many digital designers, it is not to be recommended and alternative design techniques, as discussed in Section 11.3.1, should be employed.

**The multiplexer hazard**

The 2-to-1 multiplexer provides an ideal circuit to demonstrate the problem of static hazards in a more complex circuit. The output of the multiplexer shown in Fig. 4.20 is:

$$Y = AC + B\overline{C}$$

where $C$ is the control line. So if $C=1$, $Y=A$ or alternatively $C=0$ and $Y=B$.

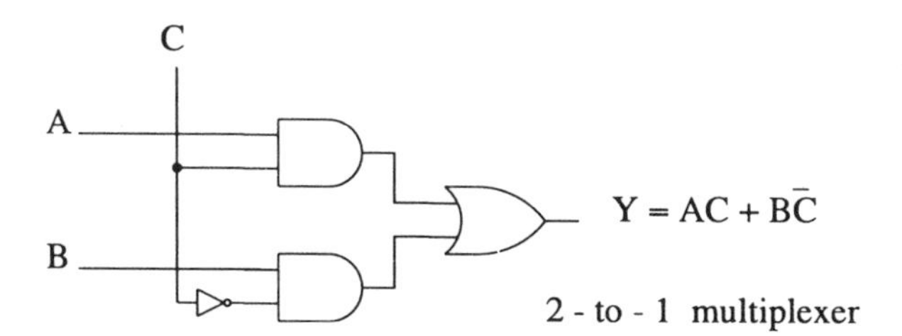


Fig. 4.20 Circuit diagram of a 2-to-1 multiplexer

We know from the simple examples above that if a hazard occurs it will be because of the two paths traced by $C$ (a race), which are then eventually OR'd together (the multiplexer is in sum of products two-level form). The error will occur because $C + \overline{C} \neq 1$.

In order to get $Y = C + \overline{C}$ we need $A=1$ and $B=1$, and so anticipate that any hazard will occur under these conditions when $C$ changes state. The timing diagram in Fig. 4.21 demonstrates that a static-1 hazard is indeed produced. (It is assumed all gates, except the NOT gate, are ideal (no delay or variable delays on any paths through them) which in no way affects the important findings.) As $C$ changes, its complement through the NOT gate is delayed, consequently when this and $C$ are OR'd together we get the negative pulse we have seen previously. It is a static-1 hazard as $C$ goes LOW.

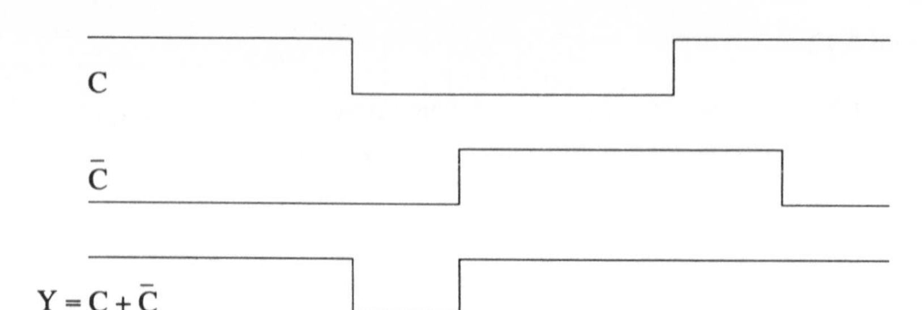


Fig. 4.21 Output from the 2-to-1 multiplexer, for inputs of $A=1$ and $B=1$ with $C$ changing state, demonstrating the occurrence of the static-1 hazard

**Hazards and Karnaugh maps**

The reason for the occurrence of this hazard in the 2-to-1 multiplexer can be seen by considering the Karnaugh map and its use in minimising the Boolean expression for the circuit. The Karnaugh map for the 2-to-1 multiplexer is shown in Table 4.8. There are three prime implicants which if all are used gives:

$$Y = AC + AB + B\overline{C}$$

Previously in this section we have not used the non-essential prime implicant $AB$.

Table 4.8 Karnaugh map for the output from a 2-to-1 multiplexer

| Y | $\overline{A}\overline{B}$ | $\overline{A}B$ | $AB$ | $A\overline{B}$ |
|---|---|---|---|---|
| $\overline{C}$ | 0 | 1 | 1 | 0 |
| $C$ | 0 | 0 | 1 | 1 |

If instead of minimising directly from the Karnaugh map, we use Boolean algebra to minimise the expression containing the three prime implicants we gain an important insight into the origin of hazards and how to prevent them.

Minimising:

$$\begin{aligned} Y &= AC + AB + B\overline{C} \\ &= AC + AB(C + \overline{C}) + B\overline{C} \quad \text{note the use of } 1 = C + \overline{C} \\ &= AC + ABC + AB\overline{C} + B\overline{C} \\ &= AC(1 + B) + B\overline{C}(A + 1) \\ &= AC + B\overline{C} \end{aligned}$$

The important point is that this minimisation depends upon the use of the Boolean identity $C + \overline{C} = 1$, which we know, because of the action of the NOT gate, is not always true and may introduce a hazard. This suggests how we may eliminate the static hazard, which is to include the non-essential prime implicant so that the expression for $Y$ will not then have been minimised using this identity.

To see the effect of including the $AB$ term, we note that if we use

$$Y = AC + AB + B\overline{C}$$

then if $((A=1)$ AND $(B=1))$ (the conditions which lead to the hazard, and define the non-essential prime implicant) this expression becomes:

$$Y = C + 1 + \overline{C} = 1$$

(since anything OR'd with 1 is 1).

So the inclusion of the $AB$ non-essential prime implicant (whose elimination relies upon the use of $C+\overline{C}=1$ during minimisation) cures the problem of the static hazard. The circuit for this form of the multiplexer is shown in Fig. 4.22; the extra AND gate forming the product AB is sometimes referred to as a blanking, or holding, gate.

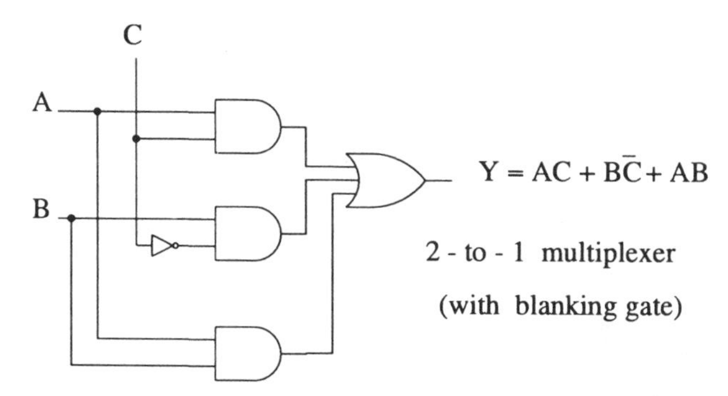


Fig. 4.22 2-to-1 multiplexer with holding gate which has no static hazards

As a general rule, static hazards can be eliminated by the inclusion of non-essential prime implicants in the 'minimised' Boolean expression. More specifically they will be non-essential prime implicants whose 'removal' relies upon the use of $X+\overline{X}=1$ where $X$ is a variable for which a race condition may exist.

## Example 4.13

Derive the product of sums form of a 2-to-1 multiplexer and then, performing the corresponding analysis to that for the sum of products form, determine whether any static hazards occur, and if they do how they may be eliminated.

### *Solution*

From Table 4.8 looping and grouping the zeros gives:

$$\overline{Y} = \overline{A}C + \overline{B}\overline{C}$$

Dualling gives the required product of sums form of:

$$Y = (A + \overline{C}) \cdot (B + C)$$

Due to the final product produced we anticipate a hazard if both racing versions of $C$ reach the AND gate. For a hazard to occur requires $C\cdot\bar{C}$ which needs both $A=0$ and $B=0$. This will be, as we saw in Fig. 4.17, a static-0 hazard produced as $C$ goes high.

Using Boolean algebra to confirm this, from the Karnaugh map using *all* 'prime implicants' for $\bar{Y}$:

$$\bar{Y}=\bar{A}C+\bar{B}\bar{C}+\bar{A}\bar{B}$$

Dualling to get the product of sums form, and using the fact that $C\cdot\bar{C}=0$:

$$\begin{aligned}
Y&=(A+\bar{C})\cdot(B+C)\cdot(A+B) \\
&=(A+\bar{C})\cdot(B+C)\cdot(A+B+\bar{C}\cdot C) && \text{using } 0=\bar{C}\cdot C \\
&=(A+\bar{C})\cdot(A+B+\bar{C})\cdot(B+C)\cdot(A+B+C) && \text{using Equation 1.15} \\
&=(A+\bar{C})\cdot(1+B)\cdot(B+C)\cdot(1+A) && \text{using Equation 1.14} \\
&=(A+\bar{C})\cdot(B+C)
\end{aligned}$$

So the minimisation process relies upon the fact that $C\cdot\bar{C}=0$ which is where the hazard arises from. The equivalent 'blanking gate' in this product of sums implementation is $(A+B)$ since for a hazard to occur both $A$ and $B$ must be 0 meaning $A+B=0$. This gate will hold the output, $Y$, low, thus preventing the static-0 hazard. Note the similarity between the sum of products and product of sums forms which is again a consequence of duality.

**A more rigorous approach**

We now consider a more rigorous analysis of the 2-to-1 multiplexer circuit, which, although it offers no more understanding of the hazards in that circuit, serves as an introduction to a method that can be applied generally to other types of hazard. We know that any hazard in the sum of products multiplexer circuit will arise because of the fact that $C+\bar{C}\neq 1$, and therefore will occur when $C$ changes state. In addition for a static hazard to occur the change in $C$ must *not* cause a change in the output, Y.[9]

We firstly draw up the truth table for the circuit including columns with the value that $C$ will change to (i.e. $\bar{C}$ as this is a digital circuit), denoted by $C^+$ and the value of $Y$ this will give, $Y^+$, as shown in Table 4.9. From this we see that there are

Table 4.9 Truth table required for the rigorous analysis of potential hazards in a multiplexer

| $A$ | $B$ | $C$ | $C^+$ | $Y$ | $Y^+$ |
|---|---|---|---|---|---|
| 0 | 0 | 0 | 1 | 0 | 0 |
| 0 | 0 | 1 | 0 | 0 | 0 |
| 0 | 1 | 0 | 1 | 1 | 0 |
| 0 | 1 | 1 | 0 | 0 | 1 |
| 1 | 0 | 0 | 1 | 0 | 1 |
| 1 | 0 | 1 | 0 | 1 | 0 |
| 1 | 1 | 0 | 1 | 1 | 1 |
| 1 | 1 | 1 | 0 | 1 | 1 |

[9] If the output does change then the only effect of the race will be to delay this change.

four input conditions for which when $C$ changes state the output, $Y$, will remain the same and so a static hazard may occur. These are $\bar{A}\bar{B}\bar{C}$, $\bar{A}\bar{B}C$, $AB\bar{C}$ and $ABC$ (i.e. the conditions when $A=B$).

When $C$ changes, one way of considering its propagation through the circuit is that it consists of two signals, $C_1$ and $C_2$, one of which forms the product $AC_1$ and the other $B\bar{C}_2$ (which are then summed). Both of these signals must eventually have the same value, but because of the race condition they may transiently be different. What this means is that if $C$, and therefore $C_1$ and $C_2$, is 0, and then changes, then either $C_1$ or $C_2$ may change first. Eventually they will both be 1 but the change may take place as either:

$$(C_1\ C_2)=(0, 0), (0, 1), (1, 1) \quad \text{or} \quad (0, 0), (1, 0), (1, 1)$$

Similarly in $C$ changing from 1 to 0 there are two possible 'routes' depending upon which of $C_1$ or $C_2$ changes first.

We can draw up a kind of truth table which allows us to see the effect of these transient values of $C_1$ and $C_2$ on the output $Y$. We do so for expressions for $Y$ using the values of $A$ and $B$ we have identified from above as likely to lead to static hazards. This table is shown in Table 4.10. The four rows hold the initial and end values (rows 1 and 4) of $C_1$ and $C_2$ when they will be the same, whilst rows 2 and 3 hold the transient conditions when they may differ. We have already considered how in changing $C$ from 0 to 1 the transition may take place via row 2 or 3 depending upon whether $C_2$ or $C_1$ changes first, whilst for $C$ changing from 1 to 0 the transition will be via row 2 if $C_1$ changes first, else via row 3.

Table 4.10 Table used to find hazards in the 2-to-1 multiplexer. The loops and arrows indicate a change in $C$ from 1 to 0 via row 2 which leads to the static-1 hazard in $Y$

| | | | A = 0<br>B = 0 | A = 1<br>B = 1 |
|---|---|---|---|---|
| | $C_1$ | $C_2$ | $Y = 0$ | $Y = C_1 + \bar{C}_2$ |
| row 1 | 0 | 0 | 0 | 1 |
| row 2 | 0 | 1 | 0 | 0 |
| row 3 | 1 | 0 | 0 | 1 |
| row 4 | 1 | 1 | 0 | 1 |

The last two columns hold the values of $Y$ showing how they depend upon $C_1$ and $C_2$ for when (($A=0$) AND ($B=0$)), and (($A=1$) AND ($B=1$)), the previously identified conditions when a static hazard may occur since the output, $Y$, should remain unchanged. For the first condition $Y=0$ for all values of $C$, therefore it does not matter which of $C_1$ or $C_2$ changes first since the output remains at 0. However, for $A=1$ and $B=1$, and hence $Y=C_1+\bar{C}_2$, we can anticipate a problem, and see that if when $C$ changes from 0 to 1, $C_2$ changes first, then the circuit transiently enters row 2 and the output will momentarily go to 0, giving a static-1 hazard. Similiarly when $C$ goes from 1 to 0, if $C_1$ changes first then row 2 will again be entered giving the same error. Using this different approach we have again found the conditions that lead to the static hazard, that is $C$ changing when both $A$ and $B$ are 1.

As the final point we note that because it is $C_2$ that goes through the NOT gate it will be this signal that is delayed and therefore $C_1$ will always change first. Consequently we will only see the possible static-1 hazard indicated by the above analysis when $C$ is changing from 1 to 0. (It is for the same reason that the inverter and OR gate circuit gives a negative going pulse on a negative, and not a positive, edge).

**More complex hazards**

That concludes our look at static hazards. However, if a hazard can be produced by a signal changing state twice when it should remain the same, could we not also have an error when a signal changes three times instead of just once? That is we get the sequence 0101 being output instead of 01. This could possibly happen for the four input combinations to the multiplexer when the output $Y$ does change as $C$ is changed (see Table 4.9) which in this case is for when $A$ and $B$ have different values.

However, further consideration of the necessary conditions to produce this type of hazard shows that it cannot happen when the signal producing the race condition only has two paths through the circuit. When there are only two paths we get a two-variable (i.e. the two paths $C_1$ and $C_2$) 'truth table' (e.g. Table 4.10) and so only one transient state can be visited before arriving at the eventual values for the two signal paths (e.g. row 1 to row 2 to row 4). To produce a hazard where there are three changes of state clearly requires a 'truth table' with more rows and hence at least three 'input variables', that is three paths for the same signal through the circuit. This brings us on to the subject of *dynamic hazards.*

## 4.3.3 Dynamic hazards

Dynamic hazards can occur when a signal has three or more paths through a combinational logic circuit. Their effect is to cause a signal which is expected to change state to do so, then transiently change back to the original state, before making the final transition to the expected state (e.g. the signal gives 1010 rather than just 10). The analysis of a circuit for a dynamic hazard is essentially a continuation of the 'rigorous approach' for static hazards described above. You should therefore be familiar with this material before continuing.

Consider the circuit shown in Fig. 4.23 together with its truth table and Karnaugh map. From this implementation we get:

$$Y=(\overline{B}+\overline{C})\cdot(AC+B\overline{C})$$

and note that $C$ has three paths through the circuit: via gates 1, 5; gates 2, 4, 5; and gates 3, 4, 5. Therefore there is the possibility of race conditions in three paths which may lead to a dynamic hazard.

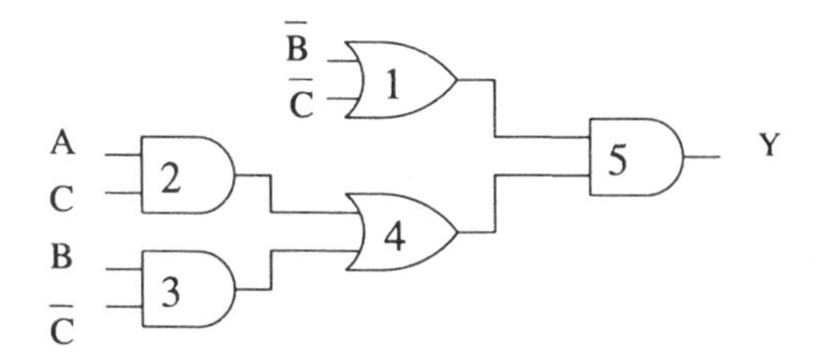


| Y | $\bar{A}\bar{B}$ | $\bar{A}B$ | $AB$ | $A\bar{B}$ |
|---|---|---|---|---|
| $\bar{C}$ | 0 | 1 | 1 | 0 |
| $C$ | 0 | 0 | 0 | 1 |

| $A$ | $B$ | $C$ | $Y$ |
|---|---|---|---|
| 0 | 0 | 0 | 0 |
| 0 | 0 | 1 | 0 |
| 0 | 1 | 0 | 1 |
| 0 | 1 | 1 | 0 |
| 1 | 0 | 0 | 0 |
| 1 | 0 | 1 | 1 |
| 1 | 1 | 0 | 1 |
| 1 | 1 | 1 | 0 |

Fig. 4.23 Combinational logic circuit, its Karnaugh map and truth table, which may possess a dynamic hazard due to the three paths for signal $C$ through the circuit

## Example 4.14

Use Boolean algebra to prove that this implementation of the circuit is functionally equivalent to the minimised expression obtained from the Karnaugh map.

### *Solution*

$$
\begin{aligned}
Y &= (\bar{B}+\bar{C})\cdot(AC+B\bar{C}) \\
&= A\bar{B}C+\bar{B}B\bar{C}+AC\bar{C}+B\bar{C}\bar{C} \\
&= A\bar{B}C+B\bar{C} \qquad \text{as from the Karnaugh map}
\end{aligned}
$$

From the truth table we see that for inputs of ($A$, $B$) of (0,1), (1,0) and (1,1), the output, $Y$, changes as $C$ changes. Therefore there is the possibility of a dynamic hazard as the output may change state, then go back to the initial value, before changing again to the final expected output.

Continuing our analysis we see that if $B=0$ then $\bar{B}=1$ and so the output, $(1+\bar{C})$, from gate 1 is always 1. This means $C$ now only has two paths through the circuit and so a dynamic hazard is not possible. Similarly, if $A=0$ then the output from gate 2 must be 0 and so again there are only two paths for $C$ through the circuit.

This leaves inputs of $A=1$ and $B=1$ to consider. (Remember it is $C$ changing that will lead to any hazard so we are looking at the effect of this whilst the other inputs are fixed.) Using a subscript, corresponding to the first gate passed through, for the three possible paths through the ciricuit, for $A=1$ and $B=1$ we write:

$$Y=(\bar{B}+\bar{C}_1)\cdot(AC_2+B\bar{C}_3)$$
$$=(0+\bar{C}_1)\cdot(1\cdot C_2+1\cdot\bar{C}_3)$$
$$=\bar{C}_1\cdot(C_2+\bar{C}_3)$$

Remember that the reason we get hazards is that transiently these three values for $C$ may not be equal (although they will be once the races have finished), which is why we are treating them as three separate variables to determine the transient values of $Y$. The truth table for the transient values of $Y$ as a function of the three values of $C$ is shown in Table 4.11.

Table 4.11 Truth table for the transient states of the circuit in Fig. 4.23

| row | $C_1$ | $C_2$ | $C_3$ | $Y$ |
|---|---|---|---|---|
| 1 | 0 | 0 | 0 | 1 |
| 2 | 0 | 0 | 1 | 0 |
| 3 | 0 | 1 | 0 | 1 |
| 4 | 0 | 1 | 1 | 1 |
| 5 | 1 | 0 | 0 | 0 |
| 6 | 1 | 0 | 1 | 0 |
| 7 | 1 | 1 | 0 | 0 |
| 8 | 1 | 1 | 1 | 0 |

This truth table shows that when the circuit has stabilised (all signals have affected the output) then all the values of $C$ will be 0 or 1. Therefore the first and last rows of this truth table correspond to inputs of $AB\bar{C}$ and $ABC$ in the truth table of Fig. 4.23.

The final stage in our analysis is to use the truth table in Table 4.11 to see if any dynamic hazards do occur. We know that for the possibility of a dynamic hazard both $A$ and $B$ must be 1 with $C$ then changing state. This will correspond to moving from either the top to the bottom (for $C$ going from 0 to 1) or the bottom to the top row of the truth table in Table 4.11.

Now, since (because of the races) there are effectively three variables $C$ which all must change, there are many possibilities as to which rows of the truth table will be visited transiently. For instance for $C$ changing from 0 to 1, then if $C_3$ changes first, then $C_2$ followed by $C_1$, then the circuit will move from row 1 to 8, transiently visiting rows 2 and 4. This will give transient outputs of 0 and 1 (for rows 2 and 4) and hence a possible dynamic hazard since the output will give 1010. This is the only possible hazard for $C$ going from 0 to 1 because of the need for the output to go to 0 first to give the incorrect transient state.

For $C$ changing from 1 to 0, we note the possible dynamic hazard if the transition is from row 8 to 1 via rows 4 and 2 (the reverse route to the above). Again this is the only possibility since we need the output, $Y$, to go HIGH first (so it can go LOW again to give the hazard). The output here will be 0101.

So we have identified two possible situations in which a dynamic hazard may occur, depending upon the relative speed of propagation of $C_1$, $C_2$ and $C_3$. For the first possible hazard $C_3$ must change first then $C_2$, and for the second $C_1$ first and then $C_2$. Now since $C_1$ only travels through two levels (gates 1 and 5) it will change faster than both $C_2$ and $C_3$. This rules out the first possible hazard as being likely, leaving only the second. This will only occur if, assuming $C_1$ does change first, $C_2$ then changes before $C_3$ (to give the transient output of 0 from row 2).

The timing diagram in Fig. 4.24 illustrates how this dynamic hazard will occur for these conditions. Note that $C$ must go from 1 to 0 and so $\bar{C}_1$ will do the opposite, as shown in the figure. This concludes our analysis of this circuit for dynamic hazards. (Note that we have *not* considered whether any static hazards are present in this circuit.)

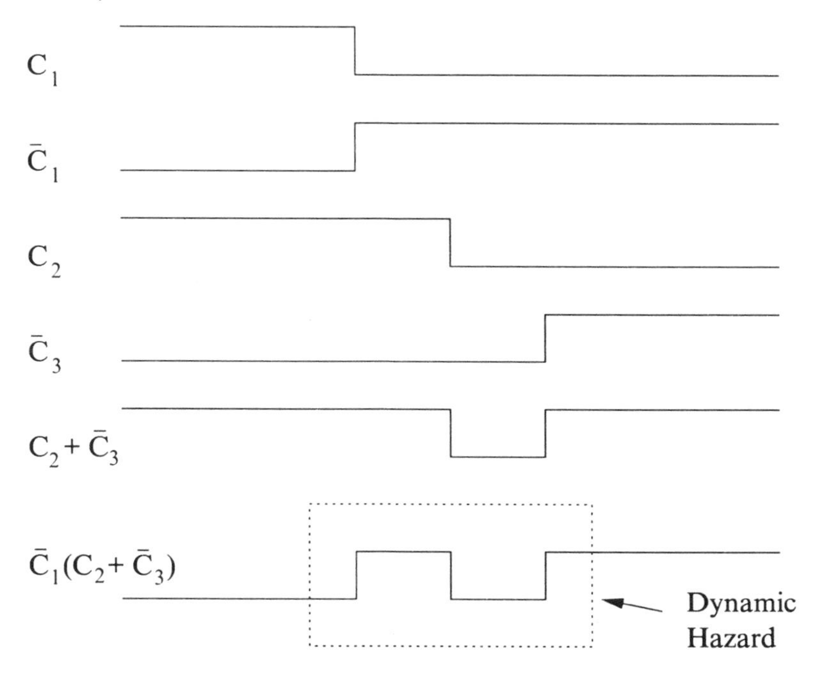


Fig. 4.24 Timing diagram illustrating the occurrence of the dynamic hazard in the circuit shown in Fig. 4.23. For the hazard to occur $C_1$ must change first (which can be assumed the case as it only has a two-gate delay), followed by $C_2$ and then $C_3$

### 4.3.4 Summary

The aim of this section was to introduce some of the problems that can be encountered when transferring combinational logic designs into practical circuits. In particular, we have seen how the finite time it takes a digital signal to pass through a logic gate (the *propagation delay*) can cause problems if a logic signal has two or more paths through the circuit along which the signals can 'race' each other (*race conditions*).

If there are two paths then *static hazards* may occur (where a signal changes state twice instead of not at all), whereas for three or more paths *dynamic hazards* may occur (where the signal changes place three times rather than once). Both types of hazard give rise to a short transient pulse (spike).

We have determined the cause of these hazards using Boolean algebra, and how this allows them to be predicted and, to a certain extent, overcome. However, in an introductory text such as this there is only room for limited examples. So although those given demonstrate clearly the principles behind the occurrence of hazards, in practice their prediction and elimination, although based upon the methods presented here, are somewhat more complex. Many of the ideas introduced here will be revisited, in later chapters, regarding the design of error-free asynchronous sequential circuits.

In practice hazards can cause particular problems if they occur in asynchronous circuits (Chapter 5) or when driving the clock lines of synchronous circuits (see Chapter 6). In other circumstances, when driving non-clock lines, the transient conditions resulting from hazards can be 'overcome' simply by delaying the sampling of such outputs until they have stabilised.

## 4.4 SELF-ASSESSMENT

4.1 What is the function of an 8-to-1 multiplexer?

4.2 Why can a multiplexer be used a a 'universal logic block'?

4.3 What does a demultiplexer do?

4.4 Why can a decoder be constructed from a demultiplexer?

4.5 What three types of combinational logic circuits can an XOR gate be used to construct?

4.6 What does a full adder do?

4.7 What type of circuit is a ripple carry adder, what basic unit is it built from, and what is its major disadvantage?

4.8 What advantage does the look-ahead carry adder have over the ripple carry adder?

4.9 What is the fundamental cause of hazards in combinational logic circuits?

4.10 What is a static hazard; what causes it and what are the two types?

4.11 What is a dynamic hazard and what causes it?

4.12 How can static hazards be overcome?

## 4.5 PROBLEMS

4.1 (a) How could a 1-of-8 multiplexer be used to generate functions $X$, $Y$ shown in Table 4.12?

(b) How could a 1-of-4 multiplexer plus one other gate be used for the same purpose?

Table 4.12 Truth table to be implemented using a multiplexer in Problem 4.1

| $A$ | $B$ | $C$ | $X$ | $Y$ |
|---|---|---|---|---|
| 0 | 0 | 0 | 1 | 1 |
| 0 | 0 | 1 | 0 | 1 |
| 0 | 1 | 0 | 1 | 0 |
| 0 | 1 | 1 | 0 | 1 |
| 1 | 0 | 0 | 0 | 0 |
| 1 | 0 | 1 | 0 | 0 |
| 1 | 1 | 0 | 1 | 1 |
| 1 | 1 | 1 | 1 | 0 |

4.2 If the data and select variables are only available in their uncomplemented forms how many two-input NAND gates would be needed to construct:
(a) a 4-to-1 multiplexer and
(b) a 1-of-4 demultiplexer.

4.3 Design the necessary combinational logic circuits to act as:
(a) an active-LOW BCD-to-decimal decoder
(b) an active-HIGH BCD-to-7-segment decoder
Assume for both decoders that the outputs for inputs greater than nine will never be input to the circuit and so can be used as 'don't care' conditions. Note that for the BCD-to-7-segment decoder some outputs are active for more than one input condition. The layout of a 7-segment display is shown in Fig. 4.25.

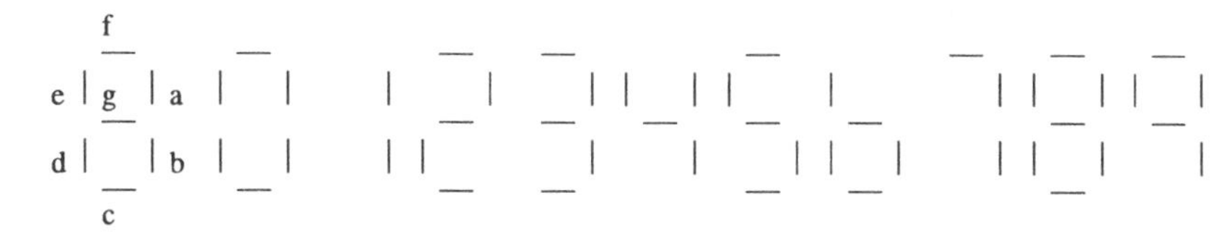


Fig. 4.25 Layout of a 7-segment display; see Problem 4.3

4.4 Assuming that only one of the inputs is ever high at any one time, give the Boolean expressions for the outputs from an active-LOW decimal-to-BCD encoder.

4.5 Design a two-level positive logic decimal-to-BCD priority encoder for decimal inputs from 0 to 4.

4.6 Fig. 4.26 shows a common implementation of a combinational logic circuit. What single gate does it represent?

4.7 How could changing a single gate in a parity checker used for a four-bit word (three data and one parity bit) constructed for two-input XOR gates be converted into a comparator for use on two-bit words?

4.8 Fig. 4.27 shows *part* of a standard design for a common digital circuit. Write down the Boolean functions produced at $D$ and $E$, and then convert them to a form which demonstrates directly the purpose of this circuit.

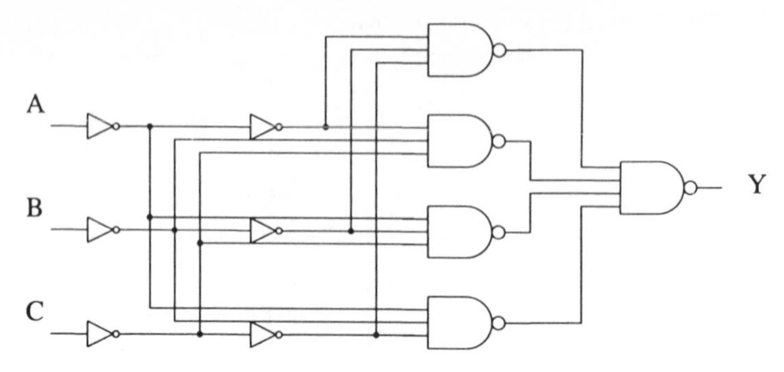


Fig. 4.26 Circuit to be analysed in Problem 4.6

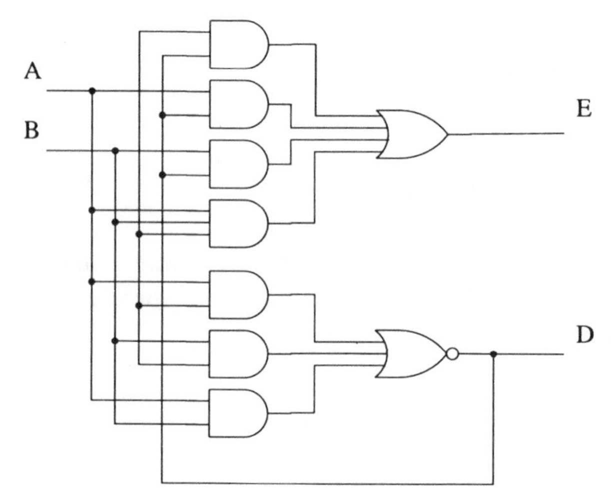


Fig. 4.27 Circuit to be analysed in Problem 4.8

4.9 How can a parallel (ripple carry adder) be converted to a parallel subtractor using XOR gates. (Hint: use two's complement subtraction.)

4.10 Table 4.13 shows part of the truth table describing the operation of the '181' Arithmetic Logic Unit (ALU) integrated circuit shown in Fig. 4.28. An input, $M$ of 0 and 1 means the ALU is operating in Boolean and arithmetic modes respectively; input $C$ is the carry bit.

Table 4.13 Truth table showing some of the functions performed by the '181' ALU (see Problem 4. 10)

| $S_3$ | $S_2$ | $S_1$ | $S_0$ | $M$ | $C$ | $F$ |
|---|---|---|---|---|---|---|
| 0 | 0 | 0 | 0 | 1 | – | $\overline{A}$ |
| 1 | 0 | 0 | 1 | 1 | – | $\overline{A \oplus B}$ |
| 1 | 0 | 1 | 1 | 1 | – | $AB$ |
| 1 | 1 | 1 | 1 | 1 | – | $A$ |
| 1 | 0 | 0 | 1 | 0 | 0 | *A plus B* |
| 0 | 1 | 1 | 0 | 0 | 0 | *A minus B* |

Confirm that for the given bit patterns in Table 4.13 the stated functions are implemented at the output, *F*. (Note that this circuit can perform further functions than these.)

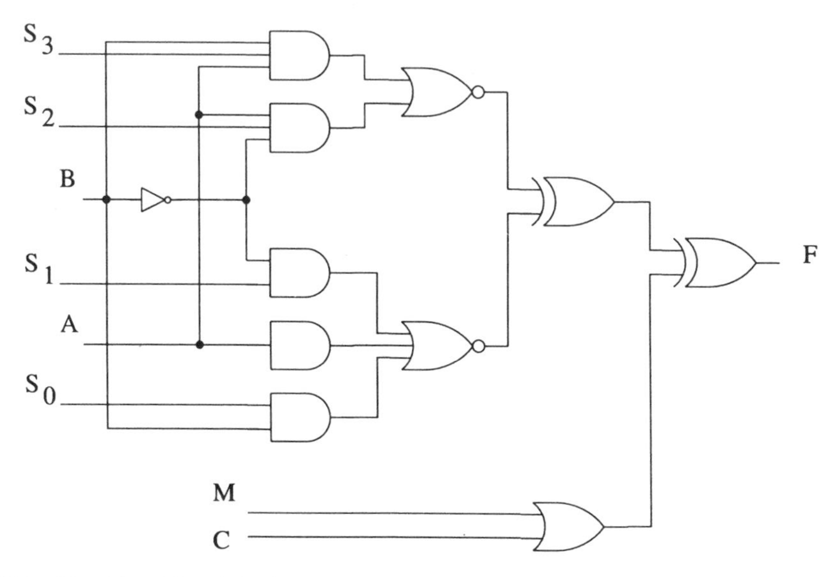


Fig. 4.28 The '181' arithmetic logic unit to be analysed in Problem 4.10

4.11 What is the output of the circuit in Fig. 4.29 for inputs *ABCD* of 1110; 1011; 0101 and 0010? What function does it perform?

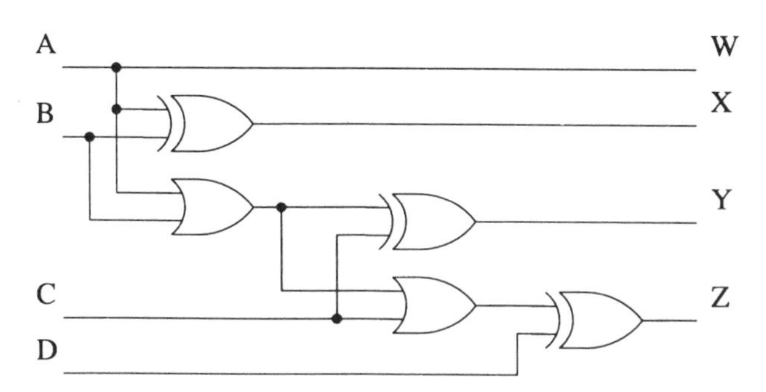


Fig. 4.29 Circuit to be analysed in Problem 4.11

4.12 The design of a circuit to perform binary multiplication could be approached by writing out the necessary truth table. Do this for the multiplication of two two-bit words.

In general if an *m*-bit and *n*-bit number are multiplied together, then (in terms of *m* and *n*) how many input and output columns, and rows, will the truth table have?

4.13 Devise a circuit which will produce a short positive going pulse on the input of a falling edge. Is this type of design to be recommended?

4.14 A function of three variables is minimised to $Y=A\bar{B}+\bar{A}C$. Draw the Karnaugh map of this function and state whether you would expect any static hazards to occur, and if so under what input conditions this would be.

If the circuit is *not* hazard free, how could it be made so?

# 5 Asynchronous sequential logic

## 5.1 SEQUENTIAL LOGIC CIRCUITS: AN OVERVIEW

All of the circuits so far considered in this book have been *combinational*. This means that their outputs are dependent *only* upon the inputs at that time. We now turn our attention to *sequential logic circuits* whose outputs are also dependent upon *past* inputs, and hence outputs. Put another way, the output of a sequential circuit may depend upon its previous outputs and so in effect has some form of 'memory'.

**General form of a sequential circuit**

Sequential circuits are essentially combinational circuits with feedback. A block diagram of a generalised sequential circuit is shown in Fig. 5.1. The generalised circuit contains a block of combinational logic which has two sets of inputs and two sets of outputs. The inputs[1] are:

- $A$, the present (external) *inputs* to the circuit;
- $y$, the *inputs* fed back from the outputs;
- $Z$, the present (external) *outputs* from the combinational circuit;
- $Y$, the *outputs* that are fed back into the combinational circuit.

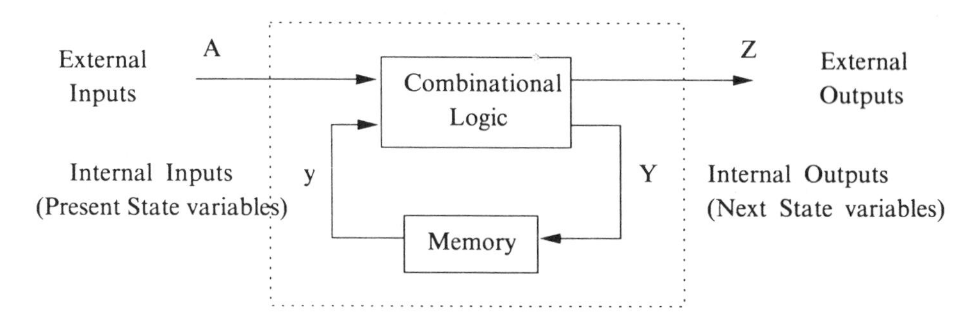


Fig. 5.1 The general form of a sequential logic circuit

Note that the outputs, $Y$, are fed back via the memory block to become the inputs, $y$, and that $y$ are called the 'present state' variables because they determine

[1]The letters (e.g. $A$, $y$) represent, in general, a number of inputs.

the current state of the circuit, with $Y$ the 'next state' variables as they will determine the next state the circuit will enter.

It is often useful to think in terms of two independent combinational circuits, one each for the two sets of outputs, $Z$ (external) and $Y$ (internal), as shown in Fig. 5.2. Both of these outputs will in general depend upon the external, $A$, and internal, $y$, (fed back) inputs.

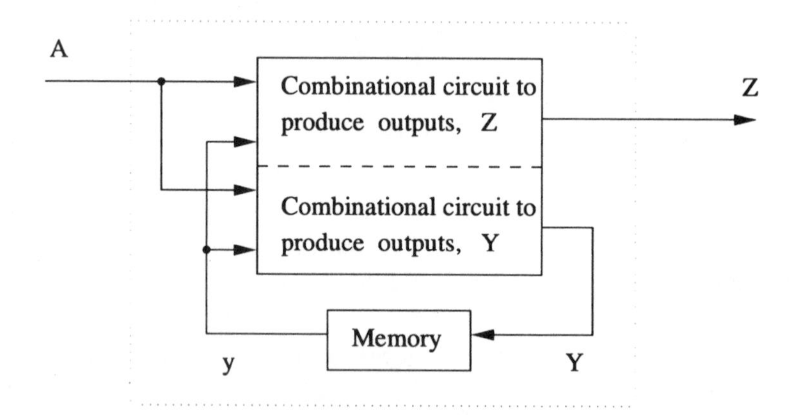


Fig. 5.2 A general sequential circuit emphasising how the outputs from the combinational logic block are functions of both the external and internal inputs

**'States' and sequential circuits**

An important concept to appreciate is that sequential circuits can be considered at any time to occupy a certain 'state'. These 'states' are dependent upon the internal feedback, and in the case of asynchronous sequential circuits, the external inputs as well. At this early stage we simply note that if the memory in a circuit has $i$ digital lines leading to and from it then it can store $2^i$ different patterns and hence the circuit possesses $2^i$ *internal states*. (We call these internal states to distinguish them from the *total states* of the circuit which are also dependent upon the external inputs.)

This idea of the circuit possessing states is fundamental to sequential circuits since they are often designed and analysed by the manner, or sequence, in which the available states are visited for given sequences of inputs. This is why in Fig. 5.1 the internal inputs, $y$, and outputs, $Y$, are labelled as the present and next state variables respectively (since they determine the current and next state of the circuit).

### 5.1.1 Asynchronous and synchronous circuits

In this introductory text we will define two broad classes of sequential circuits, namely: asynchronous and synchronous.

- The timing of the operation of *asynchronous circuits*, as the name implies, is not controlled by any external timing mechanism. Rather, as soon as changes are made at the inputs of such a circuit they take effect at the outputs. The simplest form of memory in such circuits is just a wire forming the feedback connection.[2]

- *Synchronous circuits* are those which possess a clock of some sort which regulates the feedback process. Hence the timing of changes in the outputs, in response to changes at the inputs (which may have occurred some time before), are controlled by the 'ticking' of a clock. Consequently, the timing of the operation of sequential circuits can be, and usually is, synchronised to other parts of a larger circuit. The memory in such circuits is itself made up of specialised logic circuits (called flip-flops) that essentially act as digital storage elements.

Fig. 5.3 shows the block diagrams for these two types of sequential circuits. For obvious reasons, synchronous sequential circuits are also referred to as *clocked* circuits, whilst asynchronous ones are known as *unclocked* or *free running*. Although asynchronous circuits are more difficult to design than synchronous ones they do have certain benefits. These include the fact that because they are free running their speed of operation is limited solely by the characteristics of the components from which they are built and not by the speed at which they are clocked. Consequently, asynchronous circuits have the potential to work at higher speeds than synchronous ones. Also: some systems may require a circuit to respond immediately to changing inputs (i.e. the inputs cannot be synchronised to the rest of the circuit); in very large circuits the unavoidable delays as a signal

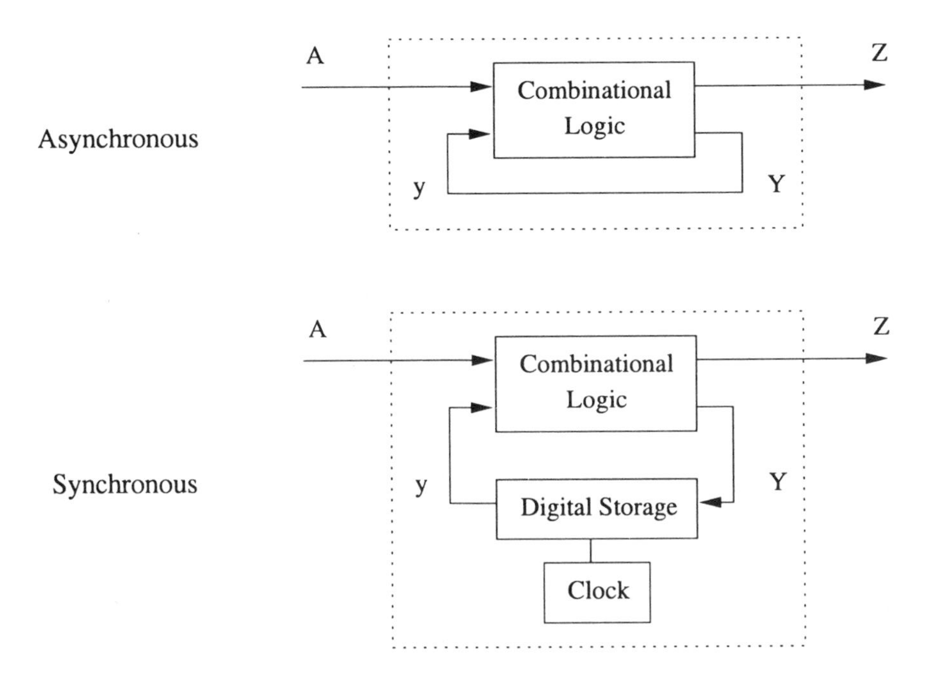


Fig. 5.3 General asynchronous and synchronous sequential circuits showing the difference between their respective 'memory'

[2]This is the only type of asynchronous circuit we will consider.

traverses the whole circuit may mean exact synchronisation is not possible; and finally flip-flops which are essential digital circuit components are themselves asynchronous circuits.

**Summary**

In this brief introduction the general properties and structure of *sequential circuits* have been introduced, together with the idea of a broad classification of such circuits as either *asynchronous* or *synchronous.*

The remainder of the chapter is split into three sections. The first is an introduction to asynchronous sequential logic circuits; the second looks at how asynchronous circuits operate via the analysis of a number of such circuits; whilst the third considers the design, and associated problems, of these circuits.

The following three chapters are also concerned with sequential logic. Chapter 6 covers flip-flops which are themselves asynchronous circuits that act as digital storage elements and are used as the memory in synchronous sequential circuits.

Chapters 7 and 8 cover synchronous sequential circuits, beginning with counters before moving on to more general examples.

## 5.2 INTRODUCTION TO ASYNCHRONOUS SEQUENTIAL CIRCUITS

The 'memory' of previous outputs in an asynchronous sequential circuit is provided by direct feedback from the internal output(s), $Y$, to the internal input(s), $y$, of the combinational logic block (see Fig. 5.3). The essence of understanding asynchronous circuits is to realise that for the circuit to be *stable* the outputs generated by the input(s) *must* be equal (i.e. $Y=y$), since these two sets of signals are connected via the feedback.

If this is not so then the circuit will be unstable with the output(s) (unmatched to the input(s)) acting as different input(s) and so producing new output(s) which will then be fed back again. This process will repeat until a stable condition is reached. This concept will become clearer as we analyse actual asynchronous sequential circuits.

The first stage of asynchronous sequential circuit analysis is to 'break' the feedback paths and treat the output(s) being fed back and the corresponding input(s), linked via the feedback, as separate variables. The circuit will only be stable when these signals have the same values.

The conditions for which the circuit is stable are known as the 'stable states'. (Note that for asynchronous sequential circuits these *total* stable states depend upon *all* of the inputs, i.e. both internal and external ones, and not just the values of the feedback, i.e. the present state, variables). Circuit *analysis* involves finding out how these stable states are both reached and related to one another, whilst *design* involves producing a circuit which enters the required stable states for the desired input patterns.

### 5.2.1 Inputs and race conditions

For an asynchronous circuit to be of use it must have more than one external input (otherwise all the circuit has to respond to is a signal alternately changing its value from 0 to 1 to 0 and so on[3]). If two of these multiple inputs are meant to change simultaneously we know that for a real circuit this can never be guaranteed and one of them will always change slightly before the other. The effect of this uncertainty, in which signal (and not always the same one) arrives first, is that the circuit may not operate as expected and actually end up in the 'wrong' state. This will make the circuit's operation unpredictable and hence render it useless.

This problem can be overcome by making sure that only *one* input to the circuit changes at a time and that there is sufficient time between changes in the inputs for the circuit to stabilise. This is called *fundamental mode operation*, which although providing a solution does inhibit the way that the circuit can be used.[4]

## 5.3 ANALYSIS

We now turn our attention to the actual analysis of asynchronous sequential circuits. The first three circuits are very simple examples whose operation can be determined intuitively. These serve to introduce some of the basic concepts of asynchronous sequential circuits such as stable states and the idea of 'memory'. The final three examples include circuits with one and two feedback signals.

### 5.3.1 Circuit 1: stable and unstable states

Consider the circuit in Fig. 5.4 which is simply an XOR gate with one of the inputs being the output which is fed back, so making it an asynchronous sequential circuit.[5] (Note that in this case, $Z$, the external output is the same as $Y$, the internal output.)

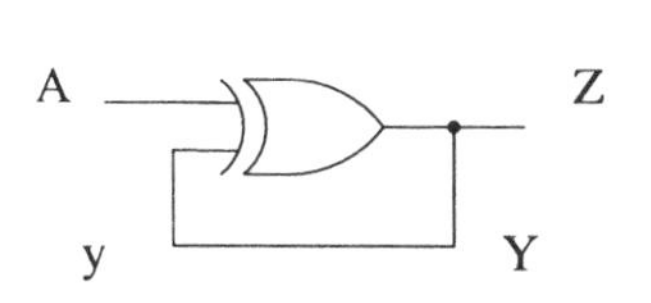


| Y | $\bar{A}$ | A |
|---|---|---|
| $\bar{y}$ | 0 | 1 |
| y | 1 | 0 |

Fig. 5.4 XOR gate based asynchronous sequential circuit. This circuit is only stable when the present, y, and next, $Y$, state variables, which are connected, are the same

[3]In an asynchronous circuit the outputs change immediately in response to a change in the inputs. Therefore it is not even possible to encode any time information, to which the circuit may respond, onto such a single signal.

[4]An alternative 'safe' means of operating such circuit is ***pulse mode***, where the inputs to the circuit are given a pulse (i.e. go from 0 to 1 and back) when active.

[5]Although it is not actually useful since it only has one external input.

If $A=0$ and $y=1$ then the output $Y=1$, so the output produced, and fed back to the inputs, matches the input (i.e. $y=Y$) and the circuit is stable. Similarly, if $A=0$ and $y=0$, then $Y=0$, and so $y=Y$ and the circuit is again stable. The fact that the circuit is stable means that all of the variables will remain unchanged until the input $A$ is changed (as this is the only variable that can be accessed, i.e. the only external input).

Now, if $A$ is changed to 1 then if $y=0$ then $Y=1$; and if $y=1$ then $Y=0$. Hence, $y \neq Y$. This will clearly lead to an unstable circuit that will continually oscillate. For example, if $y=0$ the output will be 1, which will be fed back to $y$ causing the output to go to 0 which will be fed back and so cause the output to go to 1 and so on. The speed of oscillation will be determined chiefly by the time it takes the signals to propagate through the XOR gate and back along the feedback path.

The Karnaugh map for $Y$ in terms of $A$ and $y$ is also shown in Fig. 5.4 and illustrates the operation of the circuit. For the circuit to be stable $Y$ must equal $y$, therefore for the top row, $\bar{y}$ (and hence $y=0$), the circuit will only be stable when the corresponding cell of the Karnaugh map has a 0 in it (i.e. the internal output $Y=0$). For the bottom row the circuit will only be stable when $Y=1$. We therefore see that only two stable conditions for this circuit exist. Firstly, $A=0$ and $y=0$ and secondly $A=0$ and $y=1$, that is when $A=0$ (the left hand column of the Karnaugh map) as we deduced above.

The Karnaugh map confirms the instability of the circuit when $A=1$, since nowhere in the right-hand column ($A=1$) does $y=Y$. All of the remaining circuit analyses are based upon the use of such Karnaugh maps.

### 5.3.2 Circuit 2: movement between states

The previous example introduced the idea of stable and unstable total states. Using the circuit shown in Fig. 5.5 we now look at how asynchronous sequential circuits have the potential to possess memory. This circuit has two external inputs and a single internal input which feed into a three-input OR gate.

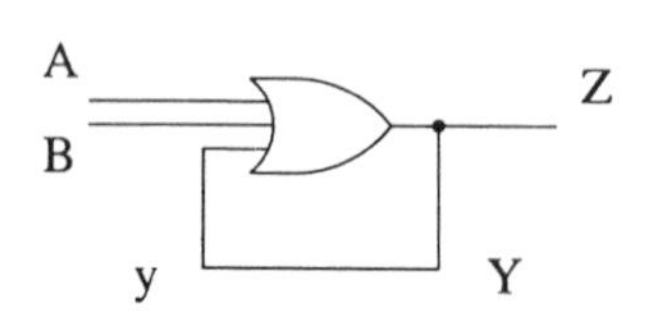


| Y | $\bar{A}\bar{B}$ | $\bar{A}B$ | $AB$ | $A\bar{B}$ |
|---|---|---|---|---|
| $\bar{y}$ | 0 | 1 | 1 | 1 |
| $y$ | 1 | 1 | 1 | 1 |

Fig. 5.5 Three-input OR gate based asynchronous sequential circuit

If either $A$ or $B$ are 1 then $Y=1$, and so $y=1$ which has no effect on the output, hence the circuit is stable. If $A$ and $B$ are both 0 then:

$$
\begin{aligned}
Y &= A+B+y \\
&= 0+0+y \\
&= y
\end{aligned}
$$

so the circuit is again stable, but whether the output is 0 or 1 depends upon what the output was *before* both external inputs went to 0. Hence the circuit has *memory* since its output under these input conditions depends upon what state it was in beforehand (i.e. whether $Y$ was 0 or 1).

However, since for any external inputs other than $A=0$ and $B=0$ the output is always 1, this means when $A$ and $B$ do both equal 0 the output will still always be 1.

Referring to the Karnaugh map in Fig. 5.5 we see that when either $A$ or $B$ are 1 the stable state is when $y=Y=1$. This is because for these inputs the 1's in the top row, $y$, indicate that the output, $Y$, is 1, which will then feed back making $y=1$ and so causing the state of the circuit to 'move' to the bottom row of the truth table where the states are stable since then $y=Y=1$. The idea that each cell of the Karnaugh map contains a state of the circuit which will be stable if $y=Y$, or else is unstable (such as when the top row contains a 1) is central to the analysis and design of asynchronous sequential circuits.

For $A$ and $B$ both 0, an output of either 0 or 1 gives a stable condition (since the output simply equals the internal input). However, when the external inputs are changed this effectively causes a horizontal movement across the Karnaugh map (as each column represents one of the possible input conditions) to a new cell and hence state. Now, since the only stable states are for $y=1$ (the bottom row) this means that horizontal movement in the map, as the inputs change, will always be within the bottom row. Consequently for $A$ and $B$ both 0, movement will always be to cell ($\bar{A}\bar{B}y$), and so the output will be 1. The lack of stable states for which $y=0$ when $A$ or $B$ are 1 means that as soon as either of the external inputs are 1 the circuit is confined to operate within the bottom row of the Karnaugh map. So the circuit will never give an output of 0.

**Summary**

The concept of the circuit moving within the Karnaugh map from state to state (cell to cell) is vital to understanding the analysis and design of these circuits. Changes in the external variables cause horizontal movement in the map (to a new column and hence input condition), whilst changes from an unstable to stable state (for fixed external inputs, i.e. within a column) cause vertical movement.

### 5.3.3 Circuit 3: memory

We have seen how asynchronous circuits can be considered to have total states (depending upon both the internal and external inputs), some stable and some unstable, and how changing the external inputs causes the circuit to move between these states. We now look at a circuit which demonstrates how this type of circuit can possess a memory of its previous states. This is shown in Fig. 5.6 together with the corresponding Karnaugh map for $Y=A+\bar{B}y$.

If $A=1$ then $Y=1$ and so the only possible stable states for this input condition

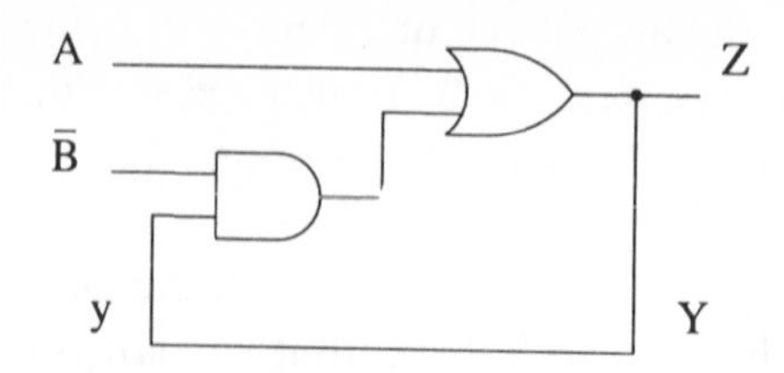


| Y | $\bar{A}\bar{B}$ | $\bar{A}B$ | $AB$ | $A\bar{B}$ |
|---|---|---|---|---|
| $\bar{y}$ | 0 | 0 | 1 | 1 |
| $y$ | 1 | 0 | 1 | 1 |

Fig. 5.6 Asynchronous sequential analysis (Circuit 3) which demonstrates the concept of memory. For inputs $\bar{A}\bar{B}$ the circuit has two stable states

will be when $y=1$, the bottom row of the Karnaugh map. When $A=0$ and $B=1$ then

$$Y=A+\bar{B}y=0$$

hence the only stable state possible is $y=Y=0$, the top row. This leaves the input condition of $A$ and $B$ both 0 to consider, which gives

$$Y=0+1\cdot y=y$$

for which $Y$ being either 0 or 1 is a stable condition. We know from the last example that which of these stable states is entered depends upon which row of the Karnaugh map the circuit was 'in' before the inputs were changed, thus causing a horizontal movement to the $\bar{A}\bar{B}$ (in this example) column.

If the inputs were previously $A=0$ and $B=1$ ($\bar{A}B$), then $y=Y=0$ and the horizontal movement left (from cell $\bar{A}B\bar{y}$) will take the circuit to the stable state with $A=0$, $B=0$ and $y=0$ (i.e. cell $\bar{A}\bar{B}\bar{y}$). However, if previously $A=1$ and $B=0$ then $y=Y=1$ (cell $A\bar{B}y$) and the horizontal movement is to the right (looping around the map) taking the circuit to stable cell $\bar{A}\bar{B}y$. So, for inputs $A$ and $B$ both 0, the output, $Y$, will depend upon what the previous inputs were and therefore from which direction the column in the Karnaugh map corresponding to these inputs was entered. Hence the circuit has *memory.*

The addition of a simple logic circuit as shown in Fig. 5.7 serves to decode state $\bar{A}\bar{B}y$ which will give an output of 1 whenever the circuit enters this state and hence

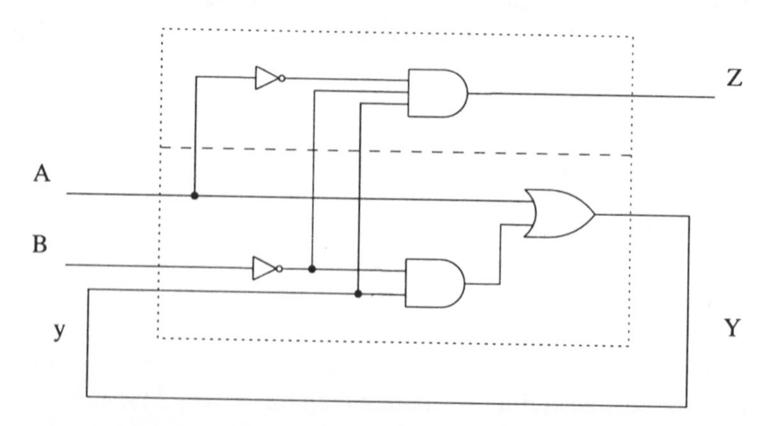


Fig. 5.7 Analysis Circuit 3 with additional output circuitry (the three-input AND gate)

when the inputs $(A,B)$ have gone from (1,0) to (0,0) (since this is the only way to enter this state). Note that this circuit now has the form of a general asynchronous sequential circuit (see Figs 5.2 and 5.3). Remember that because we are only allowing one input to change at a time (fundamental mode operation) this means the circuit can only move one column at a time as the inputs change.

**Summary**

These three examples have served to introduce asynchronous sequential circuits. We now consider three further examples which will be analysed more rigorously. Finally, note that in the first example of the XOR circuit we did not at the time discuss how the two possible stable states could be entered. It should now be clear that if $A=1$, with the circuit's output oscillating, then switching $A$ to 0 will result in a stable output whose value is dependent upon which state the oscillating circuit was in at the time that $A$ was taken low.

## 5.3.4 Circuit 4: rigorous analysis

The largely intuitive analysis of the above three circuits has served to illustrate the majority of the basic concepts of asynchronous sequential circuit analysis. We now consider three further examples, beginning with a full and rigorous analysis which amounts largely to formalising the methods we used above. The headings below relate to the various stages of the procedure.

**The circuit**

This circuit is shown in Fig. 5.8. It has two external inputs, $A$ and $B$, and one feedback signal, labelled $y$. The basic AND-OR circuit forms the combinational logic for the signal, $Y$, which is fed back, whilst the output circuit producing $Z$ is simply an AND gate. We will assume that we are operating in fundamental mode, that is that only one of the external inputs can change at a time.

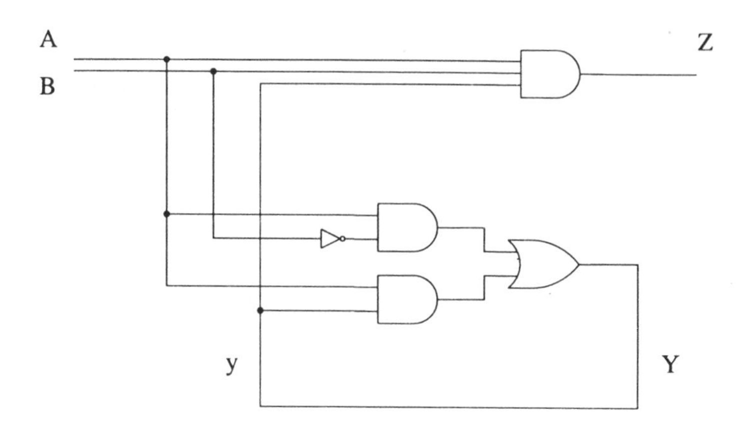


Fig. 5.8 Analysis Circuit 4

**'Breaking' the feedback path**

The first step when rigorously analysing an asynchronous sequential circuit is to 'break' all of the feedback paths. The reason for this is that we know for a stable circuit the values of the signals at each end of the feedback path(s) must be equal. By breaking the feedback path(s) and treating the signal(s) at either end as independent (e.g. $y$ and $Y$) we can begin the analysis. Here there is only one feedback path with labels $y$ and $Y$ used for the internal input and output respectively.

**Boolean expressions for the internal variables and external output**

Once the loop is broken we can consider the signal $Y$ (the internal output from the next state combinational logic) and $y$ (the input to the whole combinational logic block) to be separate variables. Hence the Boolean expressions for $Y$ and $Z$ can be written (by studying the combinational logic circuit) as:

$$Y = A\bar{B} + Ay$$

$$Z = ABy$$

**Transition table and output map**

Using these Boolean equations we next draw the Karnaugh maps for $Y$ and $Z$ for the variables $A$, $B$ and $y$. For this type of circuit analysis the convention is to use the external inputs to label the columns and the internal variables the rows. This gives the maps shown in Table 5.1. The one showing the next state variable, $Y$, in terms of the present state variable, $y$, is referred to as the *transition table*.[6]

Table 5.1 Transition table and output Karnaugh map for Circuit 4

| Y | $\bar{A}\bar{B}$ | $\bar{A}B$ | $AB$ | $A\bar{B}$ |
|---|---|---|---|---|
| $\bar{y}$ | 0 | 0 | 0 | 1 |
| y | 0 | 0 | 1 | 1 |

| Z | $\bar{A}\bar{B}$ | $\bar{A}B$ | $AB$ | $A\bar{B}$ |
|---|---|---|---|---|
| $\bar{y}$ | 0 | 0 | 0 | 0 |
| y | 0 | 0 | 1 | 0 |

**Flow tables**

The next, crucial, stage is to decide which combinations of the three inputs represent stable states for the circuit. That is for which inputs does the fed back signal, $Y$, correspond to the input signal, $y$? If these differ, then if they are 'reconnected' (since we have 'broken' the loop) clearly the circuit cannot be stable and will therefore be trying to change.

The circuit will be stable (as we have deduced in the previous examples) whenever $Y=0$ in the top row of the $Y$ transition table (i.e. $y=0$) or $Y=1$ in the bottom row (i.e. $y=1$). These 'states' can be given numbers to code them and the $Y$ transition table redrawn showing the stable states (the given number to code that state within a circle) and the unstable states (just a number with no circle) which indi-

[6]This is sometimes called the excitation table or map.

Table 5.2 Flow table for Circuit 4

| Y | $\bar{A}\bar{B}$ | $\bar{A}B$ | $AB$ | $A\bar{B}$ |
|---|---|---|---|---|
| y = 0 | (1) | (2) | (3) | 5 |
| y = 1 | 1 | 2 | (4) | (5) |

cates to which stable state the circuit will progress. This is shown in Table 5.2 and is called the *flow table* since it shows how the circuit 'flows' between states.

The stable states correspond to those cells in the transition table for which the next state variable, $Y$, equals the present state variable, $y$. Note that the transition table was drawn (by using the external inputs to label the columns) such that a change in the external inputs causes a horizontal movement in the table, which because we are operating in fundamental mode, can only be one column at a time. We see that for this circuit, all columns have a stable state, so every combination of external inputs will produce a stable output of both $Y$ and $Z$.

Whilst changing the external inputs causes horizontal movements, note that a vertical movement represents a change in the feedback signal. In this example, since all columns have a stable state, if the circuit enters an unstable state it will 'move' vertically into a stable condition where it will remain. (This is sometimes indicated by drawing arrows within the flow table to show the circuit's action when it enters an unstable state.)

For example if $A=B=0$ and $y=1$ then $Y=0$ and the circuit is unstable since $y \neq Y$, but the signal $Y$ will feedback to y (after some delay) and so eventually $y=0$. For $A=B=y=0$, $Y=0$ and so the circuit is now in stable state 1. (This is the reason why all stable states in the row indexed by $y=0$ have $Y=0$ and all unstable states (e.g. 5) have $Y=1$ which causes vertical movement to a stable state (e.g. downwards to stable state 5 in cell $A\bar{B}y$).

We will now look at some examples of the operation of this circuit for different inputs. We begin by assuming we are in stable state 3 (i.e. $A=B=1$ and $y=0$) and are operating in fundamental mode. Changing an input will move us across one column in the flow table, in this case from the $AB$ labelled column. If this horizontal movement takes us to a cell with a stable state the circuit will obviously remain there (with the appropriate output for $Z$); if it moves the circuit to an unstable state then the circuit shifts vertically to the stable state (as indicated by the arrows) corresponding to the unstable state (the number *not* in a circle) in the transiently occupied cell of the flow table.

**Operation of circuit: Case 1**

If $A=B=1$ and $y=0$ then $Y=0$ and the cell is in stable state 3 (i.e. in cell $AB\bar{y}$). If input $B$ is then changed (to 0) the circuit shifts horizontally to the right to cell $A\bar{B}\bar{y}$. This cell is unstable since $y=0$ but $Y=(A\bar{B}+Ay)=1$, so there is a downward vertical movement to cell $A\bar{B}y$ (stable state 5), where $Y=1$. The circuit then remains in this stable state since $y = Y = 1$. This movement of the circuit from one stable state to another via a transient unstable state is shown in Table 5.3.

Table 5.3 Flow table for Circuit 4: Case 1

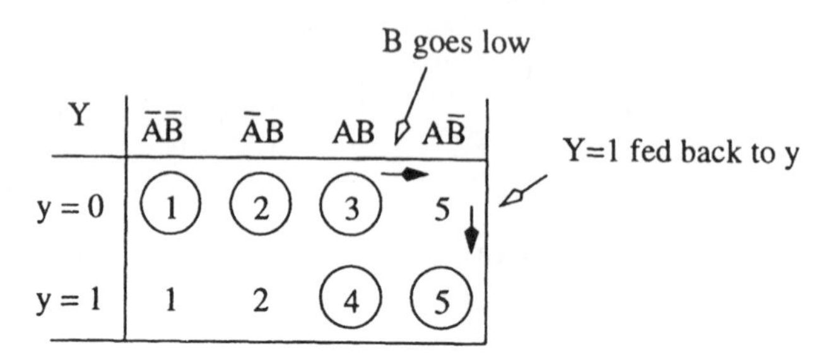


| Y | $\bar{A}\bar{B}$ | $\bar{A}B$ | $AB$ | $A\bar{B}$ |
|---|---|---|---|---|
| y = 0 | (1) | (2) | (3) | 5 |
| y = 1 | 1 | 2 | (4) | (5) |

**Operation of circuit: Case 2**

What if the circuit is again in stable state 3 and then $A$ is taken LOW? This will cause a horizontal movement to the left to cell $\bar{A}B\bar{y}$ which is another stable state, namely 2. So the circuit comes to rest in this state with $A=0$, $B=1$ and $y=Y=1$.

**Summary: use of flow tables**

To summarise the use of the flow table to determine the circuit's operation, firstly find which stable state (i.e. cell of the flow table) the circuit is in for the given inputs, and then note that:

- Changing an external input causes a horizontal movement in the flow table.
- If the circuit enters an unstable state ($y \neq Y$) there is vertical movement to a stable state (if one exists) in the column relating to the present external inputs.
- If the circuit immediately enters a stable state it will remain in it.
- If there is no stable state in a column then the circuit will simply continually enter and exit unstable states (i.e. oscillate) as $y$ and $Y$ continually follow each other.

**State diagram**

We can redraw the flow table as a state diagram which contains *exactly* the same information but for some purposes is more convenient. In the state diagram each state is represented by a node (a circle with a number in) with the states joined by arrows. These indicate how the circuit moves from state to state with the inputs causing the change indicated on the arrows. The state diagram for Circuit 4 is shown in Fig. 5.9. The number of arrows leading to each node gives the number of stable states from which that node (stable state) can be reached. (For example state 5 can be reached from states 1, 3, and 4 and so has three arrows leading to it.) There are twice as many arrows as states (nodes) since there are two input variables one of which can be changed (in fundamental mode operation).

Note that in the state diagram the states have been coded to indicate whether the ouput $Z$ is 0 or 1. Either this diagram, or the transition table, can be used to determine how changing the inputs to the circuit cause the state occupied to change.

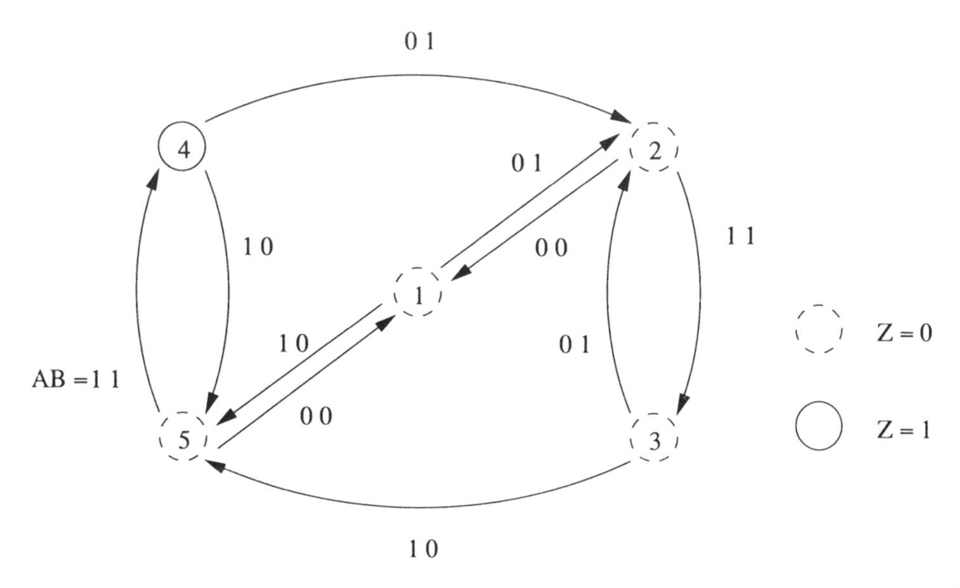


Fig. 5.9 State diagram for Circuit 4

**Use of the state diagram**

As an example, if $A=B=0$ the circuit is in state 1. Changing $B$ causes a shift to state 2 (along the arrow labelled 01), then changing $A$ causes a shift (along arrow 11) to state 3.

## Example 5.1

The inputs $(A,B)$ take the following values. Determine which states will the circuit be occupied by and illustrate this, and the output, $Z$, on a timing diagram.

$$(A,B)=(0,0),(1,0),(1,1),(0,1),(1,1),(1,0),(0,0)$$

### *Solution*

For these inputs the circuit will go into states: 1, 5, 4, 2, 3, 5, 1. The timing diagram is shown in Fig. 5.10.

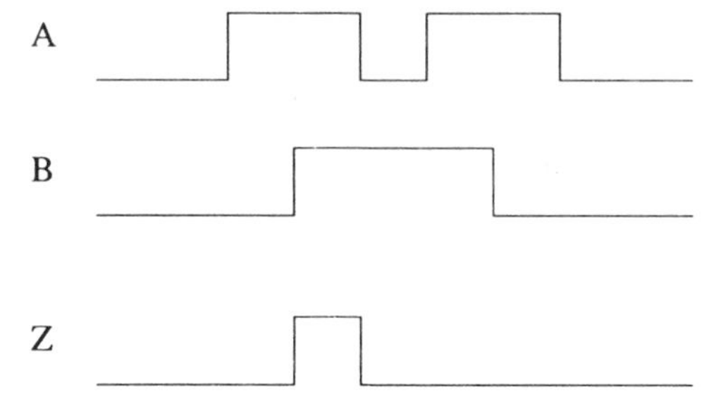


Fig. 5.10 Timing diagram for Circuit 4; see Example 5.1

**What is the circuit's function?**

We have now fully analysed this circuit and can predict what state it will be in for any given sequence of inputs. So, what is its function?

This is best investigated by considering under what conditions the external output, $Z$, becomes 1. This is when the circuit enters state 4, which is when both inputs are 1. This state can only be reached via state 5, which is entered when $A=1$ and $B=0$. So, $Z=1$ when the inputs $(A,B)$ are (1,0) and then change directly to (1,1). In other words, the circuit detects an input sequence of (1,0), (1,1).[7] This can be seen from the above example where the output goes high indicating this input sequence.

**Summary**

This completes the analysis of this circuit, which obviously possesses memory because of the way the stable state entered (when both inputs go to 1) is dependent upon the previous input conditions, and hence previous stable state.

### 5.3.5 Summary of analysis procedure

Now we have completed our first rigorous analysis of an asynchronous sequential circuit we can outline the analysis procedure. We will then use it to analyse two further circuits and then consider the design process.

Procedure:

1. Break the feedback loop(s).
2. Draw the transition tables for the next state variable(s) and Karnaugh map(s) for the external output(s) in terms of the external inputs and the present state variable(s). Remember to use the external inputs to index the columns and the present state variables to index the rows.
3. Determine which cells in the transition table give stable states, assign each of these a number and then draw the flow table.
4. Draw the state diagram if required.

### 5.3.6 Circuit 5: two internal inputs

This circuit, shown in Fig. 5.11, has two external inputs, $A$ and $B$, two internal inputs, $X$ and $Y$, and one external output, $Z$.

**Breaking the feedback paths**

This circuit has two internal outputs, $X$ and $Y$, feeding back to two internal inputs, $x$ and $y$. So these feedback paths are broken.

**Boolean expression of outputs**

From the circuit we see that:

[7] Note the similarity between the structure and function of this circuit and Circuit 3 in Section 5.3.3 which detected the sequence (1,0), (0,0).

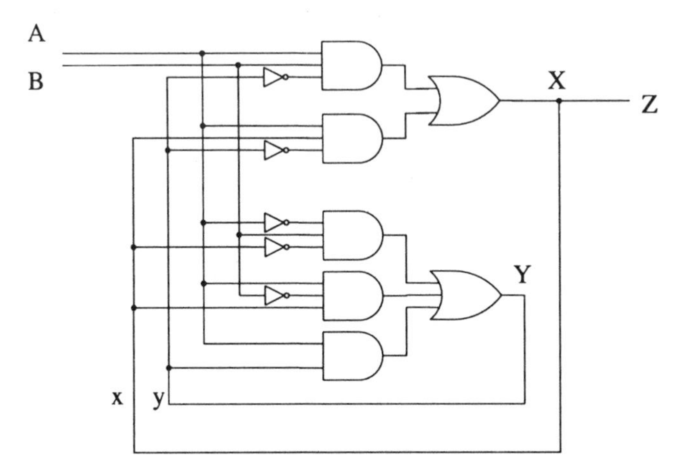


Fig. 5.11 Analysis Circuit 5, which has two external and two internal inputs

$$X = AB\bar{y} + Ax\bar{y}$$

$$Y = \bar{A}B\bar{x} + A\bar{B}x + Ay$$

with the output circuit (in this case using no logic) giving:

$$Z = x = X$$

Using these equations we can draw the Karnaugh maps for these as shown in Table 5.4.

Table 5.4 Karnaugh maps for the outputs from Circuit 5

| X | $\bar{A}\bar{B}$ | $\bar{A}B$ | $AB$ | $A\bar{B}$ |
|---|---|---|---|---|
| $\bar{x}\bar{y}$ | 0 | 0 | 1 | 0 |
| $\bar{x}y$ | 0 | 0 | 0 | 0 |
| $xy$ | 0 | 0 | 0 | 0 |
| $x\bar{y}$ | 0 | 0 | 1 | 1 |

| Y | $\bar{A}\bar{B}$ | $\bar{A}B$ | $AB$ | $A\bar{B}$ |
|---|---|---|---|---|
| $\bar{x}\bar{y}$ | 0 | 1 | 0 | 0 |
| $\bar{x}y$ | 0 | 1 | 1 | 1 |
| $xy$ | 0 | 0 | 1 | 1 |
| $x\bar{y}$ | 0 | 0 | 0 | 1 |

| Z | $\bar{A}\bar{B}$ | $\bar{A}B$ | $AB$ | $A\bar{B}$ |
|---|---|---|---|---|
| $\bar{x}\bar{y}$ | 0 | 0 | 0 | 0 |
| $\bar{x}y$ | 0 | 0 | 0 | 0 |
| $xy$ | 1 | 1 | 1 | 1 |
| $x\bar{y}$ | 1 | 1 | 1 | 1 |

**Transition table**

The next stage is to determine under what conditions the circuit is stable which, by extending the argument used in the previous examples, will be when *both* internal inputs, $x$ and $y$, match the signals being fed back to them, $X$ and $Y$. In order to see this it is helpful to combine the Karnaugh maps for $X$ and $Y$ into the single transition table shown in Table 5.5. The stable total states will be those cells where the values of $XY$ match those of the variables $x$ and $y$ labelling the rows.

**Flow table**

The stable total states can then be given numbers and circled whilst the unstable

Table 5.5 Transition table for Circuit 5

| STATE XY | $\bar{A}\bar{B}$ | $\bar{A}B$ | $AB$ | $A\bar{B}$ |
|---|---|---|---|---|
| $\bar{x}\bar{y}$ | 0 0 | 0 1 | 1 0 | 0 0 |
| $\bar{x}y$ | 0 0 | 0 1 | 0 1 | 0 1 |
| $xy$ | 0 0 | 0 0 | 0 1 | 0 1 |
| $x\bar{y}$ | 0 0 | 0 0 | 1 0 | 1 1 |

states are simply labelled with the states they will lead to, as shown in Table 5.6. Note that because this circuit has two feedback signals there are four possible *internal* states corresponding to the four possible combinations of these variables, and hence four rows in the excitation matrix. This means that when the circuit is unstable in any cell it now has three cells (in the same column) into which it can move (with one internal input there is only one other cell in each column).

Table 5.6 Flow table for Circuit 5

| STATE | $\bar{A}\bar{B}$ | $\bar{A}B$ | $AB$ | $A\bar{B}$ |
|---|---|---|---|---|
| $\bar{x}\bar{y}$ | (1) | 3 | 6 | (2) |
| $\bar{x}y$ | 1 | (3) | (4) | (5) |
| $xy$ | 1 | 3 | 4 | 5 |
| $x\bar{y}$ | 1 | 3 | (6) | 5 |

With this circuit when both $A$ and $B$ are 0 all unstable states lead to stable state 1; similarly in the second column when $A=0$ and $B=1$ there is a single stable state to which all unstable states (the other three cells in this column) lead. Note that in this column, $\bar{A}B$, rows 3 and 4 do not lead *directly* to the stable state, but rather get there indirectly via row 1 (since in these rows, $(X, Y)=(0,0)$), which leads directly to state 3 (row 2).

In the third column, which has inputs of both $A$ and $B$ being 1, there are two stable states, 4 and 6, with one each of the two unstable states leading to state 4 and 6 respectively.

The fourth column, $\bar{A}B$, also has two stable states, 2 and 5, but here the two feedback variable combinations giving unstable states both lead to state 5 (the one from row 4 indirectly via row 3). Hence the only way of entering state 2 is from state 1.

### State diagram

Finally we can draw the state diagram (for fundamental mode operation) as shown in Fig. 5.12. There is a node for each stable state, with two arrows leading from every node (corresponding to movement to the left and right in the flow table as one of the two external input variables is changed).

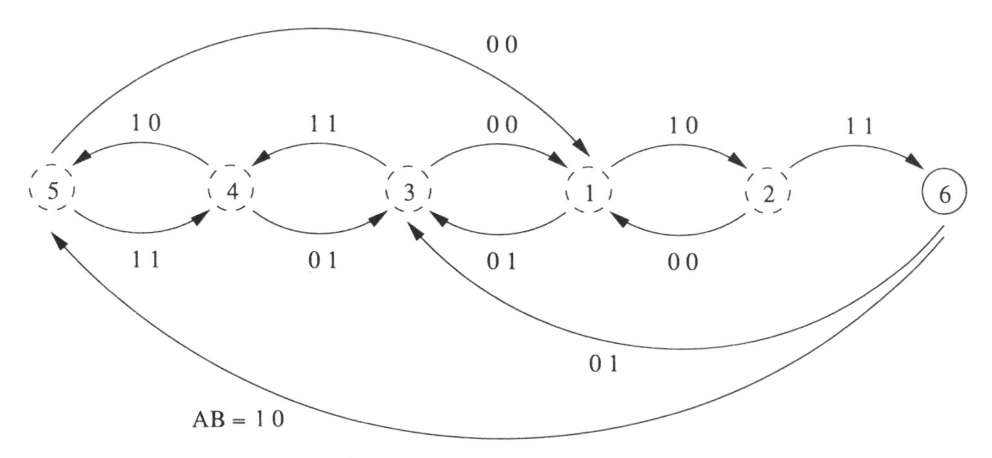


Fig. 5.12 State diagram for Circuit 5. Note the dashed states have an external output $Z=0$, and the solid $Z=1$

### Circuit operation

Having completed the basic analysis we can examine the circuit's operation under certain conditions. To do this we look at which states give an output of 1 which by comparing the flow table for the circuit (Table 5.6) and the Karnaugh map for $Z$ (in Table 5.4) can be seen to be only stable state 6. From the state diagram (Fig. 5.12) it can be seen that this state is only reached from state 2 when the inputs $(A,B)$ are (1,1). However, state 2 can only be reached from state 1 with inputs of (1,0), whilst state 1 can only be entered with inputs of (0,0). This circuit therefore uniquely detects an input sequence of $(A,B)=(0,0),(1,0)$ and then (1,1) to which it responds by sending the external output, $Z$, high. This completes the analysis, with the consequence of any input sequence capable of being found from either the flow table or state diagram.

### Example 5.2

Which states would the circuit enter for inputs of:

(a) 01, 11, 01, 00, 10, 11, 01, 11
(b) 00, 10, 00, 10, 11, 10, 11, 01
(c) 01, 00, 10, 00, 10, 11, 10, 11
(d) 00, 01, 11, 10, 00, 10, 11, 01

Draw the output waveforms for these inputs.

***Solution***

(a) 3, 4, 3, 1, 2, 6, 3, 4
(b) 1, 2, 1, 2, 6, 5, 4, 3
(c) 3, 1, 2, 1, 2, 6, 5, 4
(d) 1, 3, 4, 5, 1, 2, 6, 3
The corresponding output waveforms are shown in Fig. 5.13.

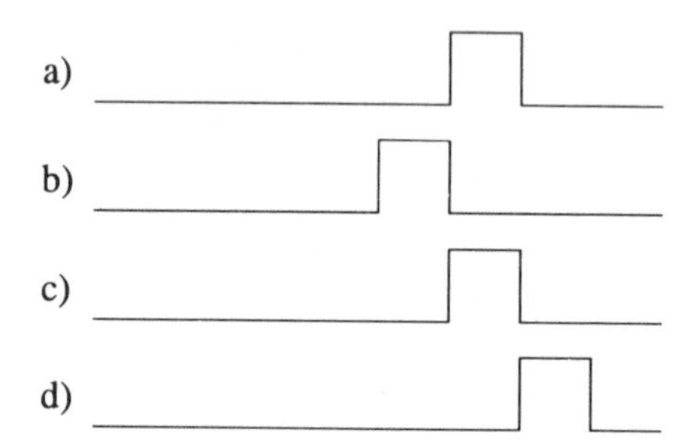


Fig. 5.13 Output waveforms from Circuit 5 for the inputs given in Example 5.2

## 5.4 CIRCUIT 6: A BINARY STORAGE ELEMENT

The final asynchronous sequential circuit we are going to analyse is shown in Fig. 5.14 and consists of two cross-coupled NOR gates. As well as analysing its operation like the previous circuits we will also use it to investigate the consequence of relaxing the restriction of fundamental mode operation.

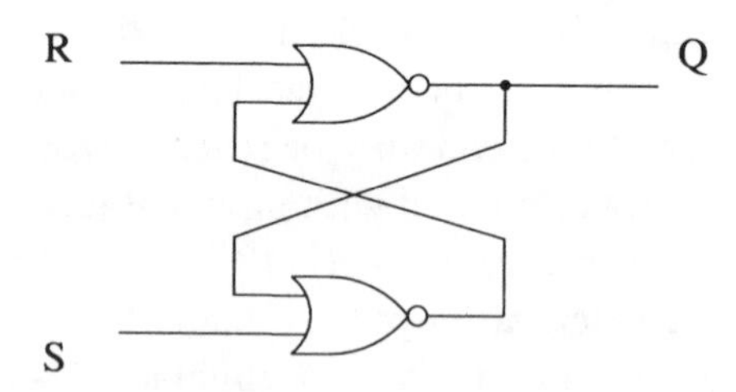


Fig. 5.14 Analysis Circuit 6: the SR flip-flop

### 5.4.1 Analysis

**Breaking the feedback path**
This circuit has two feedback paths. However, the circuit can be redrawn as shown in Fig. 5.15 which demonstrates that we only need to split one of the paths because in this circuit they are not independent.[8]

[8]The circuit can be analysed by considering two feedback paths but this will obviously be more complicated and results in exactly the same conclusions.

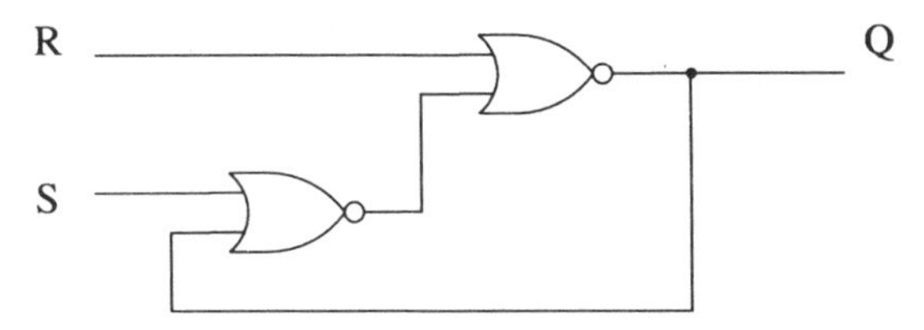


Fig. 5.15 Circuit 6 redrawn to show that only one dependent feedback path exists

**Boolean expression of outputs**

From the circuit we see that:

$$Q=\overline{R+\overline{(S+q)}}=\bar{R}\cdot(S+q)=S\bar{R}+\bar{R}Q$$

**Transition table**

We can now draw the transition table as shown in Table 5.7 which shows that all possible input combinations possess a stable state. These are shown numbered and circled in the flow table also in Table 5.7.

Table 5.7 The transition table and flow table for Circuit 6

| Q | $\bar{S}\bar{R}$ | $\bar{S}R$ | $SR$ | $S\bar{R}$ |
|---|---|---|---|---|
| $\bar{q}$ | 0 | 0 | 0 | 1 |
| $q$ | 1 | 0 | 0 | 1 |

| Q | $\bar{S}\bar{R}$ | $\bar{S}R$ | $SR$ | $S\bar{R}$ |
|---|---|---|---|---|
| $\bar{q}$ | (1) | (2) | (3) | 5 |
| $q$ | (4) | 2 | 3 | (5) |

**State diagram**

We can draw the state diagram (for fundamental mode operation) which is shown in Fig. 5.16. Having completed the basic analysis we can examine the circuit's operation under certain conditions.

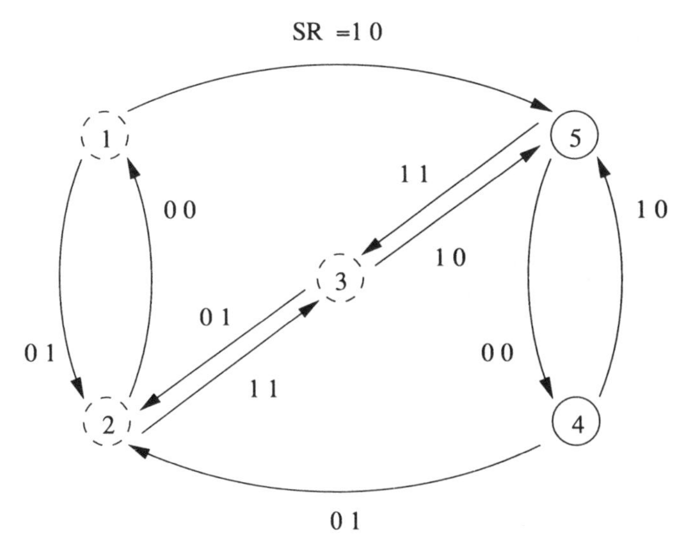


Fig. 5.16 The state diagram for Circuit 6. Solid circles indicate the states for which $Q=1$

## Example 5.3

If the inputs $(S,R)$ take the following values then what states does the circuit occupy and what happens to the output, $Q$.

$$(S,R)=(1,1),(0,1),(0,0),(1,0),(0,0)$$

### *Solution*

Changing the inputs causes horizontal movement in the maps, which will be followed by a vertical movement to a stable state should one not be entered initially. The circuit goes firstly to state 3 (inputs (1,1)):

- (0,1) causes movement to the left to stable state 2;
- (0,0) causes movement to the left to stable state 1;
- (1,0) causes movement to the 'left' (to the far right hand column) and unstable (transient) state 5, which is followed by vertical movement to stable state 5;
- (0,0) causes movement to the right, back to the first column, *but* with the circuit now occupying state 4 (due to downward movement when in column 4 (inputs (1,0)) to stable state 5).

The output, $Q$, will be 0 until state 5 is entered (since this is in the bottom row of the flow table) when it will become 1 and remain 1 as the circuit moves into state 4.

## Example 5.4

What states will the circuit occupy for the following inputs?

$$(S,R)=(0,1),(0,0),(0,1),(1,1),(1,0),(0,0),(0,1),(0,0)$$

### *Solution*

The circuit will occupy states: 2, 1, 2, 3, 5, 4, 2, 1 with corresponding outputs of 0, 0, 0 , 0, 1, 1, 0, 0.

## 5.4.2 Race conditions

In order to study some important features of asynchronous and other circuits we now relax the fundamental mode conditions and so allow more than one input to change at the same time.

**Non-critical races**

Consider Circuit 6 in the above section with inputs $(S,R)=(0,1)$ and therefore in state 2 (cell $\bar{S}R\bar{q}$), with both inputs then changing to give $(S,R)=(1,0)$. We now have the situation for a race condition since both $S$ and $R$ must change. If $S$ changes first the circuit goes transiently to state 3, cell $SR\bar{q}$ and then, as $R$ changes, on to stable state 5, cell $S\bar{R}q$ (via unstable state, $S\bar{R}\bar{q}$). However, alternatively $R$ may change first in which case the circuit will transiently enter state 1 (cell $\bar{S}\bar{R}\bar{q}$) before finally settling in state 5 (again via unstable state, $S\bar{R}\bar{q}$).

The important point to note is that whichever signal changes first the circuit eventually ends up in state 5. This is therefore an example of what is called a *non-critical race*. It is a race condition because there are two signals changing simultaneously and hence 'racing' each other to see which one effects a change first. It is a non-critical race because the final result, that is the eventual stable state, does *not* depend upon which signal arrives first at, or travels fastest through, the circuit.

**Critical races**

Now consider $(S,R)=(1,1) \rightarrow (0,0)$. Initially the circuit is in stable state 3. If $S$ changes first then the circuit goes from state 3 to 2 and finally settles in state 1 as $R$ changes. However, if $R$ is the first to change the circuit goes firstly to state 5 and then settles in state 4. Here we have an example of a *critical race* since the eventual state of the circuit (for the same inputs) is critically dependent upon which signal arrives first.

This critical race condition clearly renders this circuit's operation unpredictable, and therefore unusable, *unless* it is either operated in fundamental mode (remember we had relaxed this restriction) *or* alternatively the input conditions $S=R=1$ are not allowed since it is only by entering this state that the unpredicatability occurs. The first option demonstrates why fundamental mode operation is such a powerful restriction because of the way it helps eliminate race conditions.

### 5.4.3 The SR flip-flop

We now consider the second option which basically removes state 3, leaving four states. Since of these four states, two each have an output of 0 and 1, then we can combine these into a state diagram with two nodes, corresponding to outputs of 0 and 1 as shown in Fig. 5.17.

This state diagram shows that the output, $Q$, is *set* to 1 when $S=1$ and *reset* to 0

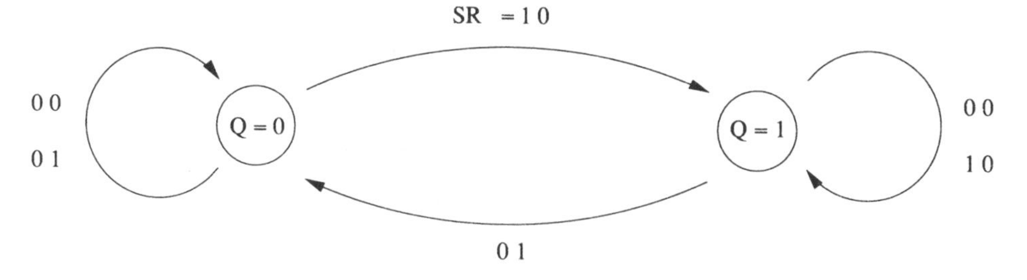


Fig. 5.17 State diagram for Circuit 6, in terms of the output, $Q$, rather than the circuit's states

when $R=1$ (remember we are not allowing both inputs to be 1). If the inputs then both become 0 the circuit retains the same output due to the existence of the two stable states (1 and 4) for these input conditions (column 1).

This circuit, operated under the conditions that both inputs must not be 1 simultaneously,[9] is what is known as a Set-Reset (SR) flip-flop. As the name implies the output of this flip-flop can be *set* to 1 by making $S=1$ or *reset* to 0 by making $R=1$. If both inputs are 0 then the circuit remains in the same state, and hence the output is unchanged.

Clearly the circuit has 'memory' since when both inputs are 0 it remembers what the output was before this input condition (since it has two stable conditions and different outputs for an input of (0,0)). A flip-flop can therefore act as a digital storage element that can store a 0 or 1 (i.e. a binary digit or *bit*) and is the basis for SRAM devices (see Section 10.3.1). As we will see in the next chapter, other types of flip-flop also exist.

The function of the SR flip-flop can be expressed in a variety of ways. Obviously the flow table and state diagrams completely describe its action, but more typically it is the truth table, shown in Table 5.8 or Boolean description:

$$Q=S\bar{R}+\bar{R}q$$

that is used. These are considered again, together with the operation of other types of flip-flop, in the next chapter.

Table 5.8 Truth table of the SR flip-flop

| S | R | Q | |
|---|---|---|---|
| 0 | 0 | unchanged | |
| 0 | 1 | 0 | RESET |
| 1 | 0 | 1 | SET |
| 1 | 1 | – | not used |

## 5.5 INTRODUCTION TO ASYNCHRONOUS SEQUENTIAL CIRCUIT DESIGN

The design of asynchronous sequential circuits follows essentially the reverse of the analysis procedure. That is from flow table, to transition table, to Karnaugh maps for the individual output variables and finally the Boolean equations. However, in order to produce reliable circuits the procedure becomes somewhat more complex than this simplified description suggests.

Consequently, what follows is a discussion of the design process, rather than its application which can be found in other texts.

**The design route**

Asynchronous circuit design usually begins with the *primitive flow table* which is similar to the flow table but only possesses one stable state per row. This is then

[9] It is worth reiterating that the reason is that if both of the inputs are 1 and then change together, there is a critical race condition which may lead to an output of either 0 or 1, i.e. unpredictable behaviour.

studied to see whether any of the resulting states are equivalent (in terms of their stable states, where their unstable states lead and the associated outputs). If any states are equivalent then they can be combined to produce the *minimum-row primitive flow table* (still with one stable state per row) which can be further reduced by *merging* equivalent rows to give the flow table we have used during circuit analysis.

The *state assignment* is then performed in which binary codes are given to the states in the flow table. Note that there are many state assignments possible (i.e. any binary code can be given to any state). This leads to the *transition table* and then on to the final Boolean equations for the output variables.

**Hazards**

Unfortunately the process is not quite this straightforward, and further thought is necessary to produce a reliable circuit. We saw in Section 4.3 that combinational circuits can contain hazards and therefore the effects of these in the combinational block of our asynchronous circuit must be taken into account. In addition asynchronous circuits possess their own potential hazards.

We have seen how relaxation of fundamental mode operation can lead to race conditions. In a similar way if a circuit possesses more than two internal variables and these change 'simultaneously' then the order in which they change may result in different eventual states being reached. (In this situation the flow table will have more than two rows, and the order in which these rows are visited within the same column, upon external inputs being changed, may differ and lead to an unexpected stable state.) Techniques for predicting and eliminating such problems exist and this is done during state assignment. It may involve the introduction of additional intermediate states which are visited transiently during the actual required transition.

Spikes in the outputs may occur due to the same cause (i.e. transiently visited states) and can be eliminated by consideration of the 'don't care' conditions when drawing up the output table.

Finally, even if neither critical races nor hazards in the combinational logic block exist an asynchronous circuit may still possess *essential hazards*. These are due to internal delays in the circuit which make it appear as if a single external input has changed three times (i.e. $0 \rightarrow 1 \rightarrow 0 \rightarrow 1$) rather than just once. These can be eliminated by adding in appropriate delays to the circuit to ensure signals propagate in the 'correct' order.

**Summary**

In *principle* asynchronous design is straightforward *but* the presence of hazards means that care must be taken for such designs to function correctly. The hazards are usually dealt with by: firstly eliminating critical races; then ensuring all combinational logic is free of hazards; and finally spotting and eliminating essential hazards by the addition of strategically placed delays.

## 5.6 SELF-ASSESSMENT

5.1 What is the basic difference between *sequential* and *combinational* logic circuits?

5.2 What is the general form of a sequential logic circuit?

5.3 How does the number of *internal* and *total* states of a sequential circuit relate to the number of outputs from the circuit's 'memory' and the circuit's inputs?

5.4 What are the basic differences between *asynchronous* and *synchronous* sequential circuits?

5.5 What conditions must be met for an asynchronous sequential circuit to be stable?

5.6 In what way can changes in the inputs to an asynchronous sequential circuit be restricted to ensure correct operation?

5.7 What is meant by 'breaking the feedback path' in the analysis of an asynchronous sequential circuit?

5.8 What are the: *transition table*, *flow table* and *state diagram*?

5.9 What movement in the flow table is caused by:
(a) a change in the inputs to an asynchronous sequential circuit
(b) movement to an unstable state?

5.10 If no stable state exists for certain input conditions what will happen to the output of an asynchronous sequential circuit when these conditions are present?

5.11 What is the difference between *non-critical* and *critical* races?

## 5.7 PROBLEMS

5.1 Analyse fully the circuit shown in Fig. 5.18 by producing the transition table and output map, the flow table and finally the state diagram. What function does the circuit perform? Comment on any similarites with Circuit 4.

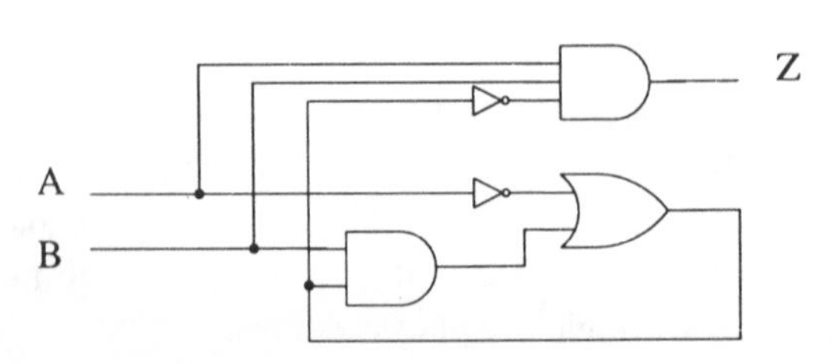


Fig. 5.18 Circuit to be analysed in Problem 5.1

5.2 Analyse fully the circuit shown in Fig. 5.19 by producing the transition table and output map, the flow table and finally the state diagram. Compare the function of this circuit with that in Problem 5.1.

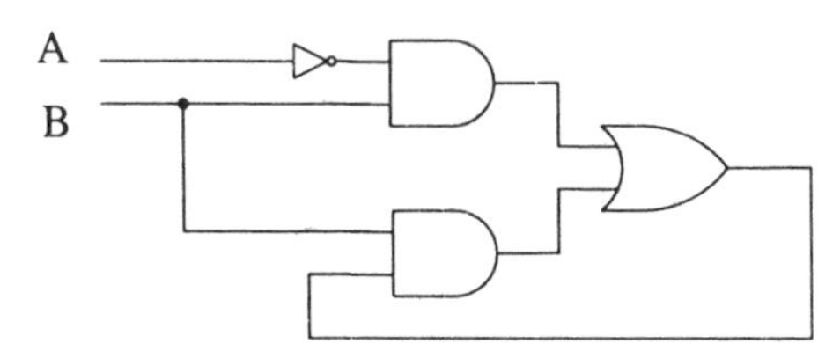


Fig. 5.19 Circuit to be analysed in Problem 5.2

5.3 Analyse fully the circuit shown in Fig. 5.20 by producing the transition table and output map, the flow table and finally the state diagram. Comment on any similarities to Circuit 4 and those analysed in Problems 5.1 and 5.2. How could this circuit be used to detect (i.e. produce an output $Z=1$) an input sequence of (0,1),(1,1)?

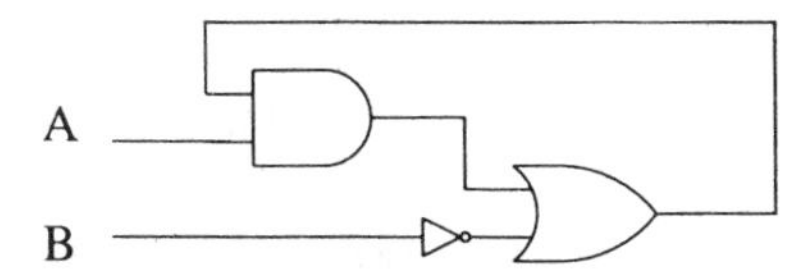


Fig. 5.20 Circuit to be analysed in Problem 5.3

5.4 Analyse the operation of a circuit constructed from cross-coupled NAND gates (see Circuit 6, Section 5.4).

5.5 Once Problems 5.1, 5.2 and 5.3 have been completed use what you have learned to design a circuit to detect the input sequence (0,1),(0,0).
Comment on any similarities between your design and Circuit 3.

5.6 Analyse Circuit 6, the cross-coupled NOR gates, as a circuit with two feedback paths (i.e. *without* combining the dependent feedback signals).

# 6 Flip-flops and flip-flop based circuits

## 6.1 INTRODUCTION

Flip-flops[1] are vital ingredients in all except purely combinational logic circuits and are therefore extremely important. The SR (Set-Reset) flip-flop was introduced in the last chapter and illustrates an important point, namely that all flip-flops are asynchronous sequential logic circuits. However, by controlling their use they can be considered as synchronous circuit elements, which is exactly the approach taken here. Rather than providing a detailed description of how flip-flops are designed and operate, they are presented as discrete circuit elements (e.g. like a multiplexer or full adder) to be used in subsequent designs.

In general, flip-flops possess data inputs (e.g. the S and R inputs), an output, $Q$ (and its complement, $\bar{Q}$), and also (as we will see) a 'clock' input which controls the activation, or clocking, of the flip-flop. That is the timing of the change in the flip-flop's output in response to its inputs.

### 6.1.1 Flip-flop types

The SR flip-flop can be set, or reset, or held in the same state via control of its inputs. However, it cannot be made to change state (i.e. its output give the complementary value) or *toggle*. Further thought reveals that if it could its operation would be unpredictable since it is an asynchronous circuit and therefore if made to toggle it would do so continuously (i.e. oscillate) until new inputs were presented.

However, by *gating* the inputs to an SR flip-flop via AND gates under control of the flip-flop's complementary outputs ($Q$ and $\bar{Q}$) it is possible to produce a flip-flop whose output can be made to toggle (i.e. go from 0 to 1 or vice versa) when activated (see Problem 6.1). This is then an example of a T-type (Toggle) flip-flop whose output either remains the same (when the input $T=0$) or toggles (when $T=1$).

Alternative input gating (again using AND gates via control of the flip-flop's outputs) allows the SR flip-flop to be transformed into a device called a JK flip-flop which combines the characteristics of the both the SR and T-types (see Problem 6.2). The JK operates as for an SR flip-flop with the addition that both of its inputs can be 1, in which case the output toggles.

[1]A flip-flop is basically a circuit capable of storing a 0 or 1.

The fourth and final type of flip-flop has the simplest operation acting only to delay the transfer of its input to its output. Clearly the activation of this D-type (Delay) flip-flop must be controlled externally since otherwise it would simply be a wire link. Hence it has three external connections: the input, $D$, and output, $Q$, and a *clock* input which controls the timing of the transfer of the input value to the output.

In fact all types of flip-flops are available in clocked form which basically means that they have an additional *clock* input, with the flip-flop's outputs only responding to the input conditions when the clock line goes active (i.e. the flip-flop is 'clocked').

To summarise, there are four types of flip-flop:

**SR** Set-Reset; must not allow both inputs to be 1 simultaneously.
**T** Toggle type; on clocking the output either remains the same or toggles depending if the input is 0 or 1.
**JK** Offering the capabilities of both the SR and T types.
**D** Delay-type flip-flop; upon clocking the output follows the input.

The operation of these four flip-flops can be described in several ways.

- A *truth table* which shows what the flip-flop's output, $Q^+$, will be for all possible input combinations. (Note the use of $Q$ and $\bar{Q}$ in the output column which respectively mean that the output either remains as it is or toggles.)
- An *excitation table* which gives the inputs that must be used to produce a given output transition.
- A *Karnaugh map* containing the same information as the truth table but in a different format. (Note there is a cell for every possible value of $Q$, and so more cells than rows in the truth table.)
- The *next state equation* which is the minimised form of the output, $Q^+$, from the Karnaugh map as a function of the flip-flop's inputs and the flip-flop's present output (state), $Q$.

These are shown for all types of flip-flop in Table 6.1.

### 6.1.2 Flip-flop operation

Being asynchronous circuits the brief description of flip-flops given above clearly cannot adequately describe their precise operation.[2] Although the SR flip-flop does find uses in its basic (unclocked) form (see Section 6.2.1), the other three types are always clocked, that is the changes in the outputs occur under the control of a clock[3] signal.

[2]For example the JK flip-flop is usually implemented as a ***master-slave*** device in order to give reliable operation. This consists of two appropriately gated SR flip-flops (see Problem 6.3). Also D-type flip-flops may not be constructed from basic logic gates at all, but rather from circuit elements called transmission gates (see Section 9.3.6) which are not logic gates.

[3]The clock input to a flip-flop is sometimes labelled as the ***strobe*** or ***enable*** input.

Table 6.1 The four types of flip-flop, and their truth and excitation tables, Karnaugh maps and next state equations

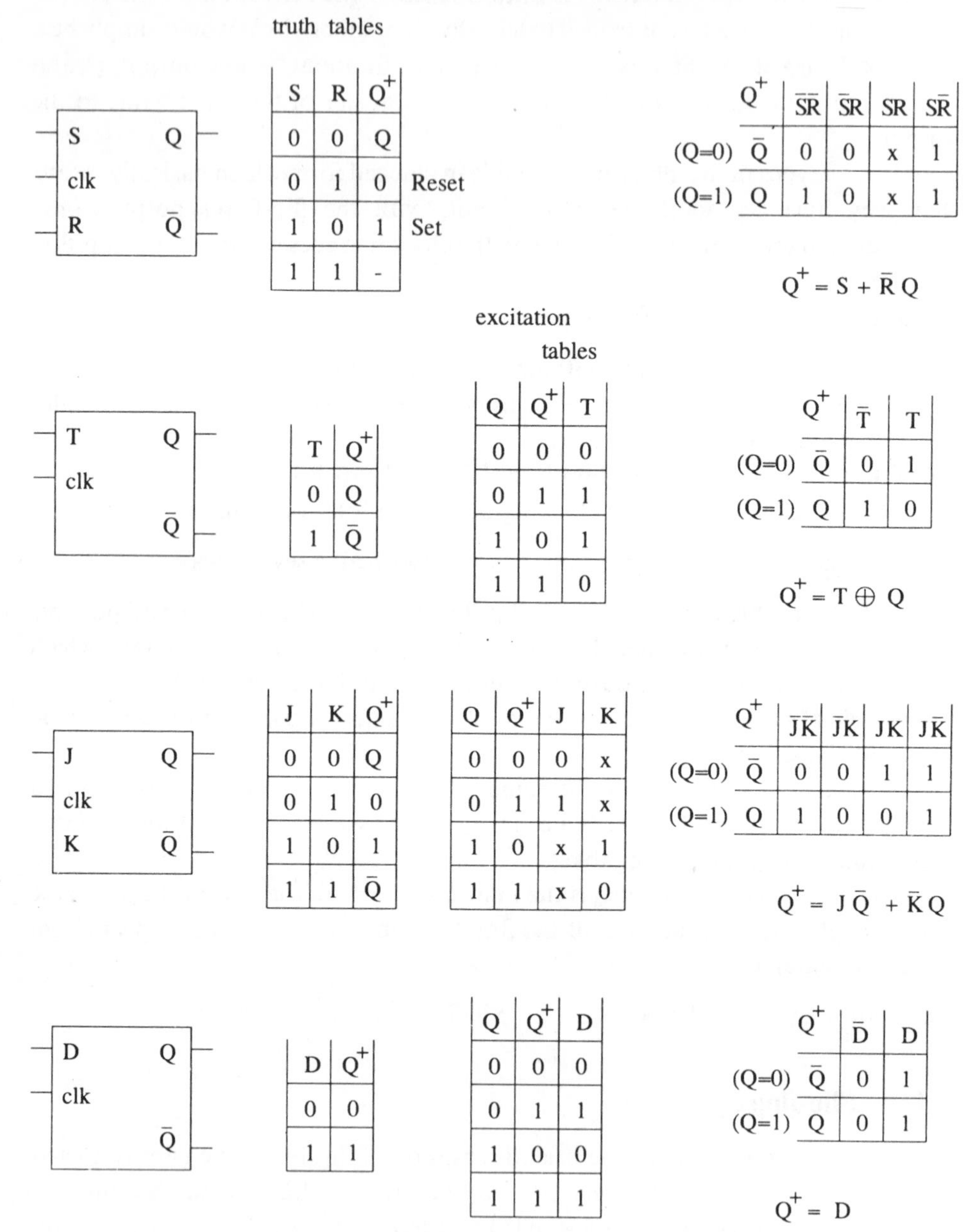


| S | R | $Q^+$ | |
|---|---|---|---|
| 0 | 0 | Q | |
| 0 | 1 | 0 | Reset |
| 1 | 0 | 1 | Set |
| 1 | 1 | - | |

| $Q^+$ | $\bar{S}\bar{R}$ | $\bar{S}R$ | $SR$ | $S\bar{R}$ |
|---|---|---|---|---|
| (Q=0) $\bar{Q}$ | 0 | 0 | x | 1 |
| (Q=1) $Q$ | 1 | 0 | x | 1 |

$$Q^+ = S + \bar{R}Q$$

| T | $Q^+$ |
|---|---|
| 0 | Q |
| 1 | $\bar{Q}$ |

| Q | $Q^+$ | T |
|---|---|---|
| 0 | 0 | 0 |
| 0 | 1 | 1 |
| 1 | 0 | 1 |
| 1 | 1 | 0 |

| $Q^+$ | $\bar{T}$ | $T$ |
|---|---|---|
| (Q=0) $\bar{Q}$ | 0 | 1 |
| (Q=1) $Q$ | 1 | 0 |

$$Q^+ = T \oplus Q$$

| J | K | $Q^+$ |
|---|---|---|
| 0 | 0 | Q |
| 0 | 1 | 0 |
| 1 | 0 | 1 |
| 1 | 1 | $\bar{Q}$ |

| Q | $Q^+$ | J | K |
|---|---|---|---|
| 0 | 0 | 0 | x |
| 0 | 1 | 1 | x |
| 1 | 0 | x | 1 |
| 1 | 1 | x | 0 |

| $Q^+$ | $\bar{J}\bar{K}$ | $\bar{J}K$ | $JK$ | $J\bar{K}$ |
|---|---|---|---|---|
| (Q=0) $\bar{Q}$ | 0 | 0 | 1 | 1 |
| (Q=1) $Q$ | 1 | 0 | 0 | 1 |

$$Q^+ = J\bar{Q} + \bar{K}Q$$

| D | $Q^+$ |
|---|---|
| 0 | 0 |
| 1 | 1 |

| Q | $Q^+$ | D |
|---|---|---|
| 0 | 0 | 0 |
| 0 | 1 | 1 |
| 1 | 0 | 0 |
| 1 | 1 | 1 |

| $Q^+$ | $\bar{D}$ | $D$ |
|---|---|---|
| (Q=0) $\bar{Q}$ | 0 | 1 |
| (Q=1) $Q$ | 0 | 1 |

$$Q^+ = D$$

Given that a flip-flop is clocked there are still several ways in which this can be performed. For instance the clock (or control) line going active may then mean that *any* changes in the inputs (during the active clock signal) take effect at the outputs. In this case the flip-flop is said to be *transparent* (since as long as the clock is active the outputs can 'see' right through to the inputs).

With such a clock signal the flip-flop is effectively a *level triggered* device with the inputs taking effect as soon as the clock line reaches, and whilst it remains at, its active level. Obviously in order for such devices to operate correctly it will usually be necessary to ensure the inputs do not change when the clock is active.

Alternatively flip-flops may be *edge triggered*. An edge refers to a rising or falling logic signal (e.g. going from 0 to 1 or 1 to 0), referred to as positive and negative edges respectively. An edge-triggered flip-flop will change its outputs in response to the inputs when an edge appears on the clock line. Therefore it is not transparent since ideally it responds to the inputs at a particular instant in time. This is the most common form of flip-flop used and the one that we will use in subsequent designs.

### 6.1.3 Timing requirements

In order for flip-flops to perform correctly (and effectively as synchronous components) they must be operated within certain timing constraints. These are the *set-up* and *hold* times, and refer respectively to the times that the inputs must be held at the required values before and after being clocked. Adhering to these times guarantees correct operation of the asynchronous flip-flops.

This requirement imposes a limit on the speed at which a flip-flop can be toggled (changed repeatedly between states), with the fundamental frequency of this toggling being one of the parameters used to describe the performance of a flip-flop.

### 6.1.4 Other inputs

In addition to the inputs and outputs shown in Table 6.1, flip-flops (available as circuit elements) usually possess *preset* (i.e. set $Q=1$) and *clear* (i.e. reset $Q=0$) inputs. These inputs may be synchronous in which case the flip-flop's outputs will change when it is next clocked, or asynchronous which means the outputs will change immediately in response to either of these inputs being active.

Figure 6.1 shows a positive edge triggered JK flip-flop with active-LOW preset, $\overline{Pr}$, and clear, $\overline{Cl}$, and a negative edge triggered D-type with active-HIGH preset

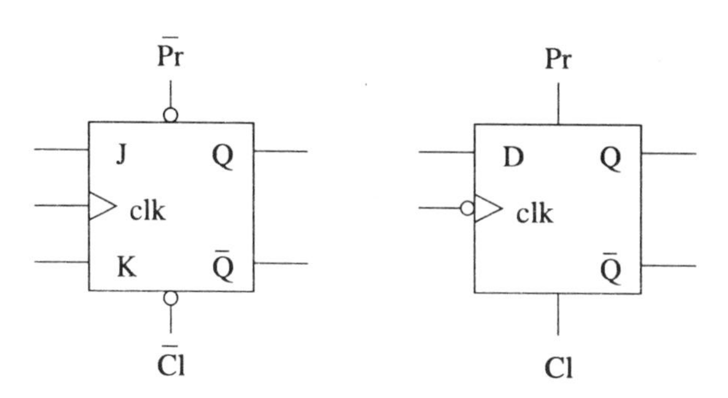


Fig. 6.1 Circuit symbols for a positive edge triggered JK flip-flop and a negative edge triggered D-type flip-flop

and clear. Remember that a bubble indicates inversion and that active-LOW means the signal must be LOW to take effect (assertion level logic). The wedge symbol on the clock input indicates edge triggering, so a bubble and a wedge means negative edge triggered.

Finally it should be noted that many elegant heuristic designs using flip-flops exist which make use of the preset and clear inputs to perform the required circuit operation in a simple manner (e.g. Problem 6.5).

## 6.2 SINGLE FLIP-FLOP APPLICATIONS

The following are two common uses of single flip-flops, one clocked and the other unclocked.

### 6.2.1 Switch debouncing

All mechanical switches will 'bounce' when closed which means the contacts do not make immediate firm contact. The consequence of this is that the actual voltage from the switch oscillates when contact is made. If this signal is to act as the input to a digital circuit then instead of the circuit seeing only a single transition it will actually see many. A common use of an SR flip-flop is to clean up this signal to ensure only a single transition is produced. The SR can do this because once set or reset it will remain in this state if both inputs are held at 0 (and so be immune to the oscillations).

**Example 6.1**

Draw the outputs that will be obtained from the circuit in Fig. 6.2 and determine how an SR flip-flop can be used to clean up these signals.

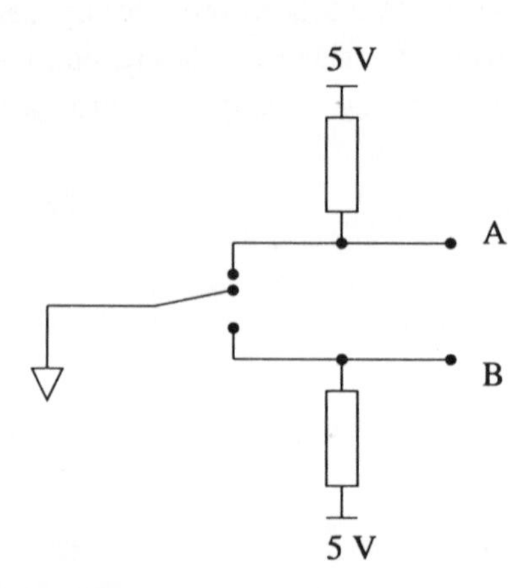


Fig. 6.2 Switch circuit whose output, exhibiting contact bounce, is 'cleaned up' as described in Example 6.1

***Solution***

Fig. 6.3 shows the signals that will be obtained from points $A$ and $B$. We note that the contact bounce gives conditions of $A$ and $B$ both 1 during which time we require the outputs to remain constant. We cannot use inputs of $S=R=1$ for an SR flip-flop but note that if both inputs to the flip-flop are inverted (to give an $\overline{\text{S}}\overline{\text{R}}$ flip-flop) then we will obtain the required clean waveform shown in Fig. 6.4.

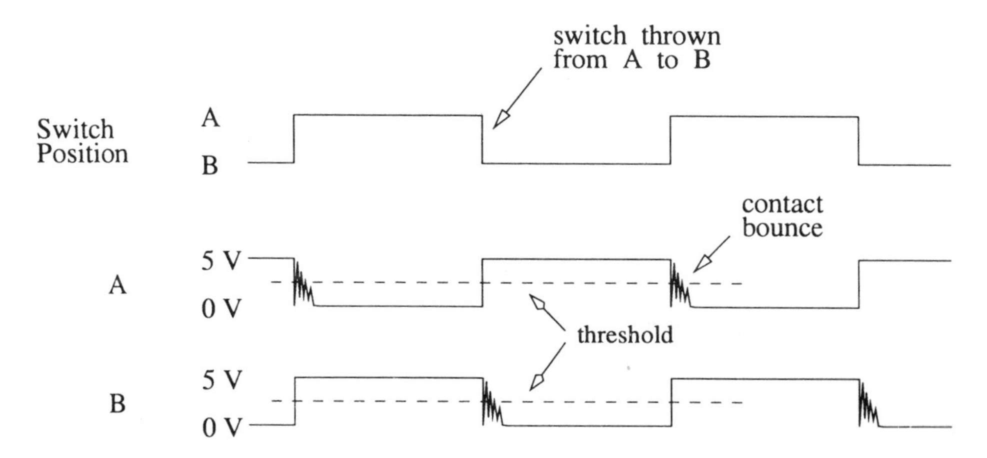


Fig. 6.3 Output from the circuit in Fig. 6.2 which illustrates the problem of contact bounce. A logic level of 1 is represented by a voltage above the illustrated threshold

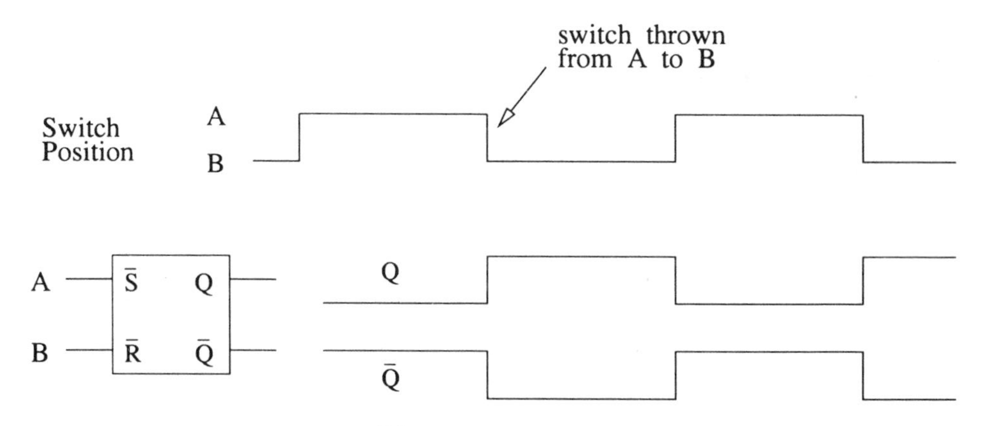


Fig. 6.4 Switch debouncing using an $\overline{\text{S}}\overline{\text{R}}$ flip-flop; see Example 6.1

Input $A$ going LOW, and so $\bar{S}$ LOW, will set $Q=1$. Then as this transition occurs the contact bounce giving $A$ and $B$ both HIGH will cause both $\bar{S}$ and $\bar{R}$ to be HIGH (i.e. $S=R=0$) and so the flip-flop will remain in the same ($Q=1$) state as required. Similarly, input $B$ going LOW will reset $Q$ to 0. Contact bounce giving $A=B=1$ will not affect $Q$ which remains LOW.

### 6.2.2 Pulse synchroniser

In synchronous circuits it is vital that events (signals) are correctly synchronised to one another. Consider an asynchronous pulse which must be synchronised to the master clock within the digital system it is entering. How can this be achieved? A D-type flip-flop provides a solution since it 'allows' the pulse to enter the system only when it has been clocked as shown in Fig. 6.5.

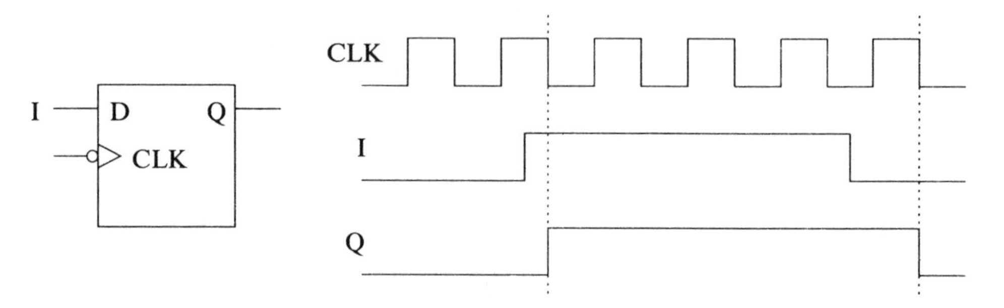


Fig. 6.5 Synchronisation of a pulse using a negative edge-triggered D-type flip-flop

**Example 6.2**

Draw the relevant waveforms showing pulse synchronisation using a positive edge triggered flip-flop.

***Solution***

These are shown in Fig. 6.6.

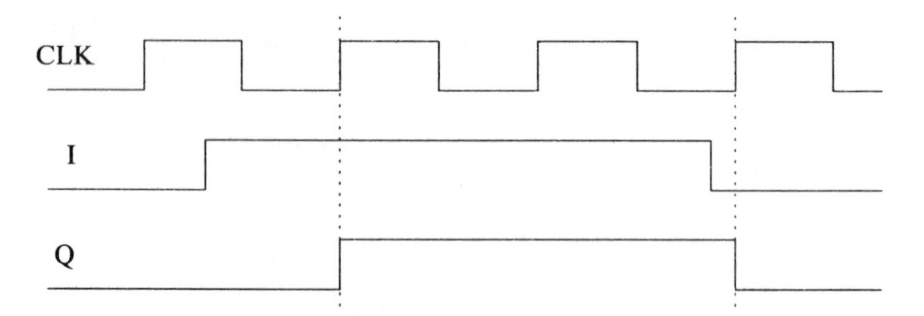


Fig. 6.6 The use of a positive edge triggered flip-flop for pulse synchronisation, as discussed in Example 6.2

Use of an AND gate allows synchronised gating of the clock by the input pulse as shown in Fig. 6.7 where $Q \cdot CLK$ is the clock signal and the output from the D-type AND'd together. Note that the propagation delay of the D-type has been included.

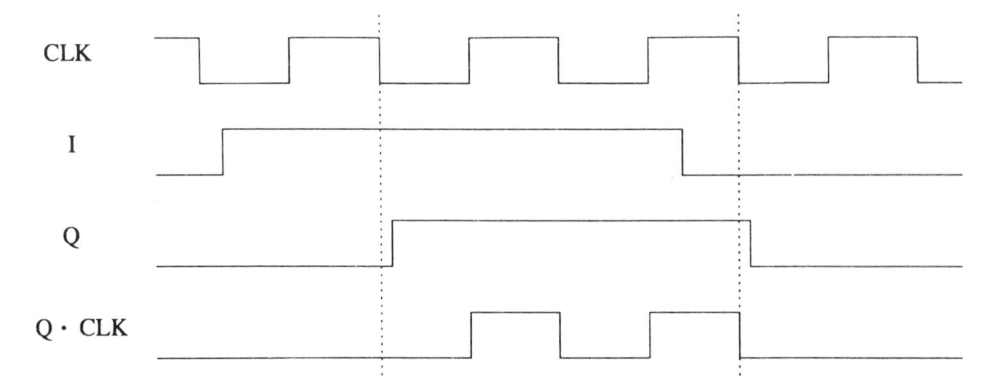


Fig. 6.7 Synchronised gating of a clock using a negative edge triggered D-type. Note that the delay of the D-type is shown (as the lag in *Q* changing state)

## Example 6.3

Consider the use of the positive edge triggered D-type synchroniser from the last example for the synchronised gating of a clock.

### *Solution*

The waveforms are shown in Fig. 6.8. The important thing to notice is that in this case the propagation delay of the D-type causes the production of spikes (and shortening of the initial gated clock pulse) following the AND'ing process which would be liable to cause incorrect circuit operation. This circuit should therefore *not* be used. Further examples of this type are discussed in Sections 11.2 and 11.3.

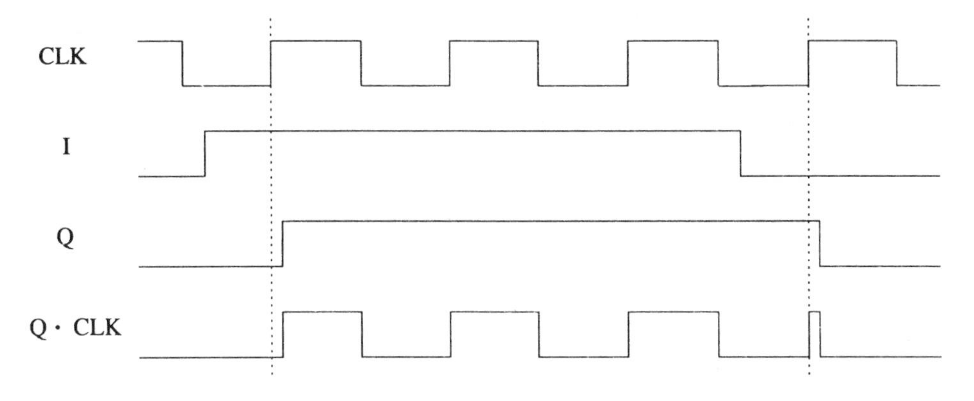


Fig. 6.8 Synchronised gating of a clock using a positive edge triggered D-type, as discussed in Example 6.3

# 6.3 REGISTERS

Because it possesses 'memory', a single flip-flop can be used to store a single binary digit (a bit), hence $n$ flip-flops can store an $n$-bit word. Such an arrangement is called a register. Considered as circuit elements in their own right registers have many applications including arithmetic operations, special types of counter and the simple storage of binary patterns. We will now look at some of the most common of these.

## 6.3.1 Shift registers

A shift register is simply a collection of clocked flip-flops linearly connected together so that information can be passed between them. That is the value stored in one flip-flop is shifted to the next in line when the register is clocked. Each flip-flop has its output(s) connected to the input(s) of the next in line. Obviously to be of use data must be able to be moved in and out of this structure, and this can be performed either *serially* or in *parallel.*

Used serially, data is passed into the flip-flop at the end of the line and then fed along from flip-flop to flip-flop. In parallel mode data is input to all flip-flops simultaneously. Fig. 6.9 shows a three-bit shift register constructed out of D-type flip-flops. Note how: the inputs and outputs of adjacent flip-flops are connected; all the clear and clock lines are tied together; and the gated preset arrangement allows the register to be parallel loaded with any desired pattern.

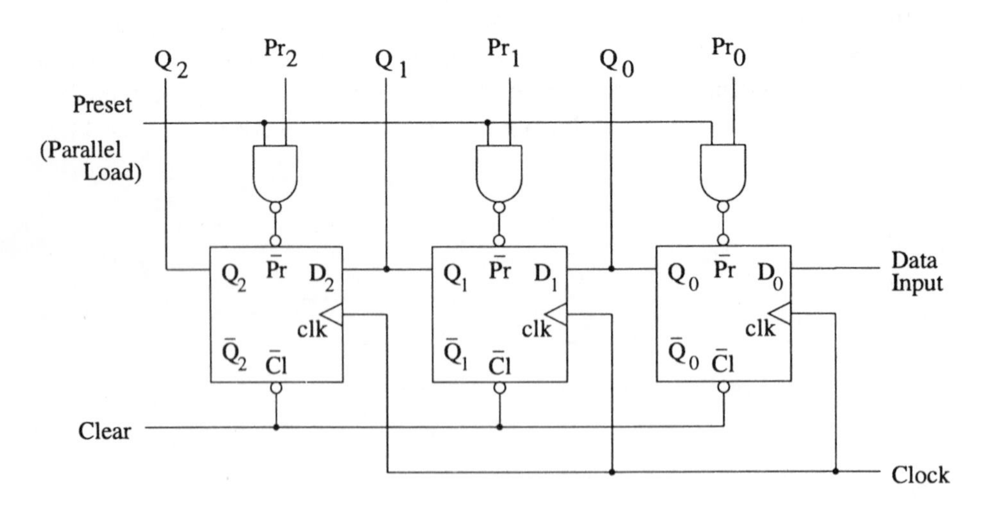


Fig. 6.9 A three-bit shift register

Because there are two ways of entering and extracting data from the shift register this gives four possible modes of operation: serial in, parallel out (SIPO); serial in, serial out (SISO); parallel in, serial out (PISO); and parallel in, parallel out (PIPO).

Although some of these forms are of more use than others we will briefly look at how the shift register shown in Fig. 6.9 would be used in each of these four modes.

**SIPO**

- Clear all flip-flops by taking clear LOW.
- Set clear and preset HIGH for normal flip-flop operation.
- Apply serial data train to the first flip-flop and clock at a rate synchronised to the data train so that one bit is entered for each clock trigger (for three bits of data for this shift register).
- Data is now residing in the shift register and can be read out in parallel from all three flip-flops simultaneously using $Q_0$, $Q_1$ and $Q_2$.

**Example 6.4**

Draw the contents of a four-bit shift register, at each clock pulse, which has the binary pattern 0110 serially loaded into it.

***Solution***

This is shown in Fig. 6.10.

| | $Q_3$ | $Q_2$ | $Q_1$ | $Q_0$ |
|---|---|---|---|---|
| Reset | 0 | 0 | 0 | 0 |
| Clock 1 | 0 | 0 | 0 | 0 |
| Clock 2 | 0 | 0 | 0 | 1 |
| Clock 3 | 0 | 0 | 1 | 1 |
| Clock 4 | 0 | 1 | 1 | 0 |

Fig. 6.10 Contents of a four-bit shift register when being serially loaded with binary pattern 0110 (see Example 6.4)

Note that in the SIPO shift register the data is transformed from being separated in time (temporally) to separated in space (spatially), within the individual flip-flops.

**SISO**

For the serial in, serial out shift register data is loaded in exactly the same way as for the SIPO but is then simply clocked out serially via $Q_2$. Obviously once loaded the data need not be accessed immediately and so can be stored. Also, it can be clocked out at a different rate, so providing a method of buffering data between two digital systems running at different clock speeds.

Note that because a SISO can be operated with only two connections (to get the data in and out) its size is not constrained by necessary access to any other inputs and outputs.

**PISO**

- Clear all flip-flops by taking clear LOW.
- Present the parallel data (a three-bit word) to the preset input lines $Pr_2$, $Pr_1$, $Pr_0$.
- Write this data into the register by taking preset enable HIGH.

This means the three-bit word is stored in the register, and can be read out as for the SISO using three clock pulses. Note that the PISO performs a spatial-to-temporal conversion of data.

**PIPO**

The PIPO shift register takes the data in and outputs it in parallel. Hence there is no shifting of data within the circuit, rather it simply acts as three memory cells.

The register discussed here can only shift data in one direction. *Bidirectional* shift registers are also available that can shift data in either direction. Note that one potential use of shift registers is for the multiplication or division by factors of two (by simply shifting data left or right respectively, see Section 2.5).

### 6.3.2 Applications of shift registers

**Digital delay lines**

If a single-bit data stream is fed serially into a shift register and then read out serially from the output of one of the register's flip-flops then the effect is that of delaying the data stream. For a clock period $T$ then if the data is read from the $n$th stage of the register, the data is delayed by $(n-1)T$.

**Sequence generator**

If a binary pattern is fed into a shift register it can then be output serially to produce a known binary sequence. Moreover, if the output is also fed back into the input (to form a SISO connected to itself) the same binary sequence can be generated indefinitely.

When a SISO shift register is connected to itself this is usually referred to as a re-entrant shift register, dynamic shift register, ring buffer or circulating memory. Variations on this type of circuit are used for data encryption, error checking and for holding data during digital signal processing.

**Ring counters**

Shift registers can be used to produce a type of simple counter whose advantage (in addition to the simplicity) is that they can operate at very high speeds since there is no need for any external control or decoding circuitry (necessary for most counters as we will see in the next chapter).

Such counters are formed by simply using a re-entrant shift register (the serial output is fed back to the serial input) which is (usually) loaded with a solitary high value. The register is then clocked and the output, taken from any one of the flip-flops, simply goes high every time the single stored bit arrives at that flip-flop (i.e. after $N$ clock cycles giving what is known as a mod-$N$ counter).

A second type of counter can be produced by connecting the $\bar{Q}$ output from the last flip-flop of a SISO back to the input and then loading a single 1. This is usually referred to as one of the following: twisted ring, switched tail, Johnson or Moebius[4] counter. A mod-2$N$ twisted ring counter requires $N$ flip-flops.

**Example 6.5**

Tabulate the contents of four-bit ring and Johnson counters which have a single bit entered and are then clocked.

***Solution***

These are shown in Fig. 6.11.

4-bit Ring

| | $Q_3$ | $Q_2$ | $Q_1$ | $Q_0$ | |
|---|---|---|---|---|---|
| Reset | 0 | 0 | 0 | 0 | |
| | 0 | 0 | 0 | 1 | |
| | 0 | 0 | 1 | 0 | |
| | 0 | 1 | 0 | 0 | |
| | 1 | 0 | 0 | 0 | |
| | 0 | 0 | 0 | 1 | mod-4 |
| | 0 | 0 | 1 | 0 | mod-4 |
| | 0 | 1 | 0 | 0 | mod-4 |
| | 1 | 0 | 0 | 0 | mod-4 |
| | 0 | 0 | 0 | 1 | |
| | 0 | 0 | 1 | 0 | |

4-bit Johnson

| | $Q_3$ | $Q_2$ | $Q_1$ | $Q_0$ | |
|---|---|---|---|---|---|
| Reset | 0 | 0 | 0 | 0 | |
| | 0 | 0 | 0 | 1 | mod-8 |
| | 0 | 0 | 1 | 1 | mod-8 |
| | 0 | 1 | 1 | 1 | mod-8 |
| | 1 | 1 | 1 | 1 | mod-8 |
| | 1 | 1 | 1 | 0 | mod-8 |
| | 1 | 1 | 0 | 0 | mod-8 |
| | 1 | 0 | 0 | 0 | mod-8 |
| | 0 | 0 | 0 | 0 | mod-8 |
| | 0 | 0 | 0 | 1 | |
| | 0 | 0 | 1 | 1 | |

Fig. 6.11 The outputs from four-bit ring and Johnson counters; see Example 6.5

## 6.4 SELF-ASSESSMENT

6.1 What are the four types of flip-flop?

6.2 Why must the T and JK flip-flops be clocked (synchronous).

6.3 In what ways can the operation of a flip-flop be described; what does the excitation table tell you?

6.4 Draw the truth table and excitation table for a JK flip-flop and relate their respective entries to each other (i.e. which entries in the two tables correspond to each other?).

[4]A Moebius strip is a loop made from a strip of paper with a single twist in, meaning it only has one side.

6.5 What does the next state equation for a flip-flop tell you?

6.6 What do the terms *transparent*; *level triggered* and *edge triggered* mean?

6.7 What happens to the output $Q$ when a flip-flop is preset or reset; and what do asynchronous and synchronous mean when referred to these operations?

6.8 What is a shift register, and why can such a circuit be used to multiply a binary number by $2^n$.

6.9 How many flip-flops are required to construct mod-12 ring and Johnson counters?

## 6.5 PROBLEMS

6.1 Determine how the circuit shown in Fig. 6.12 functions as a T-type flip-flop. What problem would there be when $T=1$ and how could it be resolved. (Hint: remember that the SR flip-flop must have a propagation delay.)

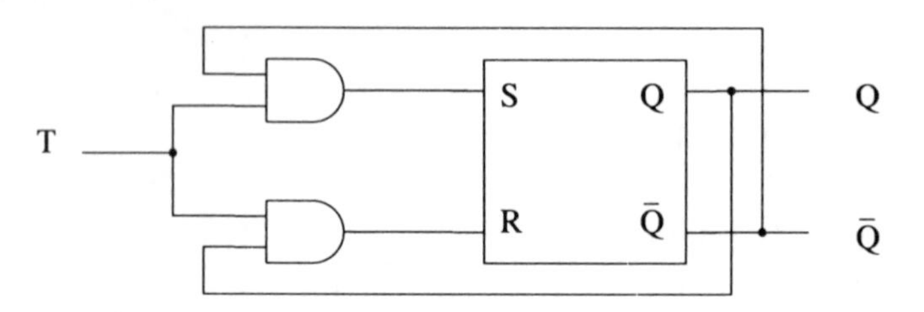


Fig. 6.12 Circuit to be analysed in Problem 6.1

6.2 Determine how the circuit shown in Fig. 6.13 functions as a JK-type flip-flop. Under what input conditions may a problem occur?

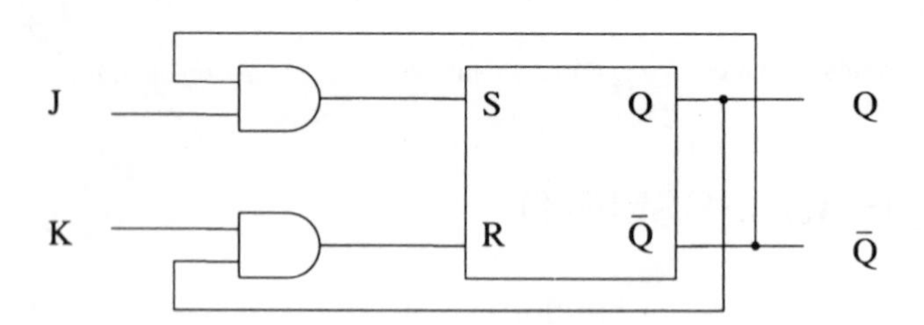


Fig. 6.13 Circuit to be analysed in Problem 6.2

6.3 Fig. 6.14 shows a *master-slave* flip-flop where the C input is a square-wave clock signal. Analyse its operation and find why it does not suffer from the problems afflicting the circuits in Problems 6.1 and 6.2.

6.4 How could:

(a) a JK flip-flop be used as a D-type?
(b) a JK flip-flop be used as a T-type?
(c) a D-type flip-flop be used as a T-type?

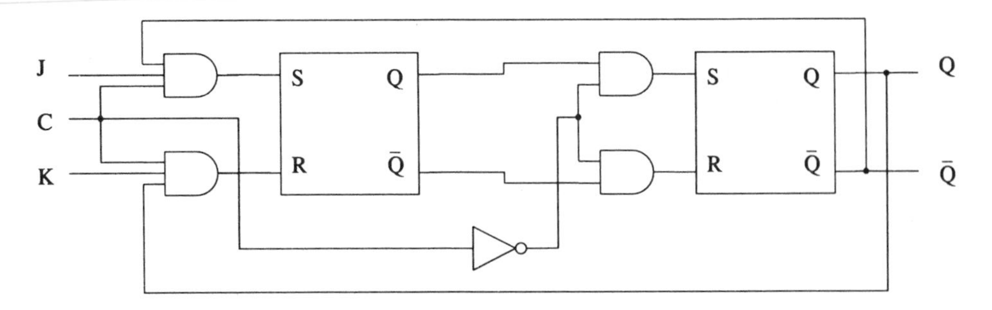


Fig. 6.14 Circuit to be analysed in Problem 6.3

6.5 What does the circuit in Fig. 6.15 do?

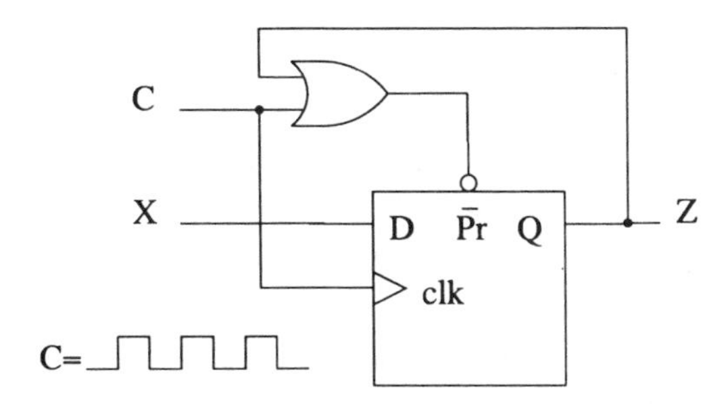


Fig. 6.15 Circuit to be analysed in Problem 6.5

6.6 Show the circuits for, and outputs from, a mod-6 ring counter and Johnson counter. What problems would arise if somehow one of the unused states (i.e. binary patterns not held by the flip-flops during normal operation of the Johnson counter) was entered.

# 7 Counters

## 7.1 INTRODUCTION

Counters are one of the most widely used logic components in digital electronic circuits. In addition to simply counting they can be used: to measure time and frequency; increment memory addresses; and divide down clock signals amongst a host of other applications. They are basically special forms of synchronous sequential circuits in which the state of the circuit is equivalent to the count held by the flip-flops used in the circuit. In this chapter we will look only at counters which count in binary sequence, although the next chapter describes how to design circuits with any required count sequence, such as Gray code for instance.

The 'mod' of the counter is the number of states the counter cycles through before resetting back to its initial state. So a binary mod-8 counter has eight count states, from $000_2$ to $111_2$ (e.g. the mod-8 counter actually counts from 0 to 7). All of the counters we will look at use flip-flops as the storage elements that hold the count state. Therefore, a mod-$N$ counter will need to contain $n$ flip-flops, where $2^n \geq N$.

### 7.1.1 Asynchronous and synchronous counters

To divide the counters we will look at into two types: *asynchronous* and *synchronous.* When used *with respect to counters*[1] these adjectives describe whether the flip-flops holding the states of the circuit are all clocked together (i.e. synchronously) by a master clock or rather asynchronously, with each flip-flop clocked by the one preceding it.

- Asynchronous counters:
  - are also known as ripple counters;
  - are very simple;
  - use the minimum possible hardware (logic gates);
  - employ flip-flops connected serially, with each one triggering (clocking) the next;
  - have an overall count which 'ripples' through, meaning the overall operation is relatively slow;

[1]Note that asynchronous counters are not asynchronous circuits as described in Chapter 5.

  - require virtually no design.
- Synchronous counters:
  - use interconnected flip-flops, but all are clocked together by the system clock;
  - use the outputs from the flip-flops, to determine the next states of the following flip-flops (rather than simply clocking them);
  - require no settling time due to rippling (as all flip-flops are clocked synchronously);
  - need designing, to determine how the present state of the circuit must be used to determine the next state (i.e. count);
  - usually need more logic gates for their implementation.

Although we will not consider such circuits it is possible to design hybrid asynchronous/synchronous counters (e.g. two synchronous four-bit counters connected asynchronously to produce an eight-bit hybrid) that possess some of the advantages of both types and which are appropriate in some applications.

Due to their universal use in logic circuits, counters are widely available as logic elements in a wide range of forms. These often include clear and preset facilities (which can be used to load a particular count state), and up-down counts available within the same device. Because the design of binary counters is largely intuitive the approach we will take is to firstly consider simple circuits that act as mod-$2^n$ counters, and then how they must be modified to produce a general mod-$N$ count.

## 7.2 ASYNCHRONOUS COUNTERS

The output of a T-type flip-flop with $T=1$ will simply toggle every time it is clocked. So the circuit in Fig. 7.1 constructed from negative edge triggered T-type flip-flops will give the waveforms shown. Note that each flip-flop is clocked by the output from the preceding flip-flop with all flip-flop inputs tied HIGH so they toggle.

The outputs of the flip-flops will only change when the output from the preceding flip-flop changes to produce a negative edge. Since this will be once every clock period of the preceding flip-flop, the effect is for each flip-flop to produce an output with twice the period of the one clocking it. In effect the clock is divided down by 2 at each stage. Two flip-flops connected like this will produce a four-bit counter, and the three in Fig. 7.1 an eight-bit counter ($2^3=8$).

Any $2^n$-bit counter can be produced like this, whilst any particular count state for a counter built from $n$ flip-flops can be decoded using an $n$-input AND gate (or equivalent circuit) connected to the $Q$ or $\bar{Q}$ from all flip-flops as required.

**Down counters**

A count down circuit can be produced by either simply replacing the negative edge triggered flip-flops for positive edge triggered ones *or* using the $\bar{Q}$ outputs to trigger the next flip-flop (see Problem 7.1).

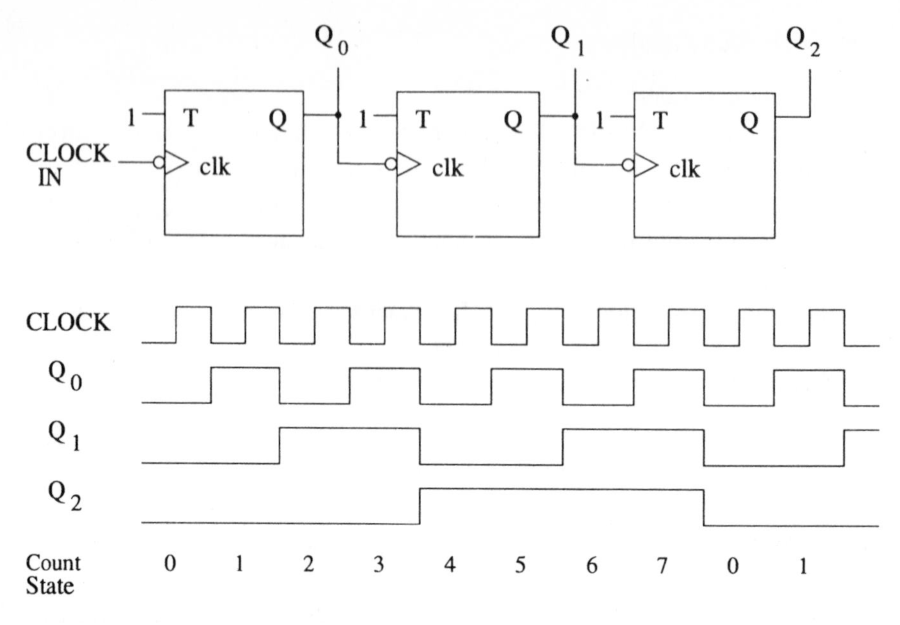


Fig. 7.1 The count action of rippled T-type flip-flops

### 7.2.1 Mod-$N$ asynchronous counters

A mod-$N$ (or divide-by-$N$) asynchronous counter, where $N=2^n$, will count up to $(N-1)$ (an output of all 1's) before resetting to all 0's and beginning the count sequence again.

A general mod-$N$ counter can be produced by using flip-flops with clear inputs and then simply decoding the $N$th count state and using this to reset all flip-flops to zero. The count will therefore be from 0 to $(N-1)$ repeated since the circuit resets when the count gets to $N$. Note that because the $N$th state must exist before it can be used to reset all of the flip-flops there is the likelihood that glitches will occur in some of the output lines during the resetting phase (since an output may go high as the reset count is reached, and then be reset to 0).

**Example 7.1**

Design a mod-10 binary up-counter using negative edge JK flip-flops with active-LOW clear.

***Solution***

Four flip-flops are required, and decimal state 10 must be decoded and used to reset all flip-flops to give a repeated count from 0 to 9 (0000 to 1001). State 10 is given by $Q_3\bar{Q}_2Q_1\bar{Q}_0$ (1010) so a four-input NAND gate (as the clear is active-LOW) could be used to decode this count and clear all flip-flops. However, since states 11 to 15 will never be entered they can be considered as 'don't care' conditions and used to simplify the logic. From the Karnaugh map in Fig. 7.2 it can be seen that the count state $Q_3Q_1$ can be used to perform the reset with the subsequent circuit also shown.

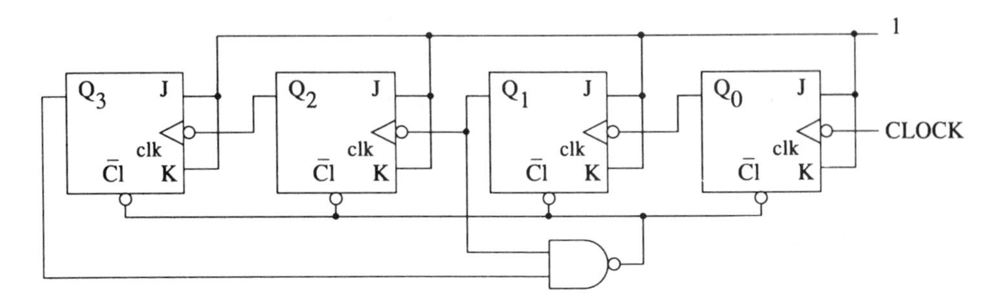


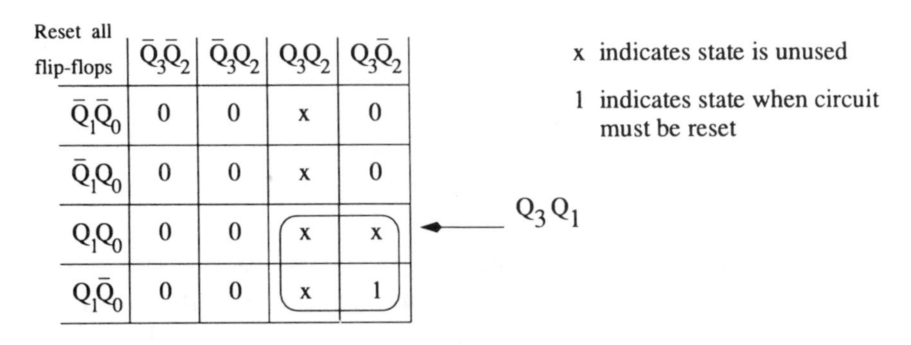

| Reset all flip-flops | $\bar{Q}_3\bar{Q}_2$ | $\bar{Q}_3Q_2$ | $Q_3Q_2$ | $Q_3\bar{Q}_2$ |
|---|---|---|---|---|
| $\bar{Q}_1\bar{Q}_0$ | 0 | 0 | x | 0 |
| $\bar{Q}_1Q_0$ | 0 | 0 | x | 0 |
| $Q_1Q_0$ | 0 | 0 | x | x |
| $Q_1\bar{Q}_0$ | 0 | 0 | x | 1 |



Fig. 7.2 The binary mod-10 asynchronous up-counter designed in Example 7.1

**Summary**
Asynchronous (ripple) counters are easy to design but, because the count has to ripple through the system, timing problems can occur and glitches can be generated. Consequently the speed of operation of this type of counter is limited.

## 7.3 MOD-$2^n$ SYNCHRONOUS COUNTERS

Synchronous binary counters are arguably the simplest *sequential synchronous circuits*. They use the flip-flops to store the circuit's count state and (usually) have no external inputs. Thus the next state (count) is determined solely by the last state (count). We again initially take an intuitive look at mod-$2^n$ counters.

The waveforms required for a mod-8 counter were shown in Fig. 7.1, being the outputs, $Q_2$, $Q_1$ and $Q_0$, from the three flip-flops. If this synchronous mod-8 counter is to be built from negative edge triggered T-type flip-flops, then since all three flip-flops will be clocked together (as this is to be a synchronous circuit) we need to determine for each clock input whether the T input for each flip-flop must be 0 (for the output to remain the same) or 1 (for it to toggle).

By inspection of the waveforms in Fig. 7.1 it is clear that:

- $Q_0$ must toggle on every negative edge of the system clock and so we need $T_0=1$;
- $Q_1$ must only toggle when $Q_0=1$, and so we need $T_1=Q_0$;
- $Q_2$ must only toggle when $Q_0=Q_1=1$ and so we need $T_2=Q_0 \cdot Q_1$.

We must therefore use an AND gate to produce the steering logic (as it is known) to enable the toggling action of the flip-flops as required. The circuit for the mod-8 synchronous counter is shown in Fig. 7.3.

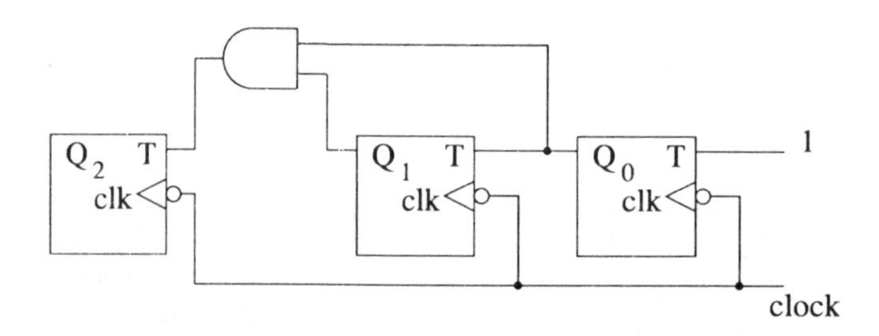


Fig. 7.3 A mod-8 synchronous counter constructed from negative edge triggered T-types

From the above analysis of the required steering logic a clear pattern emerges of how to produce any mod-$2^n$ counter (where $n$ is the number of flip-flops used). This is that the T input of each flip-flop must be the outputs from all preceding flip-flops AND'd together.

### 7.3.1 Implementation of the steering logic

Since for this type of counter in general we need the input to the $n$th T-type used to construct the circuit to be $T_n=Q_0 \cdot Q_1 \cdot Q_2 \cdots Q_{n-1}=T_{n-1} \cdot Q_{n-1}$ it is clear there are two ways in which the steering logic can be generated. One approach is to actually AND together all of the previous outputs, which for a counter with $n$ flip-flops requires AND gates with from 2 up to ($n$-1) inputs. The advantage of this *parallel* implementation is the only delay introduced is that of the single AND gate with the disadvantage that an AND gate with a prohibitively large number of inputs may be required[2] and the outputs from the flip-flops may also have to feed a prohibitively large number of inputs.

The alternative, *serial* approach, is to use only two-input AND gates and at each stage AND together the output from the each flip-flop with its input (i.e. its steering logic), for example $T_3=T_2 \cdot Q_2$.The disadvantage here is that this introduces a delay because the output of the steering logic is using a signal that must propagate (ripple) through all of the AND gates.

[2]Note also that in practice, as mentioned in Section 4.2.1, a gate with a large number of inputs will have a longer propagation delay than its two-input counterpart.

## 7.4 MOD-$N$ SYNCHRONOUS COUNTERS

We have seen that the design of mod-$2^n$ binary up-counters is straightforward, with all $2^n$ states produced, and cycled through in sequence. For general mod-$N$ counters we must begin to use (simplified) synchronous sequential circuit design techniques, with each present state of the circuit used to produce the necessary steering logic to take the circuit into the desired next state. Note that because the circuit is clocked the outputs are always stable. This is because although the next state is determined by the present state, the next state and present state variables are separated by the flip-flops, unlike the asynchronous sequential circuits studied in Chapter 5.

We begin by redesigning the mod-8 counter, which should give the circuit produced above (see the end of this section). Firstly we look at how D-type flip-flops could be used and then JK-types. This will also give us our first insight into the general differences in circuits designed using these two types of flip-flop.

### 7.4.1 Mod-8 counter using D-type flip-flops

We will need three flip-flops to give the necessary eight states from 000 (state 0) through to 111 (state 7). We begin by, as shown in Table 7.1, listing the eight possible present states of the circuit alongside the next states that must follow. The design task is to use the present state outputs from the three flip-flops to produce the required next states for the three flip-flops.

Since we are using D-type flip-flops and their outputs will equal their inputs when clocked (i.e. the next state equation is $Q^+=D$), we must simply ensure that the present states are used to produce the required next state for each flip-flop. This is easily achieved by producing a Karnaugh map for each of the three flip-flop's D-inputs in terms of the (present) outputs of the three flip-flops.

These Karnaugh maps are also shown in Table 7.1 and are (because the output of a D-type simply follows its input, i.e. $Q^+=D$) just the required next states for the circuit entered across the three maps as functions of the circuit's present states.

To complete the design we simply need to use the Karnaugh maps to simplify the required steering logic they define. This gives:

$$D_2=Q_2\overline{Q_0}+Q_2\overline{Q_1}+\overline{Q_2}Q_1Q_0$$
$$D_1=Q_1\overline{Q_0}+\overline{Q_1}Q_0$$
$$D_0=\overline{Q_0}$$

The circuit to implement this is shown in Fig. 7.4.

### 7.4.2 Mod-8 counter using JK flip-flops

The procedure for designing the counter using JK type flip-flops is fundamentally the same as for the D-type design. The major difference is that whereas the output

Table 7.1 Present and next states for a mod-8 binary up-counter, and the associated Karnaugh maps for the design of a D-type based circuit

| STATE | Present State $Q_2$ | $Q_1$ | $Q_0$ | Next State $Q_2^+$ | $Q_1^+$ | $Q_0^+$ |
|---|---|---|---|---|---|---|
| 0 | 0 | 0 | 0 | 0 | 0 | 1 |
| 1 | 0 | 0 | 1 | 0 | 1 | 0 |
| 2 | 0 | 1 | 0 | 0 | 1 | 1 |
| 3 | 0 | 1 | 1 | 1 | 0 | 0 |
| 4 | 1 | 0 | 0 | 1 | 0 | 1 |
| 5 | 1 | 0 | 1 | 1 | 1 | 0 |
| 6 | 1 | 1 | 0 | 1 | 1 | 1 |
| 7 | 1 | 1 | 1 | 0 | 0 | 0 |

Current output from flip-flops (present state); Required output from flip-flops (next state)

| $D_2$ | $\bar{Q}_2\bar{Q}_1$ | $\bar{Q}_2Q_1$ | $Q_2Q_1$ | $Q_2\bar{Q}_1$ |
|---|---|---|---|---|
| $\bar{Q}_0$ | 0 | 0 | 1 | 1 |
| $Q_0$ | 0 | 1 | 0 | 1 |

| $D_1$ | $\bar{Q}_2\bar{Q}_1$ | $\bar{Q}_2Q_1$ | $Q_2Q_1$ | $Q_2\bar{Q}_1$ |
|---|---|---|---|---|
| $\bar{Q}_0$ | 0 | 1 | 1 | 0 |
| $Q_0$ | 1 | 0 | 0 | 1 |

| $D_0$ | $\bar{Q}_2\bar{Q}_1$ | $\bar{Q}_2Q_1$ | $Q_2Q_1$ | $Q_2\bar{Q}_1$ |
|---|---|---|---|---|
| $\bar{Q}_0$ | 1 | 1 | 1 | 1 |
| $Q_0$ | 0 | 0 | 0 | 0 |

$$D_n = Q_n^+$$

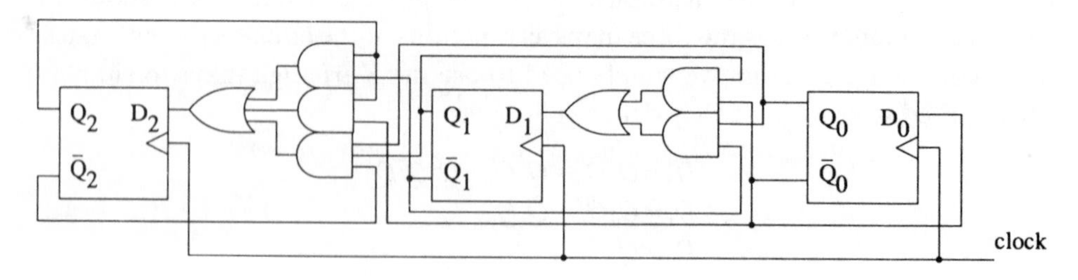


Fig. 7.4 Mod-8, D-type based, binary up-counter

from a D-type is simply its input, the output from a JK is given by the excitation function and depends upon the values on the J and K inputs (see Table 6.1).

We again write out the present and next states for the circuit (see Table 7.2) and hence the three individual flip-flops, but now also include tables for each of the

Table 7.2 The present and next states, and excitation requirements for a JK based mod-8 binary up-counter, together with the associated Karnaugh maps

| STATE | Present State $Q_2$ | $Q_1$ | $Q_0$ | Next State $Q_2^+$ | $Q_1^+$ | $Q_0^+$ | $J_2K_2$ | $J_1K_1$ | $J_0K_0$ |
|---|---|---|---|---|---|---|---|---|---|
| 0 | 0 | 0 | (0) | 0 | 0 | (1) | 0 x | 0 x | (1 x) |
| 1 | 0 | 0 | 1 | 0 | 1 | 0 | 0 x | 1 x | x 1 |
| 2 | 0 | 1 | 0 | 0 | 1 | 1 | 0 x | x 0 | 1 x |
| 3 | 0 | 1 | 1 | 1 | 0 | 0 | 1 x | x 1 | x 1 |
| 4 | 1 | 0 | 0 | 1 | 0 | 1 | x 0 | 0 x | 1 x |
| 5 | 1 | 0 | 1 | 1 | 1 | 0 | x 0 | 1 x | x 1 |
| 6 | 1 | 1 | 0 | 1 | 1 | 1 | x 0 | x 0 | 1 x |
| 7 | 1 | 1 | (1) | 0 | 0 | (0) | x 1 | x 1 | (x 1) |
| | Current output from flip-flops | | | Required output from flip-flops | | | Necessary inputs for the 3 flip-flops | | |

| $J_2$ | $\bar{Q}_2\bar{Q}_1$ | $\bar{Q}_2Q_1$ | $Q_2Q_1$ | $Q_2\bar{Q}_1$ |
|---|---|---|---|---|
| $\bar{Q}_0$ | 0 | 0 | x | x |
| $Q_0$ | 0 | 1 | x | x |

| $K_2$ | $\bar{Q}_2\bar{Q}_1$ | $\bar{Q}_2Q_1$ | $Q_2Q_1$ | $Q_2\bar{Q}_1$ |
|---|---|---|---|---|
| $\bar{Q}_0$ | x | x | 0 | 0 |
| $Q_0$ | x | x | 1 | 0 |

| $J_1$ | $\bar{Q}_2\bar{Q}_1$ | $\bar{Q}_2Q_1$ | $Q_2Q_1$ | $Q_2\bar{Q}_1$ |
|---|---|---|---|---|
| $\bar{Q}_0$ | 0 | x | x | 0 |
| $Q_0$ | 1 | x | x | 1 |

| $K_1$ | $\bar{Q}_2\bar{Q}_1$ | $\bar{Q}_2Q_1$ | $Q_2Q_1$ | $Q_2\bar{Q}_1$ |
|---|---|---|---|---|
| $\bar{Q}_0$ | x | 0 | 0 | x |
| $Q_0$ | x | 1 | 1 | x |

| $J_0$ | $\bar{Q}_2\bar{Q}_1$ | $\bar{Q}_2Q_1$ | $Q_2Q_1$ | $Q_2\bar{Q}_1$ |
|---|---|---|---|---|
| $\bar{Q}_0$ | 1 | 1 | 1 | 1 |
| $Q_0$ | x | x | x | x |

| $K_0$ | $\bar{Q}_2\bar{Q}_1$ | $\bar{Q}_2Q_1$ | $Q_2Q_1$ | $Q_2\bar{Q}_1$ |
|---|---|---|---|---|
| $\bar{Q}_0$ | x | x | x | x |
| $Q_0$ | 1 | 1 | 1 | 1 |

three flip-flops showing the necessary J and K inputs to produce the desired changes. These inputs are found from the JK excitation table. For instance, in state 0 then $Q_0$ is 0, and must go to 1 for state 1. So, the inputs to flip-flop 0 must be $(J_0, K_0)=(1, x)$. In state 7, $Q_0=1$ and must become 0 for state 0, hence we need $(J_0, K_0)=(x, 1)$.

**Derivation of steering logic: Method 1**

Once this table is complete we are at the same stage as for the D-type design except, rather than using three Karnaugh maps, for the three inputs to the D-types we now need six maps, one each for the J and K inputs to the flip-flops as shown in Table 7.2.

We can then minimise the steering logic using the Karnaugh maps to give:

$$J_2 = K_2 = Q_0 \cdot Q_1$$
$$J_1 = K_1 = Q_0$$
$$J_0 = K_0 = 1$$

**Derivation of steering logic: Method 2**

Rather than using the excitation tables of the JK flip-flops to find the required steering logic, as demonstrated in Method 1, alternatively the next state equation for the JK flip-flops can be employed. Here, Karnaugh maps for the next state outputs are firstly drawn in terms of the present state outputs, which are of course the Karnaugh maps used in the D-type design shown in Table 7.1.

However, whereas for the D-type these were the expressions that had to be sent to the inputs (since the next state equation for a D-type is $Q^+ = D$) the next state equation for a JK flip-flop is $Q^+ = J\overline{Q} + \overline{K}Q$. Hence, the J and K inputs required for each flip-flop are given by the coefficients for $\overline{Q}$ and $Q$ respectively, taken from the minimised expressions derived from the Karnaugh maps for $Q^+$ for each of the flip-flops.

The minimised expressions from the Karnaugh map give (as we used for the D-type design):[3]

$$\begin{aligned} Q_2^+ &= Q_2\overline{Q_0} + Q_2\overline{Q_1} + \overline{Q_2}Q_1Q_0 \\ &= (Q_0Q_1) \cdot \overline{Q_2} + (\overline{Q_0} + \overline{Q_1}) \cdot Q_2 \\ Q_1^+ &= Q_0 \cdot \overline{Q_1} + \overline{Q_0} \cdot Q_1 \\ Q_0^+ &= \overline{Q_0} \\ &= 1 \cdot \overline{Q_0} + 0 \cdot Q_0 \end{aligned}$$

Therefore: $J_2 = Q_0Q_1$ and $K_2 = Q_0Q_1$ (using De Morgan's theorem); $J_1 = K_1 = Q_0$; and $J_0 = K_0 = 1$.

**Circuit dependence on flip-flop type**

Note that this circuit demonstrates that in general the use of D-type flip-flops will require more logic gates, since the operation of the flip-flop itself is simpler. However, this must be offset against the fact that because the D-type is simpler it can be fabricated on a smaller area (see Section 9.3.6).

We now consider how the mod-8 counters designed using T, D and JK flip-flops relate to one another (see Problem 6.4). From the intuitive design using T-types: $T_2 = Q_1Q_0$; $T_1 = Q_0$ and $T_0 = 1$. A JK flip-flop with its inputs tied together acts

[3] Since the expression for $Q_0^+$ is minimised, then it does not contain $Q_0$. Consequently in order to obtain its coefficient (which is $\overline{K_0}$) it must be reintroduced. An alternative approach, used in the next section, is to ensure the terms whose coefficients are required are *not* minimised out.

as a T-type, which is the outcome of the JK design, since both inputs to all three flip-flops are the same as in the T-type design.

For a D-type $Q^+ = D$, so for it to act as a T-type, the input to the flip-flop must pass through some simple combinational logic together with the flip-flop's output so that the signal fed to the D-type's input is $\bar{D}Q + D\bar{Q} = D \oplus Q$. Then if $D$ is 0 the flip-flop remains in the same state, whilst if it is 1 it toggles (either the output or its complement is fed back).

For the D-type design we found that $D_0 = \overline{Q_0}$, so flip-flop 0 always toggles. Flip-flop 1 is fed $D_1 = Q_0 \oplus Q_1$ which from the above we can see means it is wired as a T-type with an input of $Q_0$, whilst

$$D_2 = Q_2\overline{Q_0} + Q_2\overline{Q_1} + \overline{Q_2}Q_1Q_0 = Q_2 \oplus (Q_1Q_0)$$

meaning this is also wired as a T-type but now with an input of $Q_1Q_0$. So all three circuits are identically equivalent, demonstrating the relationships between these three types of flip-flop.

## 7.5 EXAMPLE: MOD-6 COUNTER

Design a mod-6 binary up-counter using firstly D-type and then JK type flip-flops.

A mod-6 counter will require three flip-flops. The required relationship between the present and next states plus the required inputs to the JK flip-flops is shown in Table 7.3. Note that two states are unused.

Table 7.3 Present and next state variable for a mod-6 binary up-counter

| | Present State | | | Next State | | | | | |
|---|---|---|---|---|---|---|---|---|---|
| STATE | $Q_2$ | $Q_1$ | $Q_0$ | $Q_2^+$ | $Q_1^+$ | $Q_0^+$ | $J_2K_2$ | $J_1K_1$ | $J_0K_0$ |
| 0 | 0 | 0 | 0 | 0 | 0 | 1 | 0 x | 0 x | 1 x |
| 1 | 0 | 0 | 1 | 0 | 1 | 0 | 0 x | 1 x | x 1 |
| 2 | 0 | 1 | 0 | 0 | 1 | 1 | 0 x | x 0 | 1 x |
| 3 | 0 | 1 | 1 | 1 | 0 | 0 | 1 x | x 1 | x 1 |
| 4 | 1 | 0 | 0 | 1 | 0 | 1 | x 0 | 0 x | 1 x |
| 5 | 1 | 0 | 1 | 0 | 0 | 0 | x 1 | 0 x | x 1 |
| 6 | 1 | 1 | 0 | x | x | x | x x | x x | x x |
| 7 | 1 | 1 | 1 | x | x | x | x x | x x | x x |

### 7.5.1 D-type implementation

The Karnaugh maps for the next state variables are shown in Table 7.4. From these, and minimising taking advantage of the 'don't care' conditions:

$$Q_2^+ = D_2 = Q_2\overline{Q_0} + Q_1Q_0$$
$$Q_1^+ = D_1 = Q_1\overline{Q_0} + \overline{Q_2}\,\overline{Q_1}Q_0$$
$$Q_0^+ = D_0 = \overline{Q_0}$$

Table 7.4 Karnaugh maps for the next state variables for a mod-6 binary up-counter

| $Q_2^+$ | $\bar{Q}_2\bar{Q}_1$ | $\bar{Q}_2Q_1$ | $Q_2Q_1$ | $Q_2\bar{Q}_1$ |
|---|---|---|---|---|
| $\bar{Q}_0$ | 0 | 0 | x | 1 |
| $Q_0$ | 0 | 1 | x | 0 |

| $Q_1^+$ | $\bar{Q}_2\bar{Q}_1$ | $\bar{Q}_2Q_1$ | $Q_2Q_1$ | $Q_2\bar{Q}_1$ |
|---|---|---|---|---|
| $\bar{Q}_0$ | 0 | 1 | x | 0 |
| $Q_0$ | 1 | 0 | x | 0 |

| $Q_0^+$ | $\bar{Q}_2\bar{Q}_1$ | $\bar{Q}_2Q_1$ | $Q_2Q_1$ | $Q_2\bar{Q}_1$ |
|---|---|---|---|---|
| $\bar{Q}_0$ | 1 | 1 | x | 1 |
| $Q_0$ | 0 | 0 | x | 0 |

Using the above equations to determine the next state values for the unused states of 6 (i.e. $Q_2Q_1Q_0=110$) and 7, we find that they lead to states 7 and 4, respectively. (The states can also be found by considering the Boolean values used in the 'don't care' conditions in the Karnaugh maps during minimisation.) These are shown, together with the *state diagram* in Fig. 7.5 which illustrates the sequence of states the circuit moves through as it is clocked.

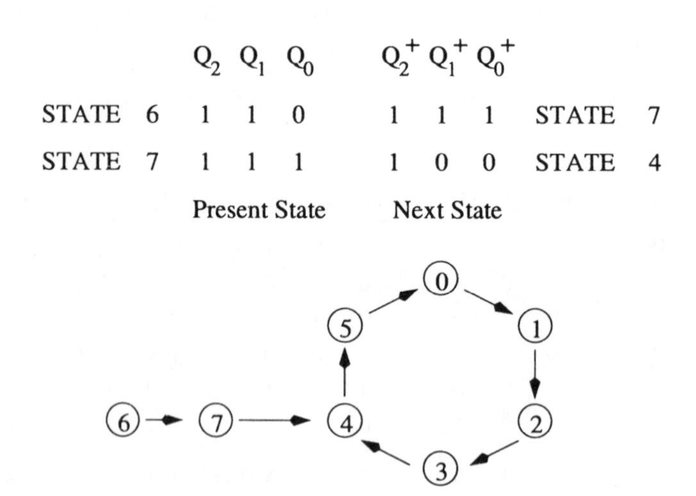

| | $Q_2$ | $Q_1$ | $Q_0$ | $Q_2^+$ | $Q_1^+$ | $Q_0^+$ | |
|---|---|---|---|---|---|---|---|
| STATE 6 | 1 | 1 | 0 | 1 | 1 | 1 | STATE 7 |
| STATE 7 | 1 | 1 | 1 | 1 | 0 | 0 | STATE 4 |
| | Present State | | | Next State | | | |



Fig. 7.5 State diagram for the mod-6 binary up-counter implemented using D-type flip-flops

Note that rather than using the 'don't care' states to aid minimisation they could have been used to ensure the unused count states led to specific states. A common choice is for them to lead to state 0 so that if either state was entered due to a circuit error then at the next clock cycle the counter would 'reset'. (To achieve this 0's are simply entered in place of the x's before the Karnaugh maps are used for minimisation.)

### 7.5.2 JK-type implementation

The Karnaugh maps for the necessary J and K inputs are shown in Table 7.5.

Table 7.5 Karnaugh maps for the J and K inputs for a mod-6 binary counter

| $J_2$ | $\bar{Q}_2\bar{Q}_1$ | $\bar{Q}_2Q_1$ | $Q_2Q_1$ | $Q_2\bar{Q}_1$ |
|---|---|---|---|---|
| $\bar{Q}_0$ | 0 | 0 | x | x |
| $Q_0$ | 0 | 1 | x | x |

| $K_2$ | $\bar{Q}_2\bar{Q}_1$ | $\bar{Q}_2Q_1$ | $Q_2Q_1$ | $Q_2\bar{Q}_1$ |
|---|---|---|---|---|
| $\bar{Q}_0$ | x | x | x | 0 |
| $Q_0$ | x | x | x | 1 |

| $J_1$ | $\bar{Q}_2\bar{Q}_1$ | $\bar{Q}_2Q_1$ | $Q_2Q_1$ | $Q_2\bar{Q}_1$ |
|---|---|---|---|---|
| $\bar{Q}_0$ | 0 | x | x | 0 |
| $Q_0$ | 1 | x | x | 0 |

| $K_1$ | $\bar{Q}_2\bar{Q}_1$ | $\bar{Q}_2Q_1$ | $Q_2Q_1$ | $Q_2\bar{Q}_1$ |
|---|---|---|---|---|
| $\bar{Q}_0$ | x | 0 | x | x |
| $Q_0$ | x | 1 | x | x |

| $J_0$ | $\bar{Q}_2\bar{Q}_1$ | $\bar{Q}_2Q_1$ | $Q_2Q_1$ | $Q_2\bar{Q}_1$ |
|---|---|---|---|---|
| $\bar{Q}_0$ | 1 | 1 | x | 1 |
| $Q_0$ | x | x | x | x |

| $K_0$ | $\bar{Q}_2\bar{Q}_1$ | $\bar{Q}_2Q_1$ | $Q_2Q_1$ | $Q_2\bar{Q}_1$ |
|---|---|---|---|---|
| $\bar{Q}_0$ | x | x | x | x |
| $Q_0$ | 1 | 1 | x | 1 |

**Method 1**
Minimising directly from the Karnaugh maps and using the 'don't care' states to aid minimisation gives:

$$J_2 = Q_1 Q_0 \quad K_2 = Q_0$$
$$J_1 = \overline{Q_2} Q_0 \quad K_1 = Q_0$$
$$J_0 = K_0 = 1 \quad \text{i.e. it is wired as a T-type.}$$

**Method 2**
Using the Karnaugh maps in Table 7.4 we again minimise using the 'don't care' states to aid this process. Note that since we are looking for the coefficients of the $\overline{Q_n}$ and $Q_n$ terms from the $Q_n$th flip-flop we do *not* minimise out these terms. This gives:

$$Q_2^+ = (Q_1 Q_0) \cdot \overline{Q_2} + \overline{Q_0} \cdot Q_2$$
$$Q_1^+ = (\overline{Q_2} Q_0) \cdot \overline{Q_1} + \overline{Q_0} \cdot Q_1$$
$$Q_0^+ = 1 \cdot \overline{Q_0} + 0 \cdot Q_0$$

On picking out the relevant coefficients these give the same Boolean expressions for the J and K inputs as obtained above.

Again we determine the results of entering states 6 and 7, which are shown in Fig. 7.6. Here state 6 leads to state 7 which leads to state 0 so the design is 'safe' in that it will reset to state 0 if an error occurs. The worst-case scenario would be if minimisation meant states 6 and 7 led to each other since then if either was entered the circuit would be stuck and simply move between these two states.

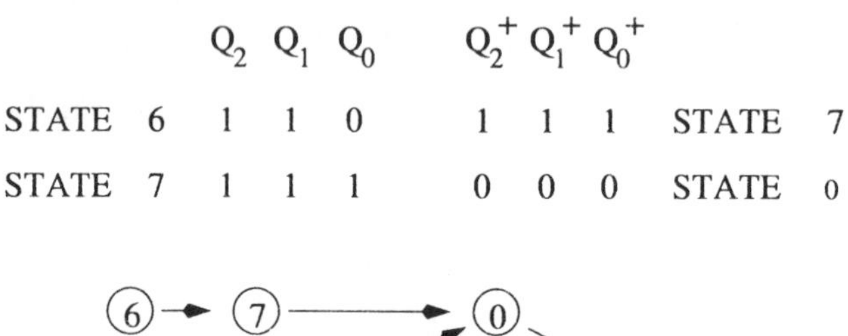

| | $Q_2$ | $Q_1$ | $Q_0$ | $Q_2^+$ | $Q_1^+$ | $Q_0^+$ | |
|---|---|---|---|---|---|---|---|
| STATE 6 | 1 | 1 | 0 | 1 | 1 | 1 | STATE 7 |
| STATE 7 | 1 | 1 | 1 | 0 | 0 | 0 | STATE 0 |

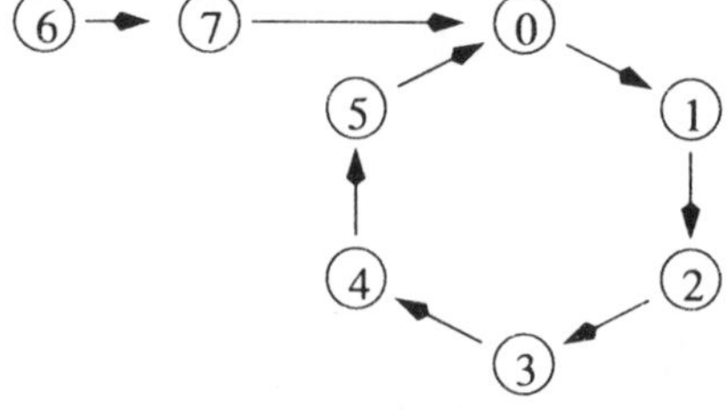


Fig. 7.6 State diagram for the mod-6 counter implemented using JK flip-flops

The reason that state 7 leads to state 0 for the JK based design (and not state 4 as when using D-types) is because a 0 rather than a 1 was used as the 'don't care' condition in cell $Q_2Q_1Q_0$ for $Q_2^+$.

**Summary**

Synchronous binary counters can be designed by simply writing out a table containing all possible states the circuit can hold as the present states, and the states that must follow these as the next states. If D-type flip-flops are used (which will generally lead to a requirement for more steering logic) then the Karnaugh maps for each of the next state variables in terms of the present state variables simply need minimising.

To use JK flip-flops the same method can be used. This requires the minimised expression to be matched with the JK's next state equation to allow the coefficients corresponding to the J and K inputs to be found. Alternatively, the required flip-flop inputs to produce the required change of output can be tabulated (using the excitation table), with this information used to produce the Karnaugh map for each of the flip-flop's inputs.

Any unused states may be used to aid minimisation or ensure that if they are entered erroneously the circuit will be eventually reset in some way. Different designs *may* lead to different state diagrams regarding the unused states.

## 7.6 SELF-ASSESSMENT

7.1 What is the 'mod' of a counter?

7.2 What are the differences between *asynchronous* and *synchronous* counters?

7.3 What do the terms *preset* and *reset* mean when referred to counters?

7.4 What design changes are necessary to turn an asynchronous up-counter into the corresponding down-counter?

7.5 What is the procedure for producing an asynchronous binary mod-N counter, and what problems may be encountered when using such a circuit in practice?

7.6 How is a synchronous binary mod-$2^n$ counter produced?

7.7 What is the procedure for producing a synchronous binary mod-$N$ counter?

7.8 In general how may flip-flops are required to produce a mod-$N$ counter, how many unused states will there be, and what is the outcome of entering these 'unused states'?

## 7.7. PROBLEMS

7.1 A logic signal is to be used to select either count-up or count-down operation from a ripple counter. What combinational logic is required between successive flip-flops to produce the required circuit?

7.2 What type of counter is shown in Fig. 7.7, and what is its *exact* function? Show how a three-input NOR gate could be used to decode count state 3, and draw the resultant output waveform.

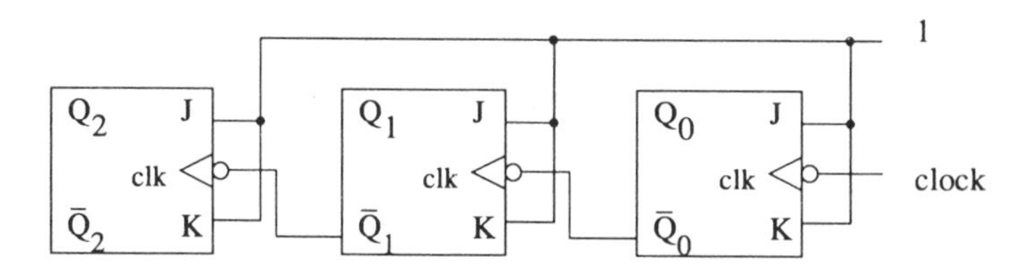


Fig. 7.7 Circuit to be analysed in Problem 7.2

7.3 What function does the circuit in Fig. 7.8 perform?

7.4 Design a mod-5 binary *ripple* counter.

7.5 Compare the use of D-type and JK-type flip-flops in the mod-6 counter designed as an example of a synchronous binary counter in Section 7.5.

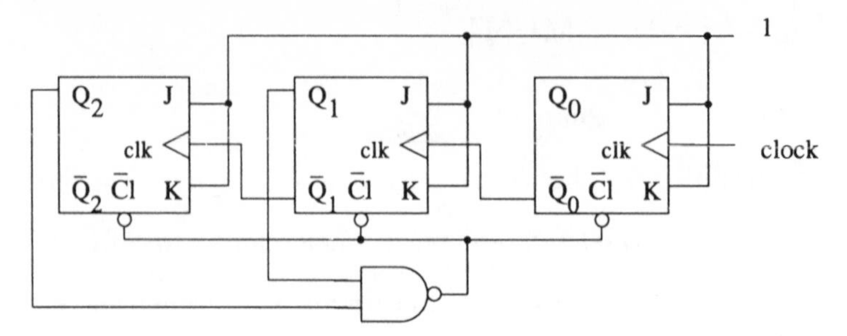


Fig. 7.8 Circuit to be analysed in Problem 7.3

7.6 Design a:
- (a) mod-5 synchronous binary counter using D-type flip-flops;
- (b) mod-5 synchronous binary counter using JK flip-flops;
- (c) mod-9 synchronous binary counter using D-type flip-flops;
- (d) mod-9 synchronous binary counter using JK flip-flops.

For all four counters use the unused states to aid minimisation. Determine what happens if the count goes into the unused state and show the results on a state diagram.

7.7 Design a mod-7 synchronous binary counter using JK flip-flops. Determine what happens if the count goes into any of the unused states and show the results on a state diagram.
How must the circuit be modified if the unused state is to lead to state 4 (i.e. outputs of 100 from the flip-flops (MSB first)).

# 8 Synchronous sequential circuits

## 8.1 INTRODUCTION

Synchronous sequential circuits were introduced in Section 5.1 where firstly sequential circuits as a whole (being circuits with 'memory') and then the differences between asynchronous and synchronous sequential circuits were discussed. You should be familiar with these ideas, and in particular the general form of a synchronous sequential circuit (see Figs 8.1 and 5.3) before continuing with this chapter.

As with asynchronous sequential circuits, the operation of synchronous sequential systems is based around the circuit moving from state to state. However, with synchronous circuits the state is determined solely by the binary pattern stored by the flip-flops within the circuit. (In Chapter 5 this was referred to as the internal state of the circuit.) Since each flip-flop can store a 0 or 1 then a circuit with $n$ flip-flops has $2^n$ possible states. Note that *all* states are stable since the present and next state variables are *not* connected directly but isolated due to the (not-transparent) flip-flops. The analysis and design of these circuits is based upon determining the next state of the circuit (and the external outputs) given the present state and the external inputs. This is therefore one application of the flip-flops' next state equations introduced in Chapter 6.

Following the introduction to sequential circuits in Section 5.1, Chapter 5 then dealt exclusively with asynchronous sequential circuits, concluding with an in-depth analysis of an SR flip-flop. Chapter 6 continued this theme of flip-flops which then meant that we could begin to look at synchronous sequential circuits since these use flip-flops as their 'memory'.

Chapter 7 looked at counters, which themselves are often considered as basic digital building blocks, and are therefore important digital circuits. The synchronous counters designed in Chapter 7 are in fact (simple types of) synchronous sequential circuits. In this chapter following a description of the way that synchronous sequential circuits can be classified, we will look at further examples of such circuits.

## 8.2 CLASSIFICATION

The general form of a synchronous sequential circuit is shown in Fig. 8.1. To recap, this has: external inputs, $A$, and outputs, $Z$; a combinational block which can be considered in two parts; and 'memory' in the form of flip-flops. The two parts of the combinational block serve to provide the internal outputs to the flip-flops, $Y$, and the external outputs, $Z$.

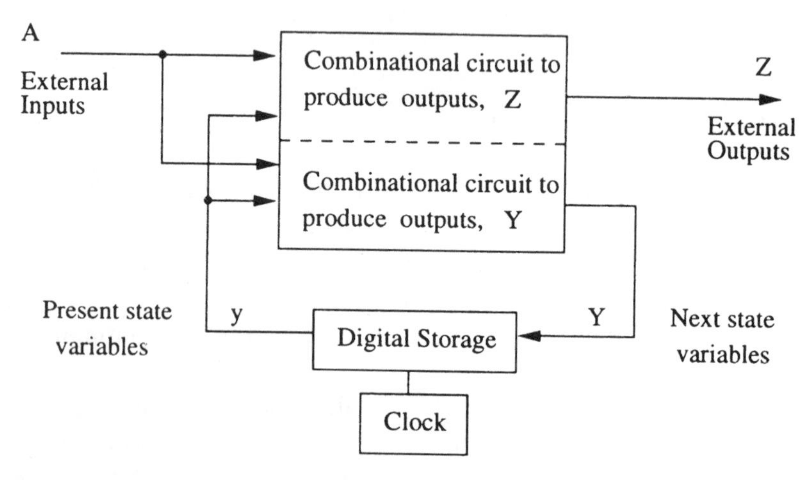


Fig. 8.1 General form of a synchronous sequential circuit

Obviously a circuit could have a simpler form and still be a synchronous sequential circuit. For instance it may have no external inputs or the external outputs may be functions of only the flip-flop's outputs (the present state variables). Consideration of such simplified circuits leads to a useful way of classifying sequential synchronous circuits.

### 8.2.1 Autonomous circuits

Autonomous circuits are those with *no* external inputs (except for the clock line) and which therefore perform independently (autonomously) of other circuits around them. Such circuits move through a set cycle of states as the circuit is clocked. The synchronous counters in the last chapter come into this category. However, the states of a general autonomous circuit obviously need not follow a binary sequence and furthermore the external ouputs need not simply be the outputs from the flip-flops (as with the synchronous counters) but could be functions of these (present state) signals. An example of an autonomous circuit is presented in Section 8.3.1.

### 8.2.2 General (Moore and Mealy) circuits

The next state of a *general* synchronous sequential circuit is dependent not only on the present state, as in an autonomous circuit, but also on the external inputs. Such general circuits can be further subdivided into two classes which are commonly referred to as Moore and Mealy models.[1]

**Moore model**

The Moore model describes a general synchronous sequential circuit where the external outputs are *only* functions of the circuit's present states (i.e. the flip-flops' outputs). Because of this in the state diagram of such a circuit the external outputs can be linked explicitly to the nodes (i.e. states). An example of such a circuit is given in Section 8.3.2.

**Mealy model**

The Mealy model is the most general since not only is the next state dependent upon the present state and the external inputs, but the external outputs are also functions of both of these sets of variables. Since the external outputs also depend upon the external inputs then in the state diagram of Mealy circuits the external outputs cannot simply be associated with a node but rather must be linked to the arrows (connecting the nodes) which are labelled with the output conditions as appropriate.

## 8.3 DESIGN EXAMPLES

### 8.3.1 Autonomous circuit

We shall design a mod-6 Gray code counter using JK flip-flops.

**Design**

A mod-6 counter has six states and therefore three flip-flips are needed. The required next states from the present states are shown in Table 8.1 together with the necessary J and K inputs to the flip-flops (obtained from the JK excitation table) and the Karnaugh maps for the six inputs. Note that the states have been labelled using their binary codes and therefore the Gray code count sequence is 0, 1, 3, 2, 6, 7, 0, 1, etc. (Design Method 1 (see last chapter) is being used.) The unused states are used to aid minimisation, the consequences of which will become clear when the state diagram is produced.

From the Karnaugh maps:

$$J_2 = Q_1\overline{Q_0} \qquad K_2 = Q_0$$
$$J_1 = Q_0 \qquad K_1 = Q_2Q_0$$
$$J_0 = \overline{Q_1} + Q_2 \qquad K_0 = Q_1$$

which completes the design.

[1]After the people who suggested such a classification.

Table 8.1 Required states for the mod-6 Gray code counter, together with the required inputs to the JK flip-flops

| STATE | Present State $Q_2$ | $Q_1$ | $Q_0$ | Next State $Q_2^+$ | $Q_1^+$ | $Q_0^+$ | $J_2$ | $K_2$ | $J_1$ | $K_1$ | $J_0$ | $K_0$ |
|---|---|---|---|---|---|---|---|---|---|---|---|---|
| 0 | 0 | 0 | 0 | 0 | 0 | 1 | 0 | x | 0 | x | 1 | x |
| 1 | 0 | 0 | 1 | 0 | 1 | 1 | 0 | x | 1 | x | x | 0 |
| 3 | 0 | 1 | 1 | 0 | 1 | 0 | 0 | x | x | 0 | x | 1 |
| 2 | 0 | 1 | 0 | 1 | 1 | 0 | 1 | x | x | 0 | 0 | x |
| 6 | 1 | 1 | 0 | 1 | 1 | 1 | x | 0 | x | 0 | 1 | x |
| 7 | 1 | 1 | 1 | 0 | 0 | 0 | x | 1 | x | 1 | x | 1 |
| 5 | 1 | 0 | 1 | x | x | x | x | x | x | x | x | x |
| 4 | 1 | 0 | 0 | x | x | x | x | x | x | x | x | x |

| $J_2$ | $\bar{Q}_2\bar{Q}_1$ | $\bar{Q}_2Q_1$ | $Q_2Q_1$ | $Q_2\bar{Q}_1$ |
|---|---|---|---|---|
| $\bar{Q}_0$ | 0 | 1 | x | x |
| $Q_0$ | 0 | 0 | x | x |

| $K_2$ | $\bar{Q}_2\bar{Q}_1$ | $\bar{Q}_2Q_1$ | $Q_2Q_1$ | $Q_2\bar{Q}_1$ |
|---|---|---|---|---|
| $\bar{Q}_0$ | x | x | 0 | x |
| $Q_0$ | x | x | 1 | x |

| $J_1$ | $\bar{Q}_2\bar{Q}_1$ | $\bar{Q}_2Q_1$ | $Q_2Q_1$ | $Q_2\bar{Q}_1$ |
|---|---|---|---|---|
| $\bar{Q}_0$ | 0 | x | x | x |
| $Q_0$ | 1 | x | x | x |

| $K_1$ | $\bar{Q}_2\bar{Q}_1$ | $\bar{Q}_2Q_1$ | $Q_2Q_1$ | $Q_2\bar{Q}_1$ |
|---|---|---|---|---|
| $\bar{Q}_0$ | x | 0 | 0 | x |
| $Q_0$ | x | 0 | 1 | x |

| $J_0$ | $\bar{Q}_2\bar{Q}_1$ | $\bar{Q}_2Q_1$ | $Q_2Q_1$ | $Q_2\bar{Q}_1$ |
|---|---|---|---|---|
| $\bar{Q}_0$ | 1 | 0 | 1 | x |
| $Q_0$ | x | x | x | x |

| $K_0$ | $\bar{Q}_2\bar{Q}_1$ | $\bar{Q}_2Q_1$ | $Q_2Q_1$ | $Q_2\bar{Q}_1$ |
|---|---|---|---|---|
| $\bar{Q}_0$ | x | x | x | x |
| $Q_0$ | 0 | 1 | 1 | x |

**State diagram**

The consequences of (accidentally) entering one of the two unused states are shown in Fig. 8.2 together with the state diagram. Note that the states are indexed by their binary rather than Gray codes. If either unused state is entered the circuit will lead back to the correct count sequence so the design is 'safe'.

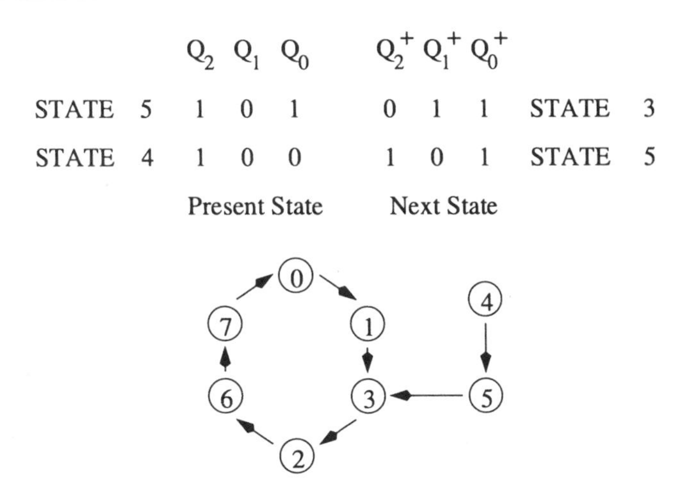


Fig. 8.2 State diagram for the mod-6 Gray code counter

## 8.3.2 Moore model circuit

Now we shall design a D-type based circuit that will count up or down, under the control of a single external input, through the first five prime numbers (in binary). We must ensure that the circuit will return to state (binary count of) 1 if an unused state is entered.

**Design**

Taking 1 as a prime number the first five primes are 1, 2, 3, 5 and 7. Therefore three flip-flops are required. States (i.e. binary counts) 0, 4 and 6 are unused and must lead to state 1 (i.e. $001_2$).

Fig. 8.3 shows: the state diagram; the necessary next states in terms of the present states and external input control, $X$; together with the Karnaugh maps for the next states, $Q^+$, of the three flip-flops in terms of the present states and $X$. By minimising the Karnaugh maps we determine that we need:

$$Q_2^+ = D_2 = Q_0 \cdot (\overline{Q}_2\overline{Q}_1X + \overline{Q}_2Q_1\overline{X} + Q_2Q_1X + Q_2\overline{Q}_1\overline{X}) = Q_0 \cdot (Q_2 \oplus Q_1 \oplus Q_0)$$

$$Q_1^+ = D_1 = \overline{Q}_1Q_0 + \overline{Q}_2Q_0X + \overline{Q}_2Q_1\overline{Q}_0\overline{X}$$

$$Q_0^+ = D_0 = Q_2 + \overline{Q}_0 + \overline{Q}_1X + Q_1\overline{X} = Q_2 + \overline{Q}_0 + (Q_1 \oplus X)$$

These give the Boolean expressions that must be implemented using combinational logic and used as the inputs to the three D-type flip-flops.

**Mealy outputs**

In this design the 'count' state can be taken straight from the flip-flops' outputs. Alternatively a particular state could be decoded using appropriate combinational logic. In both cases these would fit the Moore model since the outputs would be independent of the external input, $X$.

A further option would be to additionally use the input $X$ to decode the arrival

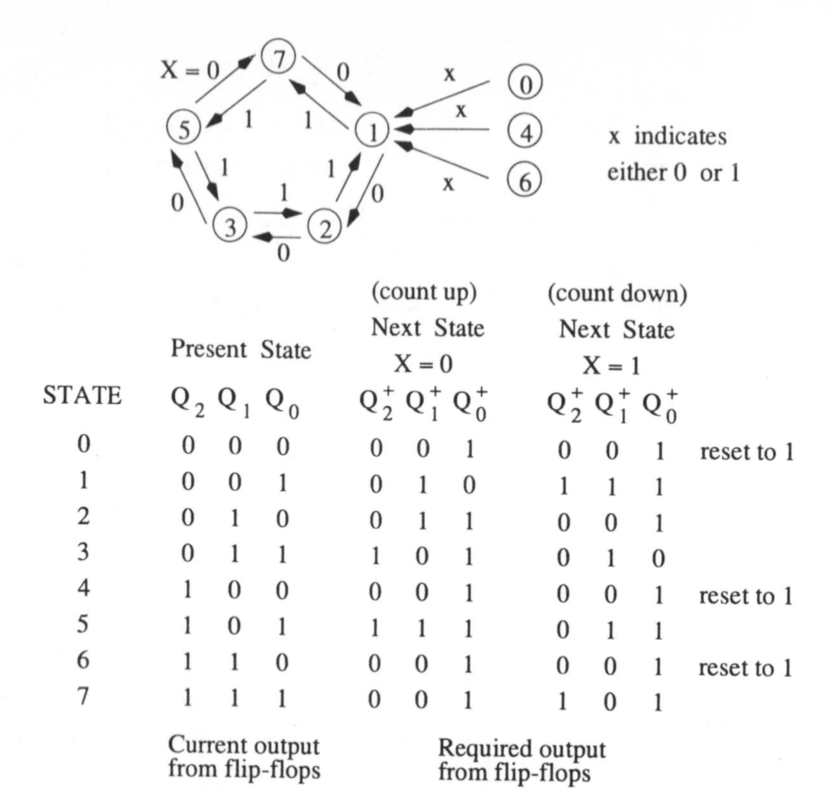


| STATE | Present State $Q_2$ | $Q_1$ | $Q_0$ | (count up) Next State X = 0 $Q_2^+$ | $Q_1^+$ | $Q_0^+$ | (count down) Next State X = 1 $Q_2^+$ | $Q_1^+$ | $Q_0^+$ | |
|---|---|---|---|---|---|---|---|---|---|---|
| 0 | 0 | 0 | 0 | 0 | 0 | 1 | 0 | 0 | 1 | reset to 1 |
| 1 | 0 | 0 | 1 | 0 | 1 | 0 | 1 | 1 | 1 | |
| 2 | 0 | 1 | 0 | 0 | 1 | 1 | 0 | 0 | 1 | |
| 3 | 0 | 1 | 1 | 1 | 0 | 1 | 0 | 1 | 0 | |
| 4 | 1 | 0 | 0 | 0 | 0 | 1 | 0 | 0 | 1 | reset to 1 |
| 5 | 1 | 0 | 1 | 1 | 1 | 1 | 0 | 1 | 1 | |
| 6 | 1 | 1 | 0 | 0 | 0 | 1 | 0 | 0 | 1 | reset to 1 |
| 7 | 1 | 1 | 1 | 0 | 0 | 1 | 1 | 0 | 1 | |

Current output from flip-flops — Required output from flip-flops

| $Q_2^+$ | $\bar{Q}_2\bar{Q}_1$ | $\bar{Q}_2Q_1$ | $Q_2Q_1$ | $Q_2\bar{Q}_1$ |
|---|---|---|---|---|
| $\bar{Q}_0\bar{X}$ | 0 | 0 | 0 | 0 |
| $\bar{Q}_0X$ | 0 | 0 | 0 | 0 |
| $Q_0X$ | 1 | 0 | 1 | 0 |
| $Q_0\bar{X}$ | 0 | 1 | 0 | 1 |

| $Q_1^+$ | $\bar{Q}_2\bar{Q}_1$ | $\bar{Q}_2Q_1$ | $Q_2Q_1$ | $Q_2\bar{Q}_1$ |
|---|---|---|---|---|
| $\bar{Q}_0\bar{X}$ | 0 | 1 | 0 | 0 |
| $\bar{Q}_0X$ | 0 | 0 | 0 | 0 |
| $Q_0X$ | 1 | 1 | 0 | 1 |
| $Q_0\bar{X}$ | 1 | 0 | 0 | 1 |

| $Q_0^+$ | $\bar{Q}_2\bar{Q}_1$ | $\bar{Q}_2Q_1$ | $Q_2Q_1$ | $Q_2\bar{Q}_1$ |
|---|---|---|---|---|
| $\bar{Q}_0\bar{X}$ | 1 | 1 | 1 | 1 |
| $\bar{Q}_0X$ | 1 | 1 | 1 | 1 |
| $Q_0X$ | 1 | 0 | 1 | 1 |
| $Q_0\bar{X}$ | 0 | 1 | 1 | 1 |

Fig. 8.3 State diagram, state tables and appropriate Karnaugh maps for the D-type based prime number up-down 'counter' discussed in Section 8.3.2

into a particular state *from a particular count direction.* This would give a circuit that would conform to the Mealy model. With such circuits care must be taken because changes in the external inputs may not be synchronised to the clock, in which case neither will changes to the external outputs. This may lead to transient states (and so spikes) in the outputs.

## 8.4 ANALYSIS

### 8.4.1 Case 1

The circuit to be analysed is shown in Fig. 8.4. From this it can be seen that it is an autonomous sequential synchronous circuit composed of two flip-flops (and therefore possessing four states) with a single output, $Z=\overline{Q_0}$. The next state equations are:

$$Q_1^+=D_1=Q_1\oplus Q_0$$

$$Q_0^+=D_0=\overline{Q_1}$$

From these the state table can be written (Fig. 8.5). We then assign letters to the four states and draw the state diagram using the state table. The timing diagram for the circuit is shown in Fig. 8.6.

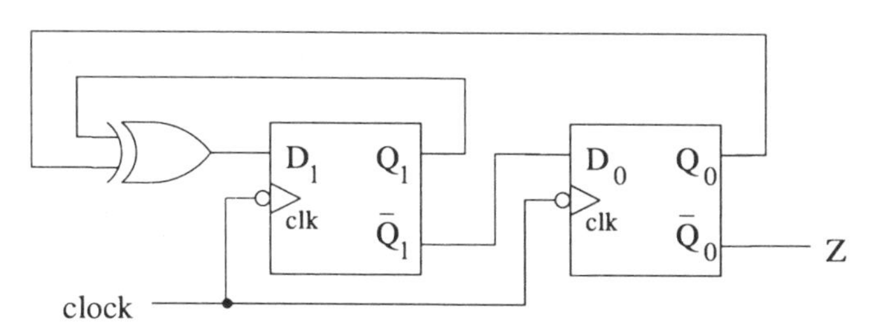


Fig. 8.4 Circuit analysed in Case 1

| | Present State | | | Next State | | |
|---|---|---|---|---|---|---|
| STATE | $Q_1$ | $Q_0$ | Z | $Q_1^+$ | $Q_0^+$ | STATE |
| A | 0 | 0 | 1 | 0 | 1 | B |
| B | 0 | 1 | 0 | 1 | 1 | C |
| C | 1 | 1 | 0 | 0 | 0 | A |
| D | 1 | 0 | 1 | 1 | 0 | D |

A

C B

D

Fig. 8.5 Next and present state table and state diagram for the circuit in Fig. 8.4

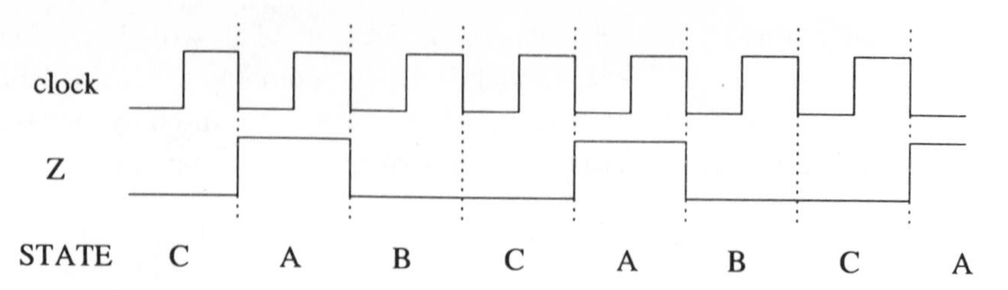


Fig. 8.6 Timing diagram for the circuit in Fig. 8.4

From these it is clear that this is a mod-3 counter with the output giving a pulse every third clock cycle. In effect it is therefore a divide-by-3 circuit. Note that if the circuit somehow entered state D it would remain there. This is therefore a poor design.

### 8.4.2 Case 2

The circuit for this example is shown in Fig. 8.7. This is another two flip-flop autonomous circuit with in this case:

$$Q_1^+ = \overline{Q_0}$$

$$Q_0^+ = Q_1$$

This gives the state table and state diagram shown in Fig. 8.8. From this, the timing diagram in Fig. 8.9 can be drawn which shows that the circuit produces four waveforms out of phase with each other by 1/4 of their period, which is four times that of the clock.

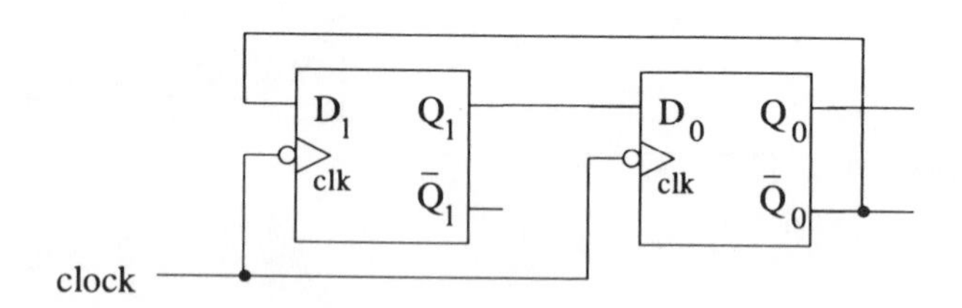


Fig. 8.7 Circuit analysed in Case 2

| | Present State | | Next State | | |
|---|---|---|---|---|---|
| STATE | $Q_1$ | $Q_0$ | $Q_1^+$ | $Q_0^+$ | STATE |
| A | 0 | 0 | 1 | 0 | D |
| B | 0 | 1 | 0 | 0 | A |
| C | 1 | 1 | 0 | 1 | B |
| D | 1 | 0 | 1 | 1 | C |

A → D → C → B → A

Fig. 8.8 State table and diagram for the circuit analysed in Fig. 8.7

Fig. 8.9 Timing diagram for the circuit in Fig. 8.7

## 8.5 SUMMARY

In this chapter we have seen how synchronous sequential circuits can be classified into different types (with the synchronous counters described in Chapter 7 being autonomous). Examples of a further autonomous design, together with a more general design (fitting the Moore model) in which the circuit's operation is also dependent upon an external input have also been given. In addition, two simple autonomous synchronous sequential circuits have been analysed.

Although these examples use only simple circuits, they demonstrate the principles underlying more complex ones. Additional complexity could come via the Mealy type circuit described in Section 8.2.2, via additional external inputs and a greater number of flip-flops.

The analysis of the circuits in the last two chapters has ultimately led to the state diagram. This is the usual *starting* point in the *design* of general synchronous sequential circuits. At this initial stage of a design the important feature is that the circuit 'moves' between states as required (possibly under the control of external inputs), which is what the *state diagram* describes. From this the *state table* can be produced and then *state assignment* performed. This is the assignment of the codes (the bit patterns held by the flip-flops) to particular states. Note that there may be more codes available than states (and hence unused states).

Although for counters the state assignment to use is obvious this is *not* generally so and many possible assignments will exist, all of which will give *functionally* (in terms of the state diagram) identical circuits. Once states have been assigned codes it is a relatively straightforward process to produce the Karnaugh maps for the next states, in terms of the present states and any external inputs, and so complete the design.

Such general design is beyond the scope of this book, but can be found in more advanced texts, and should be readily accessible to the reader with a firm grasp of the material presented in this and the preceding chapter. Some of the concepts regarding state diagrams and synchronous sequential circuit design are taken up in the following problems.

## 8.6 SELF-ASSESSMENT

8.1 What is a synchronous sequential circuit?

8.2 How can synchronous sequential circuits be classified? Illustrate your answer by drawing the modified general forms for these classes.

8.3 What is the basic design process for an autonomous synchronous sequential circuit, and how must this be amended for a general design (i.e. one with external inputs)?

8.4 What are, and happens to, the unused states in synchronous sequential circuits?

## 8.7 PROBLEMS

8.1 Redesign the mod-6 Gray code counter from Section 8.3.1 using D-type flip-flops and compare the result with the JK design.

8.2 Design a circuit using three D-type flip-flops which goes through the binary count sequence of 0, 2, 4, 6, 5, 3, 1, 0, 2, 4, 6 etc., with count state 7, if entered, leading to state 0. To what class of circuit does this design belong?

8.3 Modify the mod-7 counter designed in Problem 7.7 so that it is able to count either up and down under the control of an external input, $I$ (count up for $I=1$), ensuring all unused states lead to state 0.

8.4 Modify the mod-5 D-type flip-flop synchronous binary counter designed in Problem 7.6(a) so that it counts either up or down under the control of an external input, $I$ (count up if $I=1$).

8.5 What function does the circuit in Fig. 8.10 perform? (This should include production of the state diagram.) State what problems there could be with this circuit and produce a solution.

8.6 Analyse the operation of the circuit in Fig. 8.11. Compare this with Case 1 in Section 8.4.1.

8.7 A circuit contains two D-type flip-flops with inputs:

$$D_1 = Q_0\overline{I} + \overline{Q}_1 Q_0 + Q_1\overline{Q}_0 I \qquad D_0 = \overline{Q}_1\overline{I} + \overline{Q}_0 I$$

Determine its state diagram.

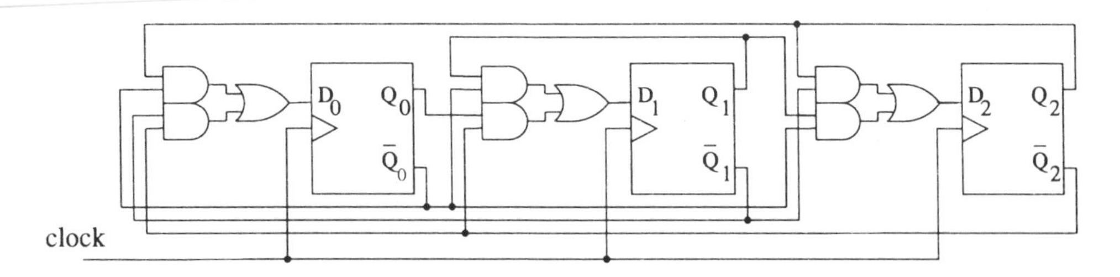


Fig. 8.10 Circuit to be analysed in Problem 8.5

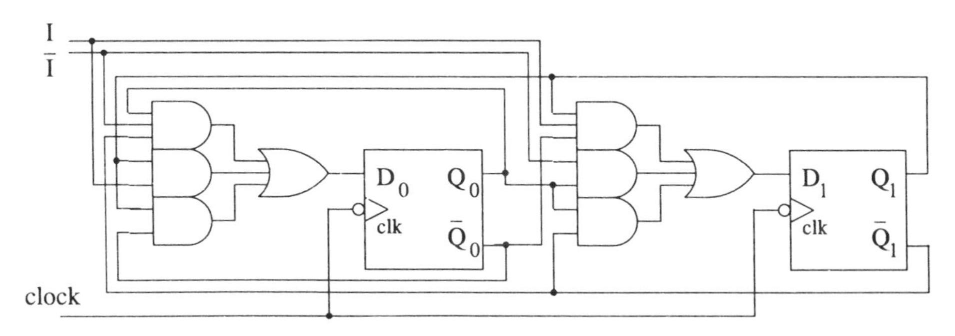


Fig. 8.11 Circuit to be analysed in Problem 8.6

8.8 Draw the state diagram for a JK flip-flop by considering it as a synchronous sequential circuit in its own right. (See Fig. 5.17 for the equivalent state diagram for an SR flip-flop.) Also draw the state table with the J and K inputs labelling the columns and the internal input (present state) labelling the rows. To what class circuit does this belong?

8.9 Produce Karnaugh maps for the internal state (i.e. the state table) and *Z* for the circuit in Fig. 8.12. Use the state table to derive the state diagram (which will be similar in form to the one in Problem 8.8). What variables is *Z* dependent upon, and therefore to what class does this circuit belong? Add values

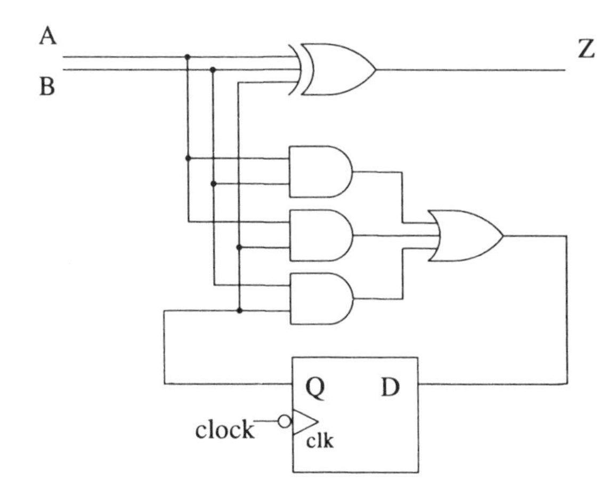


Fig. 8.12 Circuit to be analysed in Problem 8.9

for $Z$ to the state diagram (see Section 8.2.2). State what arithmetic function this circuit performs and compare this implementation with the combinational equivalent discussed in Chapter 4.

8.10 Draw the state diagrams for circuits that are to act as: (i) a parity checker and (ii) a comparator. Hints: Binary words to be operated upon are input to the circuit one bit at a time for (i) and two bits at a time for (ii); the state of the circuit should indicate odd or even parity for (i) and whether the words are equal for (ii).

8.11 Synchronous sequential circuits can be used to detect specific binary sequences entered one bit at a time. This is achieved by the circuit moving through different states (as bits are entered) until the desired pattern is received (and hence the final state is reached). Design a circuit that will detect the self-contained input sequence 1101. (That is, the final 1 in the sequence may *not* be taken as the first 1 in a following sequence.)

# 9 Choosing a means of implementation

## 9.1 INTRODUCTION

So far we have seen how to design both combinational and sequential circuits. These will, on paper, successfully perform many different functions but may well fail if the practicality of the hardware implementation issues are ignored. Ten years ago the choice of hardware options was limited; however, nowadays many choices exist for the designer, some of which are more accessible than others. The aim of this chapter is to introduce the technology options that are available so that the appropriate selection can be made from a sound engineering basis.

As far as technology is concerned designers must choose the balance they require between the circuit speed of operation and its power consumption. The two choices available are typically either bipolar or Complementary Metal Oxide Semiconductor (CMOS). However, other more exotic high-speed options are available such as Emitter Coupled Logic (ECL) and Gallium Arsenide (GaAs). CMOS offers low power consumption with moderate speeds. Alternatively, bipolar offers high speed but high power consumption. A combination of both is the ideal but was not available until only a few years ago. A mixed bipolar and CMOS technology (called BiCMOS) is now available and has an excellent combination of high speed and low power with the exception that this involves a more complex manufacturing procedure and hence is currently more expensive. As with most aspects of electronics technology the cost will certainly fall and BiCMOS may well be a low-cost technology option for the future.

The most common technology 10–15 years ago was bipolar (i.e. TTL (Transistor Transistor Logic) or ECL) but now CMOS is the preferred choice. Table 9.1 provides a comparison of logic families for various technology options. This table will provide a useful reference throughout this chapter. We shall start with a description of bipolar logic so that its limitations can be appreciated before moving to the more popular CMOS technology.

Table 9.1 Comparison of logic families

| Device | Description | Technology | Delay(ns) | Pstatic | Vohmin Volmax @Iomax | Vihmin Vilmax | Iihmax Iilmax | Iohmax Iolmax |
|---|---|---|---|---|---|---|---|---|
| 74 | Standard TTL | TTL | 10 | 10 mW | 2.4/0.4 | 2/0.8 | 40 μA/–1.6 mA | –0.4 mA/16 mA |
| 74S | Schottky clamped TTL – transistors do not enter saturation | TTL | 3 | 20 mW | 2.7/0.5 | 2/0.8 | 50 μA/–2 mA | –1 mA/20 mA |
| 74LS | Low power Schottky – as 74S but larger resistor values | TTL | 10 | 2 mW | 2.7/0.5 | 2/0.8 | 20 μA/–0.4 mA | –0.4 mA/8 mA |
| 74AS | Advanced Schottky – same as 74S but improved processing | TTL | 2 | 8 mW | 2.7/0.5 | 2/0.8 | 20 μA/–0.5 mA | –2 mA/20 mA |
| 74ALS | Advanced low power Schottky – low power version of 74AS | TTL | 4 | 1 mW | 2.7/0.5 | 2/0.8 | 20 μA/–0.1 mA | –0.4 mA/8 mA |
| 74F | Fast – compromise between S and ALS | TTL | 3 | 4 mW | 2.7/0.5 | 2/0.8 | 20 μA/–0.6 mA | –1 mA/20 mA |
| 74C | Standard CMOS – first CMOS parts in TTL pinout | CMOS | 30 | 50 μW | 4.2/0.4 | 3.5/1 | ± 2 μA | ± 4 mA |
| 74HC | High speed CMOS – improved CMOS | CMOS | 9 | 25 μW | 4.3/0.33 | 3.5/1 | ± 0.1 μA | ± 4 mA |
| 74HCT | High speed CMOS with TTL i/p voltage levels | CMOS | 10 | 25 μW | 4.3/0.33 | 2/0.8 | ± 0.1 μA | ± 4 mA |
| 74AC | Advanced high speed CMOS (1.5 μm CMOS) | CMOS | 4 | 25 μW | 4.3/0.44 | 3.5/1.5 | ± 0.1 μA | ± 24 mA |
| 74ACT | Advanced high speed CMOS with TTL i/p voltage levels | CMOS | 6 | 25 μW | 4.3/0.44 | 2/0.8 | ± 0.1 μA | ± 24 mA |
| 74(A)BCT | High speed BiCMOS for line drivers | BiCMOS | 3.5 | 600 μW | 2/0.55 | 2/0.8 | 0.07 mA/0.65 mA | –15 mA/64 mA |
| 74LVC | Low voltage (2.7–3.6 V) 1 μm CMOS | CMOS | 5 | 50 μW | 2/0.55 | 2/0.8 | ± 1 μA | ± 24 mA |
| 74LV | Low voltage (2.7–3.6 V) 2 μm CMOS | CMOS | 9 | 50 μW | 2.4/0.4 | 2/0.8 | ± 1 μA | ± 6 mA |
| 74LVT | Low voltage BiCMOS (optional 5 V inputs, 3 V outputs) | BiCMOS | 4 | 400 μW | 2/0.5 | 2/0.8 | ± 1 μA | ± 32 mA |
| 74ALVC | Advanced low voltage 1 μm CMOS | CMOS | 3 | 50 μW | 2/0.55 | 2/0.8 | ± 5 μA | ± 24 mA |
| 4000B | Early CMOS, not TTL pin compatible, 5–12 V supply | CMOS | 75 | 50 μW | 2.5/0.4 | 3.5/1.5 | ± 0.1 μA | 0.6 mA/2.3 mA |
| F100K | 100K ECL series – very fast but poor noise margins | ECL | 0.75 | 20 mW | –0.9/–1.7 | –1.2/–1.4 | 240 μA, 0.5 μA | ± 40 mA |

# 9.2 THE BIPOLAR JUNCTION TRANSISTOR

## 9.2.1 The BJT as a switch

The bipolar junction transistor or BJT as it is more commonly known can be considered in digital terms as a simple single-pole switch. It physically consists of three layers of semiconductor (which can be either N-type or P-type) of which two transistor types exist – NPN or PNP. We shall consider the operation of the NPN device since this device is used mainly in bipolar digital switching circuits.

The symbol for the NPN transistor is shown in Fig. 9.1 and is connected as a simple switch. The transistor consists of three terminals: base (b); emitter (e); and collector (c). Notice that the arrow on this type of transistor is pointing out from the emitter which indicates the direction of current flow. For the PNP the arrow points in. A simple rule for remembering the direction of the arrow is that with an NPN transistor the arrow is Not Pointing iN!

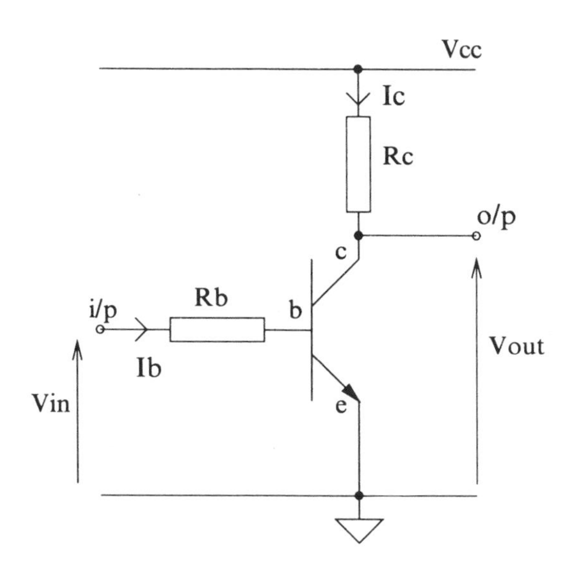


Fig. 9.1 A transistor switch

The input to the circuit in Fig. 9.1 is connected to the base terminal via the resistor $R_b$ whilst the output is taken from the collector. Several text books are available that discuss the operation of a bipolar transistor in detail.[1] However, for this simple BJT switch, and other BJT applications to follow, we just need to know the following.

1. To turn the transistor on a voltage at the base with respect to the emitter of greater than 0.7 V is needed. Under this condition a large collector current, $I_c$, flows through the transistor. The amount of current that flows is related to the

[1] B. Hart, *Introduction to Analogue Electronics*, in this series.

base current, $I_b$, by $I_c = h_{fe} I_b$, where $h_{fe}$ is called the **current gain** and is typically 100. In this condition the transistor is in the on state, called **saturation**, and the voltage across the collector to emitter is approximately 0.2 V and is called $V_{cesat}$.
2. To turn the transistor off the voltage at the base with respect to the emitter has to be less than 0.7 V. The collector current that flows is now zero (or more accurately a very small current called the leakage current). The transistor in this off state is called **cut-off** and the voltage across it is the supply voltage, $V_{cc}$, which is usually 5 V.

**Example 9.1**

Determine the value of $R_b$ needed in Fig. 9.1 to place the transistor in the saturated state when the input is 5 V. Assume that $h_{fe} = 100$ and $R_c = 1\ \text{k}\Omega$.

***Solution***

In the saturated state the voltage across the transistor is 0.2 V. We need to work back from the collector side to the base input. Using Kirchhoff's Voltage Law (KVL)

$$V_{cc} = I_c R_c + V_{cesat}$$

$$5 = I_c \times 1 \times 10^3 + 0.2$$

Hence $I_c = 4.8$ mA and so $I_b = I_c / h_{fe} = 48\ \mu\text{A}$. Using KVL on the input side:

$$V_{in} = I_b R_b + V_{be}$$

To turn the transistor on requires the base-emitter voltage to be at least 0.7 V. Hence

$$R_b = (5 - 0.7)/48\ \mu\text{A} = 89.6\ \text{k}\Omega$$

Consequently, when the input is 5 V the transistor is turned on (saturated) and hence the output is $V_{cesat}$ or 0.2 V. With the input at 0 V the transistor is turned off (cut-off) and the output is 5 V. If we let 5 V represent a logic '1' and 0.2 V a logic '0' then[2] the circuit in Fig. 9.1 performs the operation of an inverter or a NOT gate.

We shall now look at how other logic gates are implemented with BJT devices.

### 9.2.2 The BJT as a logic gate

**Diode transistor logic**

The DTL or diode transistor logic first became available commercially in 1962. The circuit diagram for a two-input DTL NAND gate is shown in Fig. 9.2(a) and although it is no longer available it does provide a useful introduction to the TTL

[2]As we shall see both logic levels are assigned to a voltage *range* rather than a single voltage.

logic family which follows. Before proceeding we need to point out that when the voltage across a diode equals 0.7 V (anode (A) voltage with respect to cathode (K)) then current will flow and this is called *forward bias*. Any voltage less than 0.7 V will result in negligible current flow. The two conditions are shown on a current/voltage plot in Fig. 9.2(b).

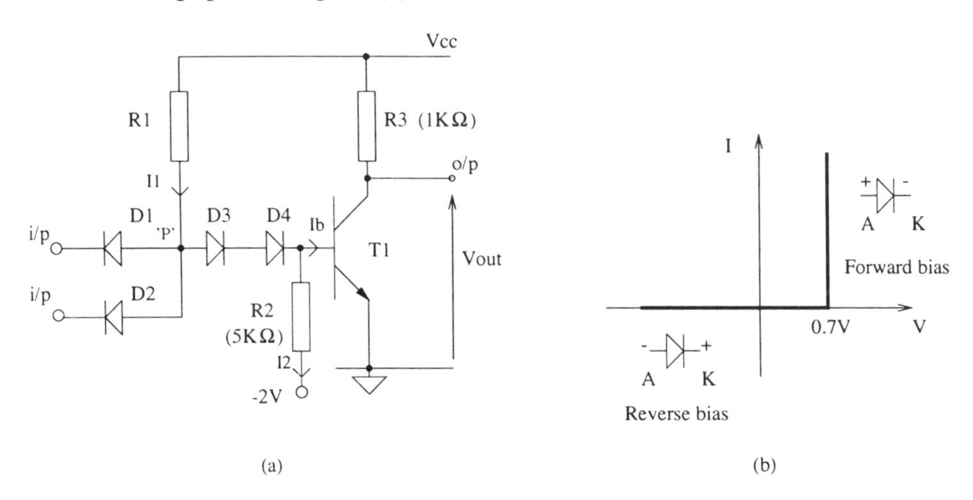


Fig. 9.2 Diode transistor logic circuit and ideal diode *I*/*V* characteristic

The circuit in Fig. 9.2(a) is actually a two-input AND gate followed by a NOT gate (i.e. a NAND gate) and functions as follows. If one input is low (less than 0.2 V) then the corresponding diode is forward biased and the voltage appearing at point 'P' is 0.9 V (since 0.7 V exists across a forward biased diode).This voltage, however, is insufficient to turn on diodes D3 and D4 and so the voltage appearing at the base of T1 is insufficient to turn on transistor T1. The current flowing through T1 is small and so the voltage dropped across R3 is also small and the output voltage is therefore close to 5 V i.e. a logic 1. Thus when either or both of the inputs are low the output is high.

Now if both inputs are high then the diodes D1 and D2 are turned off and the voltage at point 'P' rises to turn on diodes D3 and D4. Hence the voltage appearing at the base of T1 is dictated by the values of resistors R1 and R2. If R1 and R2 are chosen carefully then transistor T1 can be turned on and the output will be $V_{cesat}$ or 0.2 V.

So to summarise: if either or both inputs are low the output transistor, T1, is turned off and hence the output is high; if, however, both inputs are high ('1') then the transistor T1 is turned on and the output is low ('0'). The circuit thus operates as a two-input NAND gate.

**Example 9.2**

For Fig. 9.2(a) what value of R1 should be chosen to turn on T1 when both inputs are high? Assume that the $h_{fe}$ of T1 is 100.

***Solution***

To turn on T1 we need a base–emitter voltage of 0.7 V. Hence using KVL from R1 to R2 reveals:

$$V_{cc}=I_1R_1+V_{D3}+V_{D4}+V_{b1}=I_1R_1+0.7+0.7+0.7=I_1R_1+2.1$$

Since we know $V_{cc}$ then to find $R_1$ we need to know $I_1$, which from Kirchhoff's Current Law (KCL) is equal to the sum of $I_2$ and $I_b$ since D1 and D2 are off. Calculating each of these currents gives:

$$I_2=(V_{b1}-(-2))/R_2$$

$$I_2=(0.7-(-2))/(5\times 10^3)=0.54\,\text{mA}$$

$$I_b=I_c/h_{fe}=((V_{cc}-0.2)/R_3)/h_{fe}=4.8/(1\times 10^3\times 100)=48\,\mu\text{A}$$

Therefore $I_1=0.588$ mA and substituting this into the above equation for $V_{cc}$ to find $R_1$ gives:

$$5=(0.588\times 10^{-3})R_1+2.1\Rightarrow R_1=4932\,\Omega$$

**Standard TTL (Transistor Transistor Logic) – 74 series**

The standard TTL (short for Transistor Transistor Logic) logic gate was first marketed in 1963 under part numbers 74XXX. For example the 7400 is a quadruple (i.e. it contains four) two-input NAND gate in one package whilst the 74174 is a Hex D-type (i.e. six D-types in one package). The circuit diagram for a single two-input NAND gate implemented in TTL is shown in Fig. 9.3. Although it is not immediately obvious it does build on the DTL design of Fig. 9.2(a). The diodes D1, D2 and D3 have been replaced by a single transistor (T1) that has a multiemitter (two emitters in this case). The diode D4 and resistor R2 is replaced by the R2, T2 and R3 configuration. Finally the output stage has been replaced by a circuit that is called a *totem pole*[3] output. The multiemitter input transistor is quite simply an NPN transistor with more than one emitter which mimics the operation of the two diodes D1 and D2. The circuit operates as follows.

If at least one input is low (0.2 V) then that emitter is forward biased and the transistor is turned on (with current flowing out of the input that is low). A voltage of $V_{cesat}$ (0.2 V) appears across T1 and hence the voltage at the base of T2 is 0.4 V (i.e. 0.2+0.2). This is insufficient to turn on transistor T2 and hence the current through R2 and R3 is negligible. Consequently, the voltage at the base of T3 is 0 V and at the base of T4 is approximately 5 V. Hence T4 is turned on but not quite saturated and the output is high – but how high?

**Example 9.3**

What is the output high voltage when at least one of the inputs is low and what current would flow *out* of the input?

[3]The term *totem pole* is used simply because the components are arranged above each other.

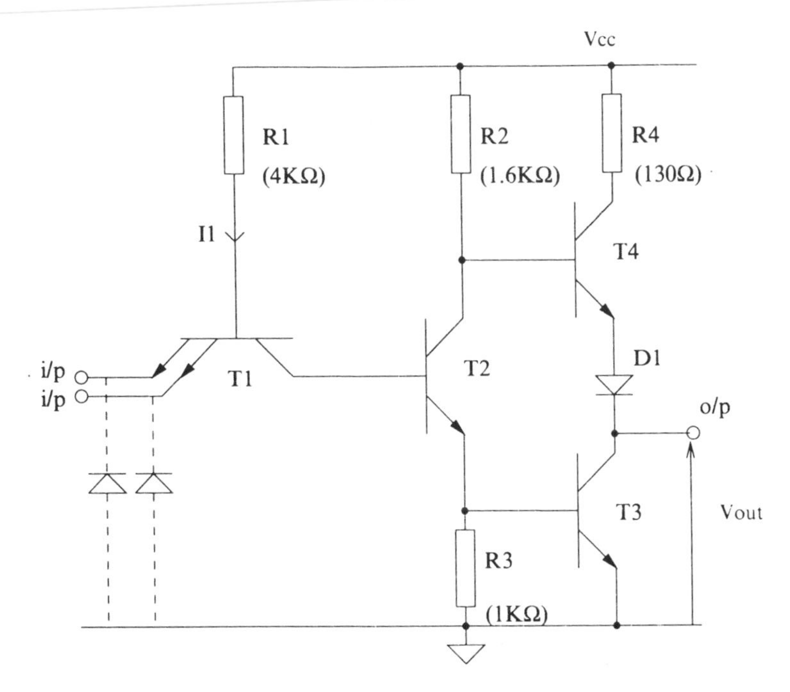


Fig. 9.3 Standard TTL two-input NAND gate – 74XXX series

***Solution***

From the above analysis we know that the voltage at the base of T4 is approximately 5 V when at least one input is low. Hence the output voltage is:

$$V_{out} = V_{b4} - (V_{be4} + V_{D1}) = 5 - (0.7 + 0.7) = 3.6\,\text{V}$$

This is classed as a TTL logic high voltage under no load and is called $V_{OH}$. Note that if a load is added which draws current through T4 then the output voltage will fall. This is caused by the voltage dropped across R2 as the base current through T4 increases and an additional volt drop across D1 and $V_{be4}$ due to their internal resistances. A minimum value for $V_{OH}$ for the 74 series is consequently set below 3.6 V at 2.4 V and is called $V_{OHmin}$.

Since one input is low then the emitter–base of T1 is forward biased and current will flow out of the emitter. Since T2 is off then the current must be supplied via R1. Thus:

$$V_{cc} = I_1 R_1 + V_{be1} + V_{in}$$

Substituting:

$$5 = I_1 \times 4 \times 10^3 + 0.7 + 0.2$$

Therefore:

$$I_1 = (5 - 0.7 - 0.2)/(4 \times 10^3) = 1.025\,\text{mA}$$

This current is referred to as the input low current (or $I_{IL}$) and any stage driving this input must be able to receive (or sink) this current and still maintain an input low of 0.2 V. Precise control of resistance values from chip to chip is difficult and hence $I_{IL}$ can vary considerably. To account for the wide tolerance in resistor values the maximum value quoted for $I_{IL}$ is 1.6 mA and is called $I_{ILmax}$. It is possible to apply a larger value of input voltage than 0.2 V and for it still to be recognised as a logic '0'. The maximum input low voltage for the 74XXX series is quoted at 0.8 V and is called $V_{ILmax}$. This will provide 0.6 V immunity (0.8 – 0.2) against a noise signal appearing at the input which would corrupt a logic '0' – such a safety tolerance is called the *noise margin* and is discussed later in this chapter.

Now, with both inputs high (3.6 V, from Example 9.3) the two emitter-bases of T1 are reverse biased and the current through R1 falls thus increasing the voltage at the base of T1 until its base–collector is forward biased. This will provide base current to turn on T2 which then turns on T3 and hence the output will be equal to $V_{cesat}$ or 0.2 V – which is sufficiently low to drive other TTL inputs. When the output drives other TTL loads then this output transistor (T3) must be able to sink $I_{IL}$ (1.6 mA) and still maintain a valid logic zero. In fact the output of any TTL gate may well drive more than one TTL input and hence the output must have sufficient current drive to drive several loads without the voltage at the output rising above 0.8 V ($V_{ILmax}$). The capacity for the output to drive more than one TTL input is called its *fan out*. Now, if the output is at 0.8 V then any slight noise will result in the output no longer providing a valid logic zero. Thus a safety margin is allowed of 0.4 V and the output low voltage must not be allowed to rise above 0.4 V – called $V_{OLmax}$.

So in summary if at least one of the inputs is low then the output is high, whilst if both inputs are high then the output is low. The circuit thus operates as a two-input NAND gate.

The two diodes (shown by dotted lines in Fig. 9.3) at the input are protection diodes to protect the gate against negative going voltages at the input caused by ringing of fast signals on the inputs. The presence of a negative voltage at the input will turn on the diodes and hence limit the input to a maximum negative voltage of –0.7 V.

**Example 9.4**

The output of a standard TTL gate is low when both inputs are high. What is the minimum value of input voltage that can be classed as a logic '1'?

***Solution***

Using Fig. 9.3, when the output is low, T3 is turned on and we can work backwards from here. The base of T3 will be at 0.7 V and in order to generate this voltage across R3 then T2 must be turned on, i.e. its base must be at 1.4 V. To turn on T2 we require a base current from T1 into T2 and hence the base-collector of T1 must be forward biased, i.e. 2.1 V. In other words, the base of T1 will be at three forward biased diode voltage drops. So that the input voltage does not influence the base of T1 we must reverse bias the emitter bases of T1. In order to achieve this we need to have a voltage at the input of greater than 2.1 V. In fact it is found for TTL that input voltages slightly less than 2.1 V (i.e. 2.0 V) are sufficient to turn on the output transistor.

The valid voltage levels for a 74 series standard TTL are thus summarised as: $V_{IHmin} = 2.0$ V; $V_{ILmax} = 0.8$ V; $V_{OHmin} = 2.4$ V; and $V_{OLmax} = 0.4$ V as can be seen in Table 9.1.

**The totem-pole output stage**

At the output of all gates there is a capacitive load ($C_L$) caused by the input capacitance of the next stage. This could be a printed circuit board interconnect or quite simply an oscilloscope lead. With the DTL circuit of Fig. 9.2(a) when the output changes from low to high, this capacitance ($C_L$) has to be charged through the collector resistor R3. Hence the delay time for the output to charge from low to high (i.e. '0' to '1') is limited by the time constant $R3 \times C_L$. To reduce this we could just reduce the value of R3 but then the power consumption will increase when the output transistor T1 is on.

The totem-pole output Fig. 9.3 gets around these problems. When the output is charging, the time constant is now dependent upon the resistance of the transistor T4, diode D1 and R4. Since R4 is only 130 Ω and both T4 and D1 are on then the time constant is much smaller than the DTL output circuit and hence the low-to-high delay is greatly reduced. In this case the transistor T3 is off and power consumption is low. This type of circuit is called an *active pull-up*.

The presence of both D1 and R4 are essential for the reliable operation of the TTL output stage. When the output is low, i.e. T3 on, the base of T4 is at a potential of: $V_{be3} + V_{cesat2} = 0.7 + 0.2 = 0.9$ V. Since the output is 0.2 V then this is insufficient to turn on the combination of T4 and D1 which results in no current being drawn from the supply. However, without the diode D1, then T4 will turn on and current will flow into T3 thus consuming power and the output voltage will rise (due to the resistance of T3) to a level between a 'low' and a 'high' (i.e. an illegal state). Hence D1 is inserted to keep T4 turned off when T3 is on. Resistor R4 is present so as to limit the current when the output is high and thus provides a short circuit protection if the output is inadvertently tied to 0 V.

**Example 9.5**

Given that the input and output currents for a standard 74 series TTL gate are as shown in Table 9.1 then how many standard 74 series TTL inputs will a single standard 74 series device drive?

***Solution***

This is called the fan out and is equal to the lesser of $I_{OLmax}/I_{ILmax}$ or $I_{OHmax}/I_{IHmax}$.

In both cases this is 10. Hence a 74 device can drive 10 other 74 series devices, i.e. it has a *fan out* of 10.

## Example 9.6

Many other logic gates can be implemented with the standard 74 series. What function does the circuit in Fig. 9.4 perform? Assume $V_{IL}=0.2$ V and $V_{IH}=3.6$ V.

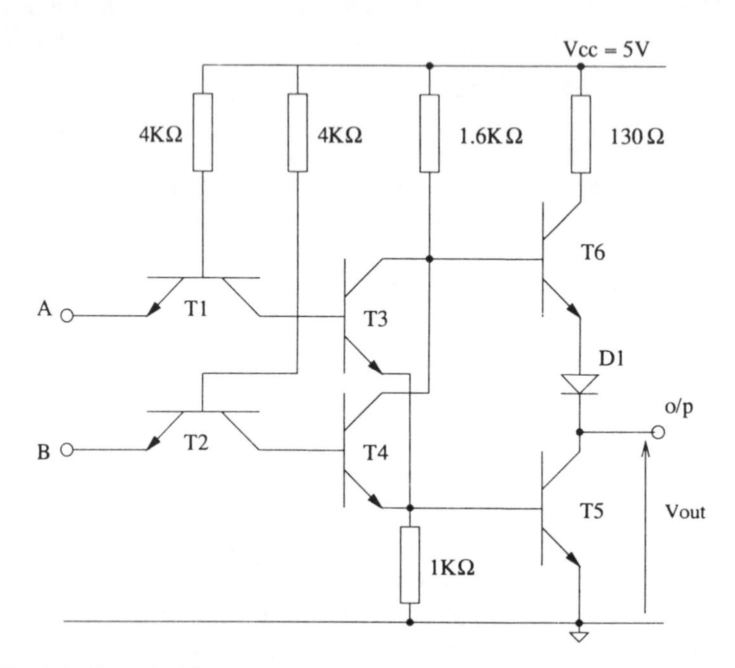


Fig. 9.4 Circuit for Example 9.6

### *Solution*

$A=0.2$ V, $B=0.2$ V. Both input transistors T1 and T2 are on and thus the bases of both T3 and T4 are at 0.4 V. This is insufficient to turn on the output transistor T5 and the collectors of both T3 and T4 are high which turns on transistor T6, thus pulling the output high.

$A=0.2$ V, $B=3.6$ V. T1 is on but T2 is off. With T1 on then the voltage at the base of T3 is 0.4 V and so this is insufficient to turn on both T3 and T5. However, since T2 is off then the base of T4 can rise so as to turn on T4 and then T5. The output is thus low.

$A=3.6$ V, $B=0.2$ V. This time T2 is on and T1 off and the transistors T3 and T5 are on, forcing the output low.

$A=3.6$ V, $B=3.6$ V. Both T1 and T2 are turned off and so both T3 and T4 are on which therefore turns on T5 and the output is low.

Since the output is only high when both inputs are low then the circuit functions as a two-input NOR gate.

### Schottky clamped TTL – 74S series

The standard TTL series has a typical propagation delay of 10ns (the term propagation delay was introduced in Chapter 4 and will be covered in more detail later in this chapter). By this we mean that when an input change occurs it takes 10 ns for the effect to propagate to the output. In the early 1970s it was found that this propagation delay could be decreased by replacing those transistors that saturate with *Schottky* clamped transistors.

So far we have seen that when a transistor turns on it enters *saturation*. This name is given to this condition because the base is saturated with charge. Before a saturated transistor can be turned off this charge must first be removed. This can take a considerable amount of time and thus slows down the switching speed of the device. Preventing the transistor from entering saturation will therefore increase the switching speed. The Schottky TTL series uses such a technique by connecting a Schottky diode between the base and collector of the transistor to stop the device entering saturation. Hence these circuits are much faster than the non-Schottky clamped series.

A Schottky diode is a metal-semiconductor diode that has a forward volt drop of only 0.3 V as opposed to the standard PN junction diode that has a forward voltage drop of 0.7 V. The Schottky diode is connected as shown in Fig. 9.5(a) between the base and collector. Without the Schottky diode, when the transistor is turned on, the base is 0.7 V above the emitter, the collector-emitter voltage is 0.2 V and this is called saturation. However, addition of the Schottky diode (which will be forward biased due to the base at 0.7 V) clamps the collector at a voltage this time of 0.4 V with respect to the emitter. Remember, a forward biased Schottky diode has 0.3 V across it. Hence the transistor is on but is not in saturation. When such a transistor is turned off it will now not take as long to change state since the base is not saturated with charge. The resulting Schottky transistor is represented by the symbol illustrated in Fig. 9.5(b).

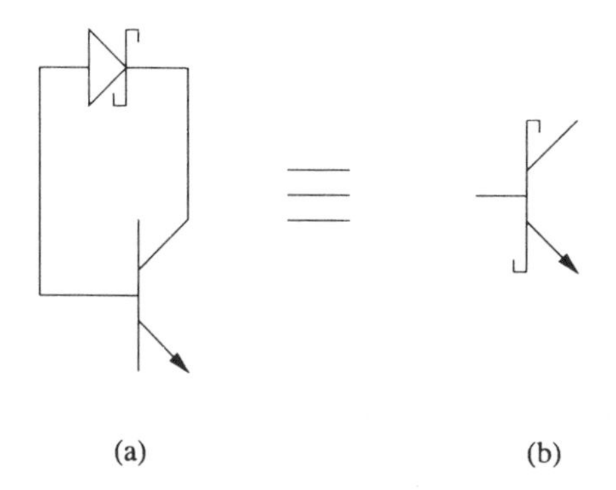


Fig. 9.5 Schottky clamped transistor and its associated symbol

The Schottky series (labelled the 74S series) thus emerged as a high-speed replacement for the standard 74 series. The circuit diagram of a two-input NAND gate implemented with Schottky clamped transistors is shown in Fig. 9.6. Apart from the use of Schottky transistors the circuit also has other improvements. The output diode D1 has been replaced with two transistor stages T3 and T5. This again provides two 0.7 V drops between the T2 collector and the output so that T5 is not turned on at the same time as T6. However, the two-transistor arrangement without the diode also improves the output current drive when the output is in the high state. In addition, Schottky transistor T4 is included so as to improve the switch-off time of transistor T6.

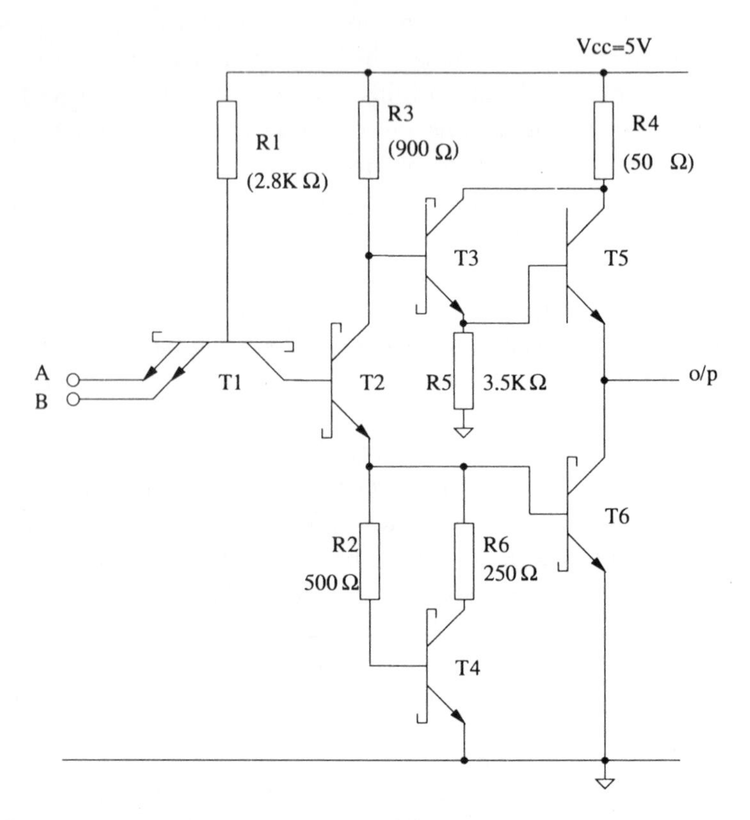


Fig. 9.6 Schottky clamped TTL NAND gate – 74SXXX series

The 74S series has a typical propagation delay of 3 ns and a power dissipation of 20 mW. The voltage levels at the output are changed slightly. In the high state since the diode has been removed then the minimum output high voltage is increased to 2.7 V. In the low state since the output transistor does not saturate then the maximum output low voltage has been increased to 0.5 V.

Due to these circuit changes the input and output currents for a Schottky clamped 74S series are different from the 74 series. From Table 9.1 we can see that the fan out is 10 when the output is low but 20 when the output is high. However, it is the lower value (i.e. 10) that indicates the number of loads that can be driven.

**Low-power Schottky – 74LS series**

The low-power Schottky clamped TTL logic family (74LS series) was released in 1975. This, as the name suggests, has a lower power dissipation than the 74S series. A circuit diagram for a two-input NAND gate is shown in Fig. 9.7. The main differences from the 74S series are:

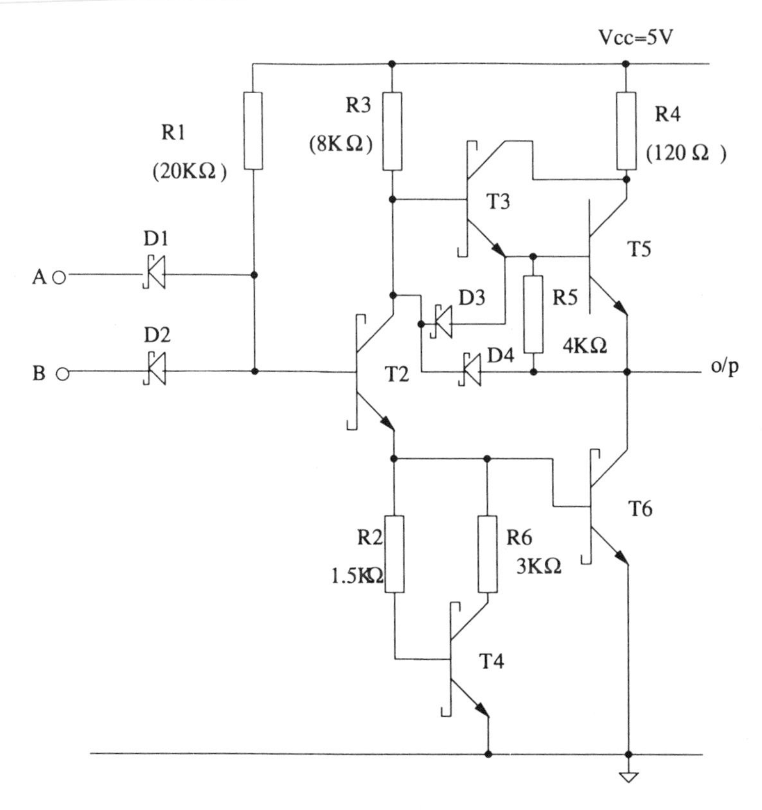


Fig. 9.7 Low-power Schottky clamped TTL NAND gate – 74LSXXX series

1. larger resistors are used throughout to reduce power consumption at the expense of longer propagation delays;
2. the multiemitter input transistor is replaced by Schottky diodes D1 and D2. This is because these diodes can take up a smaller area on chip;
3. the Schottky diodes D3 and D4 assist in the removal of charge from T5 thus speeding up the high-to-low propagation delay.

Typical delays for the 74LS series are 10 ns with a power dissipation of 2 mW. The logic levels are the same as the 74S series whilst the input and output currents for a 74LS series are such that the fan out is now 20.

**Advanced Schottky TTL – 74AS, 74ALS and 74F series**

In approximately 1980 advancements in manufacturing of the 74S and 74LS series resulted in the release of the Advanced Schottky (74AS) and Advanced Low-power Schottky series (74ALS), respectively. As can be seen in Table 9.1, the ALS and AS series provide a much faster propagation delay time than the LS and S series, respectively. Also the ALS and AS series have a significant reduction in power consumption when compared to their associated LS and S series. These improvements have been achieved by implementing the design with smaller transistors (due to improvements in manufacturing), by increasing resistor values slightly and by using subtle circuit modifications.

Another family appeared at this time and that was the 74F series (sometimes referred to as Fast). This family is a compromise between the AS and ALS series having a typical delay of 3 ns and a power dissipation of 4 mW.

# 9.3 THE MOSFET

## 9.3.1 The MOSFET as a switch

The Metal Oxide Semiconductor Field Effect Transistor (or MOSFET) has proved over the past 15 years to be a very attractive alternative to the BJT. In recent years the MOSFET has become the preferred technology mainly because manufacturing improvements have advanced further with FET processes compared to bipolar processes. A cross-section of an N-channel MOSFET is shown in Fig. 9.8(a). We shall study the device at this level since this will help in our understanding of how memory devices operate – see Chapter 10. The transistor has four terminals: gate(G); source(S); drain (D); and substrate. Just as in the BJT the MOSFET is composed of three semiconductor layers. However, for the FET the middle terminal (the gate) is separated from the P-type semiconductor substrate by a thin gate oxide of approximately 0.05 μm in thickness. The drain and source are connected to the N-type regions either side of the gate. The original MOSFETs used a metal gate but now all MOSFETs are manufactured with a polysilicon gate.[4] One symbol for an N-channel MOSFET is shown in Fig. 9.8(b).

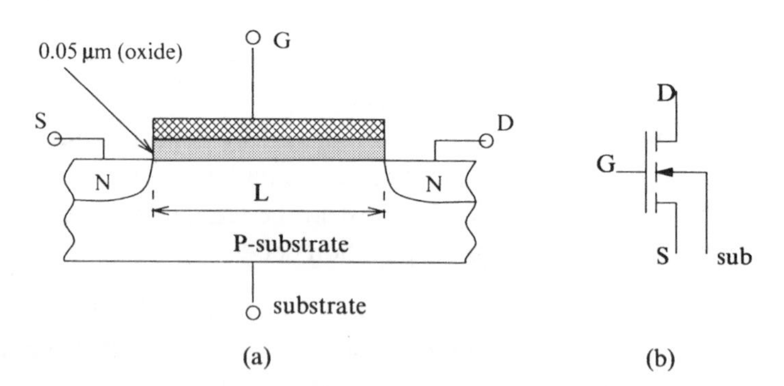


Fig. 9.8 Metal Oxide Semiconductor Field Effect Transistor

The device operates by using the voltage on the gate to control the current flowing between source and drain. When $V_{GS}$ is zero, application of a positive voltage between the drain and source ($V_{DS}$) will result in a negligible current flow since the drain to substrate is reverse biased. When $V_{GS}$ is increased in a positive direction electrons are attracted to the gate oxide-semiconductor interface. When $V_{GS}$ is greater than a voltage called the *threshold voltage* ($V_T$) the P-type material close to the gate oxide changes to N-type and hence the source and drain are

[4] L. Ibbotson, *Introduction to Solid State Devices*, in this series.

connected together by a very thin channel. Now, when a positive voltage, $V_{DS}$, is applied a current, $I_{DS}$, will flow from drain to source and the transistor is said to be turned on. The transistor can be turned on even more by further increasing the gate voltage. This is because more electrons are attracted to the oxide–semiconductor interface and the depth of the channel increases. Consequently the resistance between source and drain reduces thus increasing the current $I_{DS}$.

The current–voltage relationship of the MOS transistor can be modelled approximately with two equations depending upon the value of $V_{DS}$:

If $V_{DS} < V_{GS} - V_T$ then the device is in the *linear*[5] region and

$$I_{DS} = K[(V_{GS} - V_T)V_{DS} - V_{DS}^2/2] \tag{9.1}$$

If $V_{DS} > V_{GS} - V_T$ then the device is in the *saturation* region and

$$I_{DS} = [K/2][V_{GS} - V_T]^2 \tag{9.2}$$

where $K = (W/L)\mu C_{ox}$ and $W$ and $L$ are the width and length of the gate; $\mu$ is the mobility of carriers (this is a measure of the ease at which a carrier can pass through a semiconductor material); and $C_{ox}$ is the oxide capacitance per unit area of the thin gate oxide region. Physically, the length of the gate is the distance between the drain and source and is marked as $L$ in Fig. 9.8(a), whilst the width $W$ is the dimension into the page. Since increasing $K$ increases $I_{DS}$ then $K$ is sometimes referred to as the gain of the transistor even though it has dimensions of $\mu A\ V^{-2}$. These MOS equations thus allow the voltages around the transistor to be calculated.

Let us look at a simple N-channel MOSFET inverter (illustrated in Fig. 9.9(a)) as a means of illustrating the application of this type of transistor. Here, the transistor can be thought of as a switch such that when $V_{GS}$ is greater than the threshold voltage, $V_T$ (typically 1 V), then the transistor will turn on. Therefore

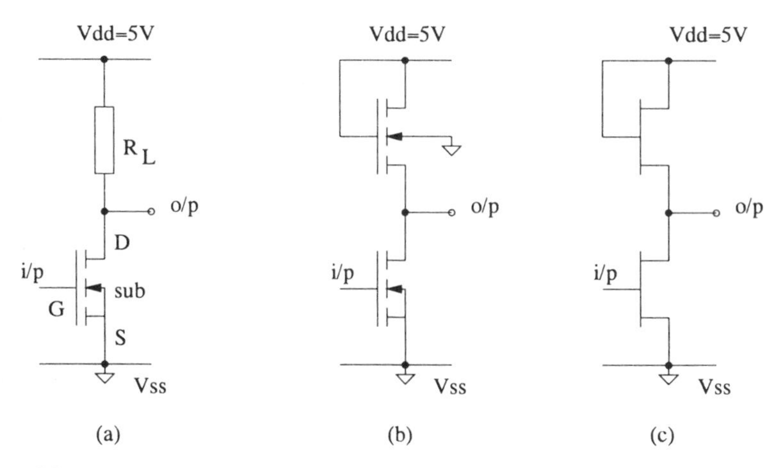


Fig. 9.9 MOS inverter

[5]The term *linear* is used because for small values of $V_{DS}$ the current $I_{DS}$ is linearly related to $V_{DS}$.

current flows from the supply through the load resistor $R_L$, through the transistor to ground ($V_{ss}$). As $V_{GS}$ increases the current flowing will increase and by choosing the appropriate value of $R_L$ then the voltage at the output will be pulled down towards 0 V. If on the other hand the gate–source voltage is less than $V_T$ then the transistor is turned off (i.e. the switch is open) and the output is pulled up to $V_{dd}$ (usually 5 V). This circuit thus operates as an inverter or a NOT gate. From the bipolar section we can see that the NMOS device operates as an *active* pull-down, whilst the resistor $R_L$ is called a *passive* pull-up.

## Example 9.7

For the circuit shown in Fig. 9.9(a) what value should be chosen for $R_L$ such that the output will be 0.5 V when the input voltage is 5 V. Assume that $K$ for the transistor is 128 μA $V^{-2}$.

### *Solution*

We need to decide which of the two Equations 9.1 and 9.2 to use. Since the input voltage ($V_{GS}$) is 5 V and the output voltage is 0.5 V ($V_{DS}$) then $V_{DS} < V_{GS} - V_T$ and hence the device is in the linear region, i.e. we use Equation 9.1:

$$I_{DS} = K[(V_{GS} - V_T)V_{DS} - V_{DS}^2/2] = 128 \times 10^{-6}[(5-1)0.5 - 0.25/2] = 0.24\,\text{mA}$$

Since we want an output voltage of 0.5 V then the voltage across the load resistor, $R_L$, will be 4.5 V and so:

$$4.5\,\text{V} = I_{DS} \times R_L = 0.24\,\text{mA} \times R_L \Rightarrow R_L = 18\,750\,\Omega$$

To create a resistor of this size would require a large area on an integrated circuit. Hence the resistor is replaced with an MOS transistor which is wired by attaching the gate to the drain and hence the device is always in saturation. The MOS transistor wired in this manner is shown in Fig. 9.9(b) and is called an *active resistor*. The area taken up by the transistor is approximately 1/200 of that of an equivalent value resistor.

There are two points to notice about the symbol for the N-channel device. Firstly the arrow pointing in on the substrate terminal indicates that the device is N-channel. Secondly, the dotted line indicates that this device is an *enhancement* mode device. With enhancement mode devices a current will only flow when a voltage above the threshold voltage is applied to the gate. Depletion mode devices exist where a current will flow even if the gate is at zero volts. The symbol for these has a continuous line between source and drain. Since we shall only be using enhancement mode devices and to simplify the drawing of the transistor symbol we shall use the symbol shown in Fig. 9.9(c) to represent an $n$-channel enhancement mode MOSFET which is easier to draw. Also it will be assumed that the substrate for all $n$-channel devices is always connected to 0 V. PMOS enhancement mode devices exist and these have the substrate connected to $V_{DD}$.

## 9.3.2 The MOSFET as a logic gate

Apart from the inverter shown in Fig. 9.9(b) and (c) it is possible to use the NMOS transistor to form other logic gates as we shall see in the next example.

### Example 9.8

For the circuits in Fig. 9.10(a) and (b) determine the functions implemented.

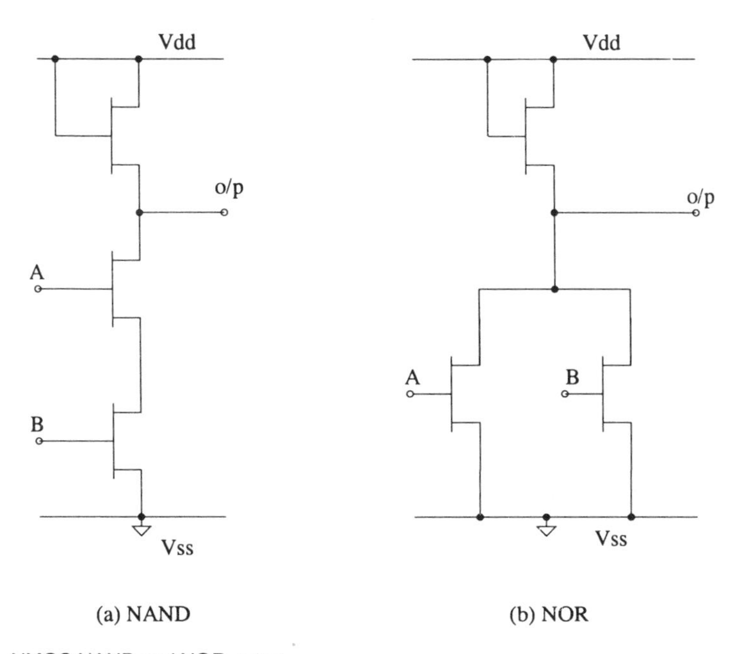


Fig. 9.10 NMOS NAND and NOR gates

### *Solution*

*Fig. 9.10(a):* if either or both of the inputs $A$ and $B$ are low (i.e. less than $V_T$) then one of the NMOS transistors will be off and hence the output voltage will be pulled up to $V_{dd}$. The only way for the output to go low is for both $A$ and $B$ to be high. The circuit thus operates as a two-input NAND gate.
*Fig. 9.10(b):* if either $A$ or $B$ or both are high then the output is pulled to ground. The only way for the output to go high is for both $A$ and $B$ to be low. The circuit thus operates as a two-input NOR gate.

This type of logic is called NMOS logic. Historically the first MOS logic that appeared was in 1970 and used PMOS transistors. It was not possible at the time to produce NMOS devices due to problems with processing. However, in 1975 these problems were remedied and NMOS logic gates were manufactured taking advantage of the higher mobility of the N-channel carriers in NMOS transistors compared to the P-channel carriers in PMOS devices. We can see from $K = [W/L]\mu C_{ox}$ that a higher mobility will result in a higher value of $K$ allowing a larger current to be passed within the same size transistor. In addition the higher the mobility, the faster the switching speed. In fact N-channel mobility is 2–3 times that of P-channel carriers and hence the NMOS logic operates at 2–3 times the speed of PMOS.

One problem of the NMOS gates (and for that matter PMOS) is that the upper transistor load is just acting as a resistor. When the lower transistor is on then current will flow from $V_{dd}$ to $V_{ss}$ and hence these types of devices consume a moderate amount of power. Consequently in 1978 both PMOS and NMOS devices were combined on to the same chip to produce the Complementary Metal Oxide Semiconductor family or CMOS as it is more commonly known.

### 9.3.3 CMOS inverter

A CMOS inverter is shown in Fig. 9.11. It consists of one NMOS and one PMOS transistor. The PMOS device is indicated by the negation sign (i.e. a bubble) on its gate and has a negative threshold voltage of typically –1 V. To turn on a PMOS device we require a voltage, $V_{GS}$, more negative than –1 V. Notice that the two drains of the two MOS transistors are connected together and form the output whilst the two gates form the single input. Due to the difference in the mobilities of the two devices the PMOS device is made with its *W*/*L* ratio 2–3 times larger than the NMOS device. This results in the two transistors having the same value of *K* so that both will have the same electrical performance.

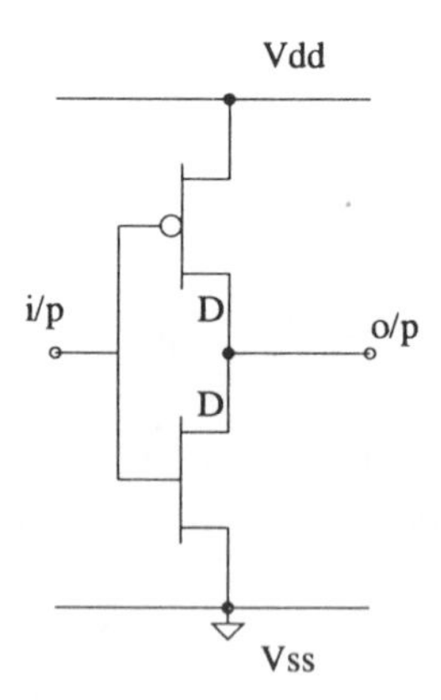


Fig. 9.11 CMOS inverter

The circuit operation depends upon the individual gate-source voltages. When the input voltage is 5 V then the NMOS $V_{GS}$ is 5 V and hence this device is on. However, the PMOS $V_{GS}$ is 0 V and so this device is turned off. The output voltage is thus pulled down to 0 V. Now with the input at 0 V the NMOS $V_{GS}$ is 0 V and hence is turned off. However, the PMOS $V_{GS}$ is –5 V and is thus turned on (remember a voltage more negative than the threshold voltage is needed to turn on a PMOS device). With the PMOS device on, the output voltage is pulled up to $V_{dd}$. The circuit thus operates as an inverter or a NOT gate.

**CMOS inverter power dissipation**

You should notice that when the input is steady at either a high or a low voltage (static condition) then one transistor is always off between $V_{dd}$ and $V_{ss}$. Hence the current flowing is extremely small – equal to the leakage current of the off transistor which is typically 100 nA. As a result of this the static power dissipation is extremely low and it is this reason that has made CMOS such a popular choice of technology.

For input voltages between $V_T$ and $V_{dd} - V_T$ then the individual MOS transistors will be switched on by an amount dictated by Equations 9.1 and 9.2 and thus current will flow from $V_{dd}$ to $V_{ss}$. When the input voltage is $V_{dd}/2$ both transistors will be turned on by the same amount and hence the current will rise to a maximum and power will be dissipated. On many integrated circuits, several thousand gates exist and hence this power dissipation can be large. It is for this reason that the input voltage to a CMOS circuit must not be held at $V_{dd}/2$. When the inputs are switching the power dissipated is called dynamic power dissipation. However, as long as the input signals have a fast rise and fall time then this form of dynamic power dissipation is small. The main cause of dynamic power dissipation, however, in a CMOS circuit is due to the charge and discharge of capacitance at each gate output. The dynamic power dissipation of a CMOS gate is therefore dependent upon the number of times a capacitor is charged and discharged. Hence as the frequency of switching increases so the dynamic power dissipation increases. The dynamic power dissipation for a CMOS gate is equal to

$$P_{\text{dynamic}} = C_L \times V_{dd}^2 \times f \tag{9.3}$$

where $f$ is the switching frequency and $C_L$ is the load capacitance.

The total power dissipated in a CMOS inverter is thus the sum of the static and dynamic components.

**Example 9.9**

Compare the power dissipated by a CMOS inverter driving a 50 pF load at (a) 10 kHz and (b) 10 MHz. What average current flows in each case. Assume a 5 V power supply.

***Solution***

(a) 10 kHz:

$$P_{\text{dynamic}} = C_L \times V_{dd}^2 \times f = 50 \times 10^{-12} \times 25 \times 10 \times 10^3 = 12.5\,\mu\text{W}$$

Also:

$$P_{\text{dynamic}} = V_{dd} \times I_{\text{average}} \Rightarrow I_{\text{average}} = 12.5 \times 10^{-6}/5 = 2.5\,\mu\text{A}$$

(b) 10 MHz:

$$P_{\text{dynamic}} = C_L \times V_{dd}^2 \times f = 50 \times 10^{-12} \times 25 \times 10 \times 10^6 = 12.5\,\text{mW}$$

Also:

$$P_{dynamic} = V_{dd} \times I_{average} \Rightarrow I_{average} = 12.5 \times 10^{-3}/5 = 2.5\,\text{mA}$$

**Example 9.10**

Calculate the output voltage and the current $I_{DS}$ flowing between $V_{dd}$ and $V_{ss}$ when the input to the CMOS inverter in Fig. 9.11 is 2.5 V. Assume that $K_N = K_P = 128\,\mu\text{A V}^{-2}$.

***Solution***

When the input voltage is 2.5 V then $V_{GSN} = -V_{GSP} = 2.5$ V. Hence both devices will be turned on by the same amount. Since $K_N = K_P$ then the output voltage will equal $(V_{dd} - V_{ss})/2 = 2.5$ V.

The current, $I_{DS}$, is determined by using one of the two Equations 9.1 or 9.2. Since $V_{DS} > V_{GS} - V_T$ for both the NMOS and PMOS transistors then both devices are in saturation and Equation 9.2 is used. Thus:

$$I_{DS} = K_N(V_{GS} - V_T)^2/2 = 128 \times 10^{-6}(2.5 - 1)^2/2 = 0.144\,\text{mA}$$

**CMOS inverter delay**

The delay for a CMOS inverter depends upon the rate of charge or discharge of all capacitors at the output. This load capacitance is due to two components called the inherent capacitance and the external load capacitance. The inherent capacitance is due to the drain regions of each transistor and the wiring connecting these two drains together. The external capacitance is due either to the input capacitance of the next stage or any parasitic off-chip capacitance. The propagation delay ($\tau_p$) of a CMOS inverter, and for that matter all CMOS gates, is approximately equal to

$$\tau_p = 2C_L/KV_{dd} \tag{9.4}$$

**Example 9.11**

A CMOS inverter has a total inherent drain capacitance at the output of 1 pF before any external load is added. What is the propagation delay for this inverter unloaded? Also, plot a graph of inverter propagation delay versus external load capacitance. Assume that $K_N = K_P = 64\,\mu\text{A V}^{-2}$.

***Solution***

Before any load is added (i.e. with 1 pF inherent capacitance) the inherent propagation delay of this inverter can be calculated from Equation 9.4. Now, since $K_N = K_P$ then the high-to-low delay will equal the low-to-high delay and it does not matter which of the two we use. Hence

$$\tau_{p(inherent)} = 2 \times 1 \times 10^{-12}/64 \times 10^{-6} \times 5 = 6.25\,\text{ns}$$

As external load capacitance is added the propagation delay will increase linearly at a rate of 6.25 ns/pF. A graph of propagation delay versus external load capacitance can be plotted and is shown in Fig. 9.12. The graph does not pass through the zero delay point since the intercept on the $y$-axis is the inherent delay before any external load is added. If we wish to decrease the delay of a CMOS gate then we must do one of two things. Either decrease the capacitance or increase $K$. The capacitance is decreased by reducing the size of the devices but this is limited to the minimum linewidth[6] achievable with the process. Hence if the designer is already at the limit of the process then all that remains is to increase $K$ which is implemented by increasing the $W/L$ ratio.

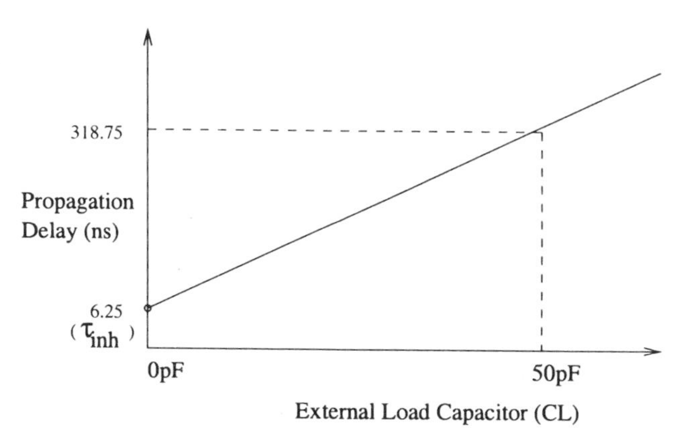


Fig. 9.12 Propagation delay versus load capacitance

Note: It is also possible with some CMOS processes to reduce delays by either increasing $V_{dd}$ (you should check the data sheet before doing this!) or by reducing the temperature (this results in an increase in mobility and hence an increase in $K$).

## 9.3.4 CMOS logic gates

We have seen how to implement the logic gates NAND and NOR using NMOS technology. In CMOS the process is just the same except that the complementary PMOS transistors are added.

### Example 9.12

What function is implemented by the circuits shown in Fig. 9.13(a) and (b)? Although not shown you should assume that the gate inputs labelled A are connected together (similarly for gate input $B$).

[6]The minimum linewidth is the narrowest feature that an IC manufacturing process can produce. The smaller the feature size the more transistors per unit area.

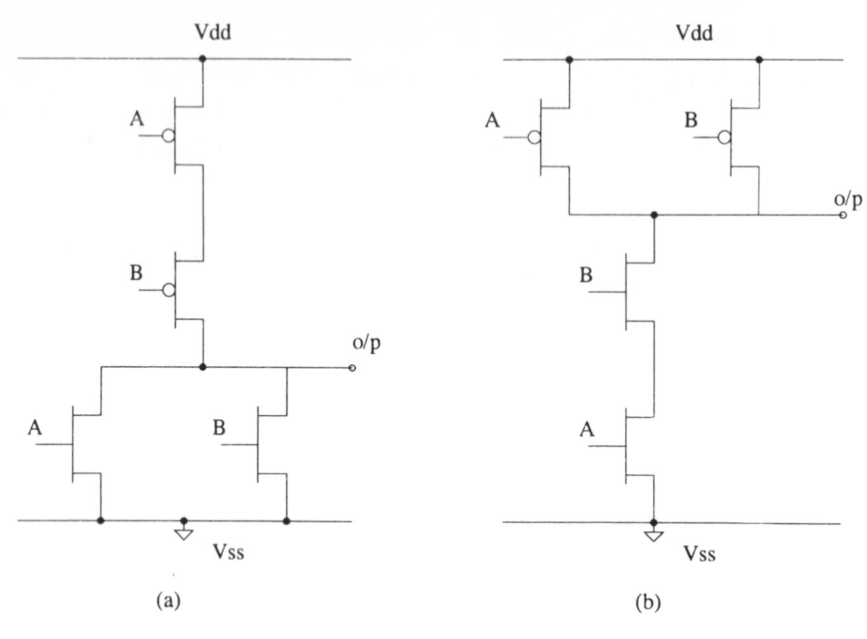


Fig. 9.13 CMOS circuits for Example 9.12

***Solution***

Fig. 9.13 (a): with either $A$ or $B$ or both high then at least one NMOS transistor is on and the output is pulled down to ground. As far as the PMOS transistors are concerned if an input is low then that PMOS transistor is turned on. Now, in this case the PMOS transistors are in series and hence only when both inputs are low will the output be pulled high. The circuit of Fig. 9.13(a) is thus a NOR gate.

Fig. 9.13(b): this time the PMOS transistors are in parallel and hence we only need one input to be low for the output to go high. Conversely, the NMOS transistors are in series and the only way for the output to go low is for both inputs to be high. The circuit of Fig. 9.13(b) is thus a NAND gate.

Note: as for the CMOS inverter when the inputs are held static at either logic 1 or logic 0 then one transistor is always off between $V_{dd}$ and $V_{ss}$ and the current flow is just due to the leakage current of the off transistor. The static power dissipation is therefore again extremely low.

### 9.3.5 Complex gates with CMOS

As we have seen in earlier chapters we can implement many complex combinational functions by connecting together the basic gates NAND and NOR. However, the result is not an efficient use of transistors. If we introduce some basic rules we can produce a more efficient CMOS transistor implementation. Consider for example Fig. 9.14(a) which shows a CMOS circuit which implements the function:

$$f=\overline{(A+B)\cdot(C+D)}$$

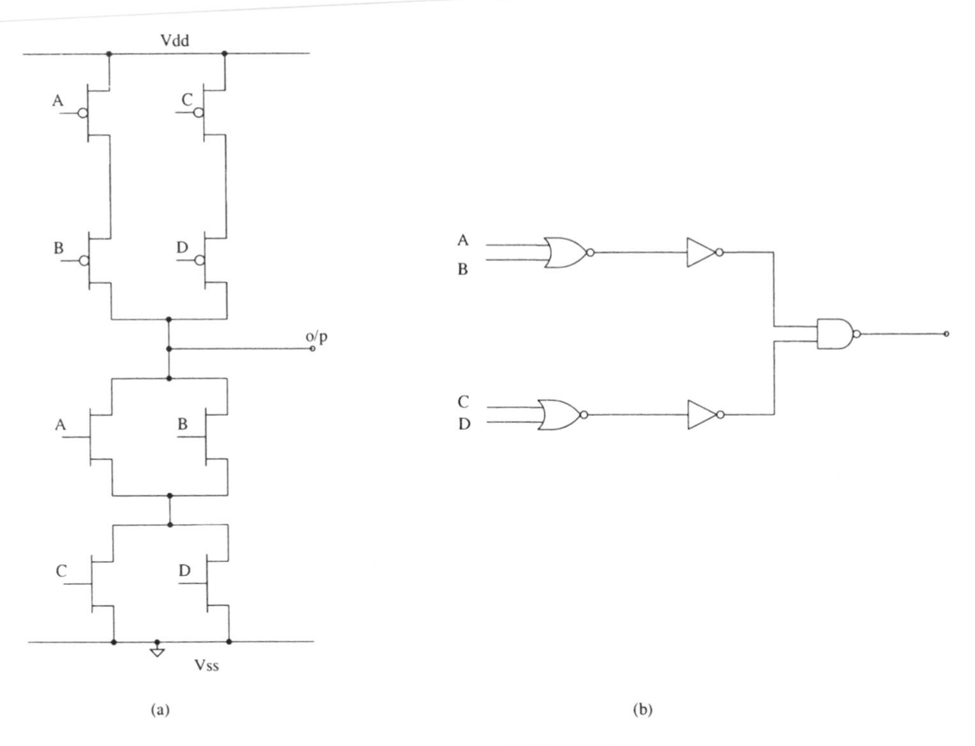


Fig. 9.14 CMOS implementation of the function $f=\overline{(A+B).(C+D)}$

The basic rules are as follows.

1. Concentrate on the NMOS network and note that from the function 'f' we can see that terms OR'd are represented as transistors in parallel and those AND'd are transistors in series, i.e. A is in parallel with B and C is in parallel with D, whilst these two networks are in series with each other.
2. To produce the PMOS network we just replace series networks with parallel networks and parallel with series.[7]

Notice that the number of transistors needed for this function is eight. If we try to implement this function directly with a NAND/NOT/NOR gate approach the circuit shown in Fig. 9.14(b) would be needed and the number of transistors required would be 16 – a rather wasteful use of silicon.

## Example 9.13

Produce an efficient CMOS transistor circuit diagram for the function:

$$f=\overline{A\cdot(B+C+D)}$$

[7]This is another illustration of the principle of duality which we introduced in Chapter 1.

### *Solution*

Using the above rules we concentrate on the NMOS network first. AND functions are networks in series whilst OR functions are networks in parallel. Since *B*, *C*, and *D* are OR'd together then they are drawn in parallel. This parallel network is in series with *A* since they are AND'd together. The network for the NMOS side is thus as shown in Fig. 9.15(a).

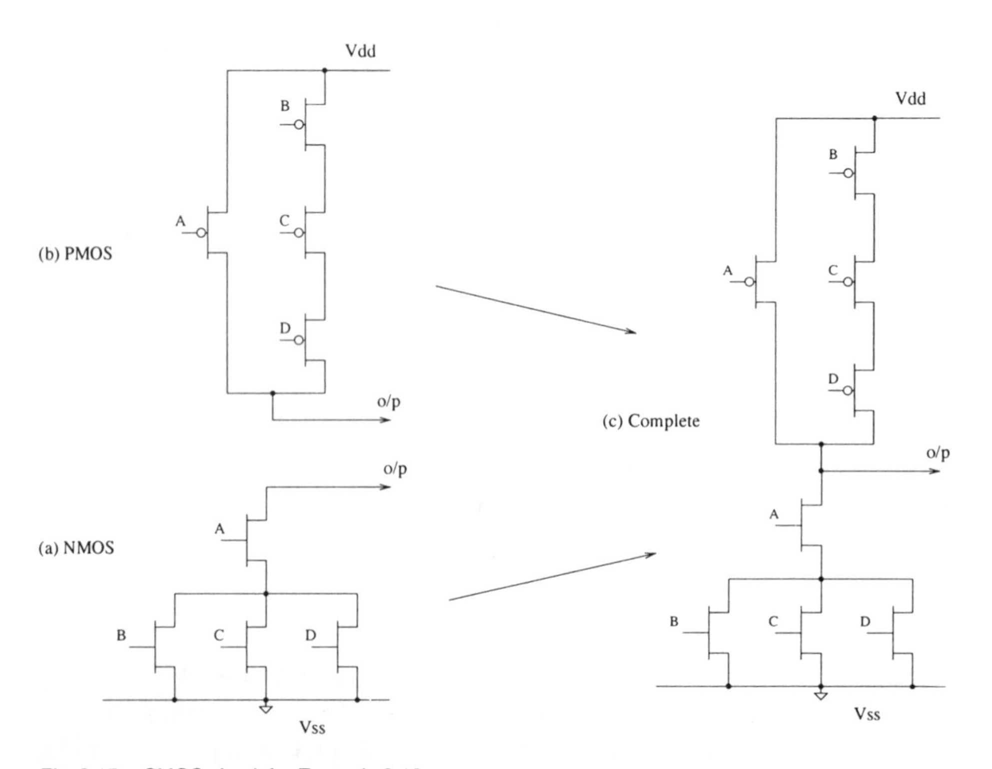


Fig. 9.15 CMOS circuit for Example 9.13

The PMOS side is the reverse of the NMOS circuit, i.e. all series networks are in parallel and all parallel are in series. The circuit for the PMOS side is thus as shown in Fig. 9.15(b). The complete CMOS circuit is as shown in Fig. 9.15(c).

We stated in Chapter 4 that if we reduce the number of gates or levels then the total delay for the circuit reduces. However, we should be careful with this technique since if we tried to produce an eight input NAND gate using a minimum number of transistors then we would have eight NMOS transistors in series and eight PMOS transistors in parallel. If $R_{ds}$ is the resistance of a transistor when turned on, then the output resistance for a high-to-low transition will be $8R_{ds}$, whilst for a low-to-high transition it will be just $R_{ds}$. Thus the high-to-low delay will be eight times that of the low-to-high delay. Hence for large input gates the minimum transistor count may not give the shortest delay. In these cases it is sometimes better to use two four-input NAND gates feeding into an OR gate.

## 9.3.6 CMOS transmission gate

The CMOS transmission gate (TG) is a single-pole switch that has a low on resistance and a near infinite off resistance. The device consists of two complementary MOS transistors back to back and is shown in Fig. 9.16(a) with its symbol in Fig. 9.16(b).The device has one input, $V_{in}$, and one output $V_{out}$. The gate of the NMOS transistor is driven from a control signal $V_c$ whilst the PMOS transistor gate is driven from $\overline{V_c}$ via an inverter (not shown).

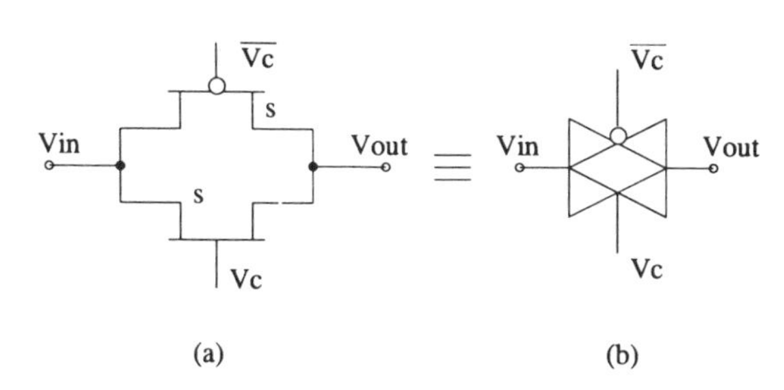


Fig. 9.16 CMOS transmission gate

Consider what happens when $V_c$ is held high (i.e. 5 V). With $V_{in}$ at 0 V then the NMOS $V_{GS}$ is 5 V and this device is turned on and the output will equal the input, i.e. 0 V. Notice that $V_{GS}$ for the PMOS device is 0 V and hence this device is turned off. The reverse is true when $V_{in}$ is held high, i.e. PMOS $V_{GS}$ is –5 V and is switched on whilst the NMOS $V_{GS}$ is 0 V and is turned off. In either case an on transistor exists between $V_{in}$ and $V_{out}$ and hence the input will follow the output, i.e. the switch is closed when $V_c$ is held high.

Now when $V_c$ is held low then the NMOS $V_{GS}$ is 0 V and the PMOS $V_{GS}$ is 5 V and so both devices are off. The switch is therefore open and the output is said to be *floating* or high impedance.

One application of this device is as a *tri-state* circuit which is discussed later in this chapter. However, many other uses have been made of this CMOS TG. Some of these are shown in Fig. 9.17(a) and Fig. 9.18(a). Fig. 9.17(a) shows a 2-to-1 multiplexer circuit. When the *select* line is high then 'bit 0' is selected and passed to the output whilst if *select* is low then 'bit 1' is passed to the output. Notice that the non-TG version of this circuit, illustrated in Fig. 9.17(b), uses many more transistors than the simple TG version.

Fig. 9.18(a) shows the use of a transmission gate as a feedback element in a level triggered D-type latch. When the clock signal is high then TG1 is closed and data at D is passed to the output (TG2 is open). When the clock goes low then TG1 is open and the data at the output is passed around the feedback loop via TG2 which is now closed. Data is therefore latched into the circuit. The equivalent non-TG version using logic gates (introduced in Problem 5.4) is shown in Fig. 9.18(b) and again uses many more transistors than the TG version. As a result of this, all CMOS flip-flops are designed using the space saving TG technique. To

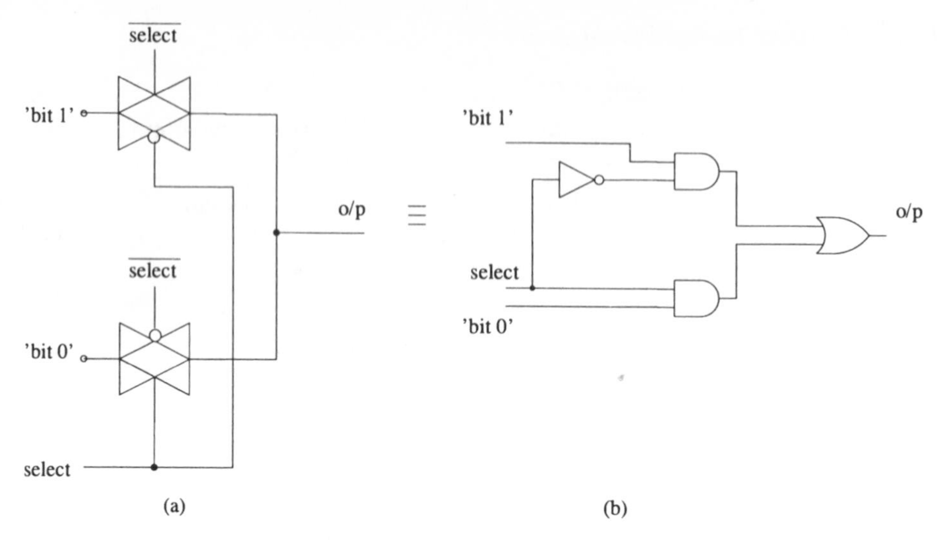


Fig. 9.17 Digital multiplexer implemented with (a) TGs and (b) logic gates

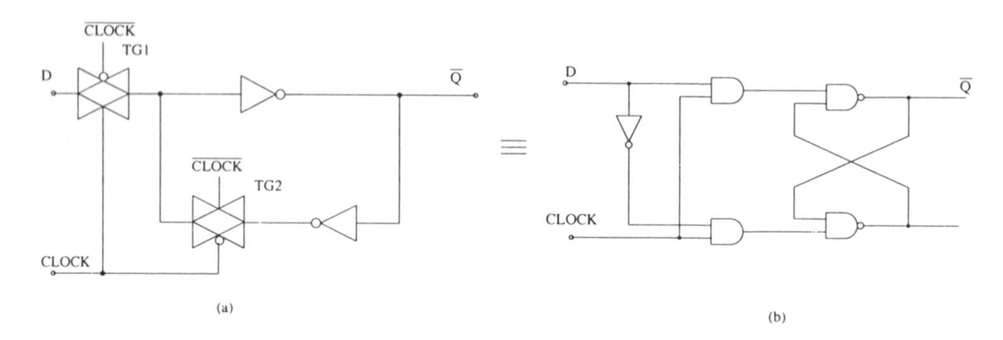


Fig. 9.18 D-type latch implemented with (a) TGs and (b) logic gates

produce a JK with CMOS TGs it is necessary to add the appropriate circuitry to a TG based D-type (see Problem 11.4). Hence CMOS JKs use more gates than D-types. It is for this reason that CMOS designs use the D-type as the basic flip-flop rather than the JK.

### 9.3.7 CMOS 4000 series logic

The 4000 series was the first CMOS logic family marketed. It was basically the raw CMOS logic gates shown in Figs 9.13–9.18 directly driving external load capacitances or other loads such as TTL gates. The circuit impedance seen looking back into the output depends upon which transistors are on or off. For example the two-input NOR gate when the output is low will present a different output impedance depending upon whether one transistor is on or both are on. This will result in differing propagation delays and variable output drive capability. Nevertheless, these devices had very low static power dissipation and with a wide power supply range of 3–15 V had good noise immunity (see later).

The 4000 series was eventually replaced by the 4000B series. This logic family is essentially the original 4000 series but with the outputs double buffered. This double buffering was quite simply two inverter stages with *W*/*L* ratios increasing at each stage so that the last stage is able to drive the off chip capacitances and other TTL loads without compromising logic levels. A transistor circuit diagram illustrating the double buffering (with approximate *W*/*L* ratios in microns) is shown in Fig. 9.19 for a two-input NOR gate in the 4000B series. These devices have a transfer characteristic which changes more abruptly from one logic level to the other compared to the 4000 series. This is due to the two extra stages at the output which also results in a much better noise margin. Delays of the order of 50–100ns are obtainable with this process.

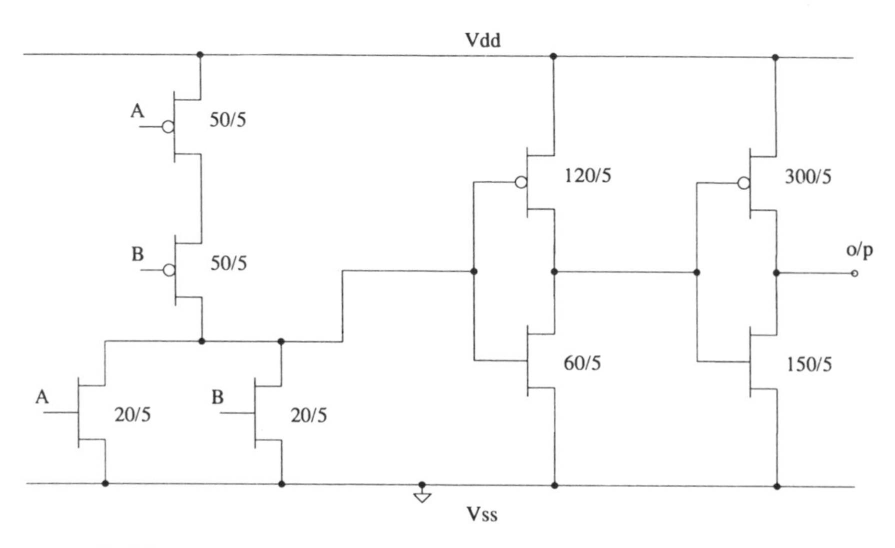


Fig. 9.19 A CMOS 4000B double buffered NOR gate

### 9.3.8 CMOS 74 series logic

Many digital electronic systems were designed at first with the 74 series TTL devices. Since the 4000B series were not pin-for-pin compatible with the TTL devices then replacement with CMOS was only possible if a complete board redesign was implemented. Hence, in order to take advantage of the low static power consumption of CMOS logic the TTL series has been gradually replaced with CMOS equivalents that have the same pin out. These CMOS logic gates all have outputs that are double buffered and buffers on the inputs which result in a good noise margin. A plethora of logic gate families now exist under the 74 CMOS series and we shall look chronologically at most of these.

**74C series**

The 74C family was the first CMOS version of the TTL 74XXX series on the market. It used 5 μm technology with all outputs double buffered, as in the

4000B series, so that they can drive other TTL logic gates as well as large off-chip capacitances. This family is now obsolete being replaced by the HC and HCT versions.

**74HC/HCT series**

The 74 HC series are fabricated with 3 μm CMOS and an increased value of $K$ (see Section 9.3.1). This results in a shorter propagation delay and increased output drive capability. These devices have a speed performance similar to the 74LS series but with a greatly reduced power consumption. Unloaded the output voltages are guaranteed to be within 100 mV of the supply. However, under a load such as driving a TTL input the voltage across the MOS output transistors will increase as current passes through them. The value of $K$ of the output stage is therefore designed such that the output voltage will still produce a legal logic 1 or 0 (i.e. a large $W/L$ ratio is used – see Problem 9.11 at the end of this chapter).

Although the HC series have the same speed as 74LS parts (see Table 9.1) they cannot be driven by LS parts. This is because the minimum $V_{IH}$ (called $V_{IHmin}$) of the HC is approximately 3.5 V whilst the minimum $V_{OH}$ for the LS part is 2.7 V and hence will not be recognised by the HC series device as a legal logic 1. To avoid this problem the 74HCT was introduced. This series again uses CMOS technology but the inputs are designed to be TTL input voltage level compatible i.e. $V_{IHmin}=2.0$ V and $V_{ILmax}=0.8$V. This is achieved by adjusting the $W/L$ ratios of the two MOS transistors in the input buffers so as to move the switching point. In the HC series the PMOS width is 2–3 times that of the NMOS (to compensate for the difference in mobilities) and the device switches at $V_{dd}/2$. However, for the HCT devices the NMOS width is approximately ten times that of the PMOS device such that the value of $V_{IH}$ is reduced to 2 V and $V_{IL}$ to 0.8 V – compatible with TTL logic levels.

**Example 9.14**

What should be the relative width-to-length ratios of the NMOS and PMOS transistors for the CMOS input buffer to create a TTL input compatible device. Assume that the mobility of NMOS carriers is three times that of the PMOS carriers.

***Solution***

The TTL input logic levels are: $V_{IHmin}=2.0$ V and $V_{ILmax}=0.8$ V. Hence the switching point should be chosen half-way between these levels i.e. at 1.4 V. The switching point will occur when both transistors are on by the same amount and from Equation 9.2

$$K_N(V_{GSN}-V_T)^2/2=K_P(V_{GSP}-V_T)^2/2$$

$$K_N(V_{in}-V_T)^2/2=K_P(V_{dd}-V_{in}-V_T)^2/2$$

$$K_N/K_P=(3.6-1)^2/(1.4-1)^2=42.25$$

From $K=(W/L)\times\mu\times C_{ox}$ and given that $\mu_n=3\mu_p$ then the NMOS $W/L$ ratio should be set at 14.08 times that of the PMOS transistor.

**74AC/ACT series**

Continual improvements in CMOS processing have led to the introduction of an improved high-speed CMOS family called the *advanced* CMOS logic designated as 74AC and 74ACT. They are direct replacements for the 74AS and 74ALS series and in some cases the 74F series. These devices use 1.5μm CMOS technology with a very thin gate oxide of approximately 400 Å (one Å $=10^{-10}$m!). This results in very high speed CMOS devices with delays of typically 5 ns. This range of devices also has a very high output current drive of 24 mA (see Table 9.1) due to the higher $K$ caused by the thin gate oxide and large $W/L$ ratios at the output. The ACT series is TTL input voltage level compatible and can be mixed with ALS and AS devices. It has the added advantage of a very low power consumption as with all CMOS devices. This range of devices is sometimes referred to as advanced CMOS logic (ACL) by Texas Instruments or FACT by National and Fairchild.

Undoubtedly more and more logic families will become available to the designer. Currently we are at the advanced, advanced stage of high-speed CMOS devices, the latest being the 74VHC series offered by National Semiconductors and the 74AHC/AHCT series marketed by Texas Instruments. We may well be approaching the limit of CMOS and the use of BiCMOS could well be the next technology choice on offer to the logic designer.

## 9.4 BiCMOS – THE BEST OF BOTH WORLDS

The advances in integrated circuit processing have led to ever decreasing transistor sizes. However, for the same quality process a MOS transistor consumes considerably less space than a bipolar transistor. Hence CMOS chips are much smaller than bipolar equivalents and hence internal capacitances are greatly reduced resulting in ever decreasing propagation delays and manufacturing costs. However, the CMOS families are limited when driving large capacitive loads such as off-chip capacitances present on data buses, and even oscilloscope leads. The bipolar transistor is much better at driving these large capacitances since for the same size device the bipolar transistor has a larger effective $K$ than the MOS device. A new technology has therefore emerged called BiCMOS that combines the best of both worlds, i.e. CMOS *and* bipolar. It contains the small CMOS logic gates but in any places where it is necessary to drive large capacitive loads then the bipolar totem-pole stage is used.

A typical BiCMOS inverter is shown in Fig. 9.20. The device operates as follows. When the input is high then the base of T1 is low and is turned off. Transistor MN is turned on and since MN2 is off then T2 turns on and the output is low. When the input switches to zero volts then the base of T1 goes high and turns on T1. Since the base of T1 is high then MN2 is turned on and the base of T2

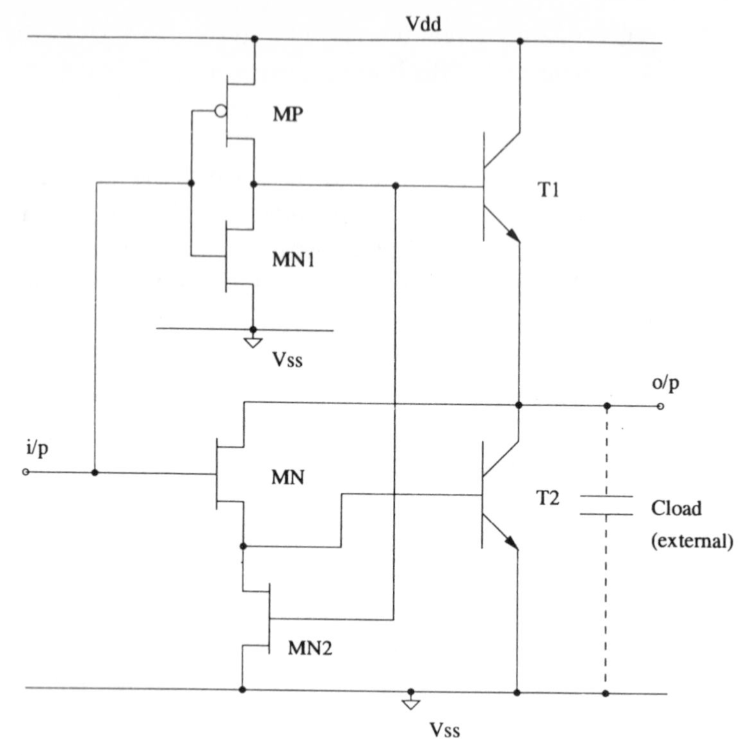


Fig. 9.20 BiCMOS inverter

is low and is thus turned off and the output goes high. Notice that when T1 turns off then MN1 provides a base discharge path, whilst when T2 turns off the base discharge path is provided by MN2.

The bipolar output thus allows large capacitances to be driven, whilst the CMOS part implements the desired function internally. A typical BiCMOS logic family is the 74BCT series which tends to have devices that are only for bus driving such as octal buffers and octal latches. These have similar speeds to the 74F family but with greatly reduced power consumption (see Table 9.1).

## 9.5 LOW-VOLTAGE OPERATION

Battery operated equipment such as lap-top computers and hand-held instruments require low-power devices. As we have seen CMOS offers extremely low power, an approximate value of which can be obtained from Equation 9.3, i.e. $C_L V_{dd}^2 f$.

As device dimensions reduce in size, the capacitance reduces leading to further reductions in the power consumed. However, a reduction in $V_{dd}$ will lead to a larger reduction in power consumption due to the square term in Equation 9.3. In addition the use of a lower $V_{dd}$ will result in fewer batteries needed and hence a lighter instrument. Consequently most portable digital equipment is made nowadays with a reduced $V_{dd}$.

Since alkaline batteries have a typical voltage of 1.35 V and NiCd have a typical voltage of 1.2 V then a multiple of this is usually needed for the power supply. A range of 2.7–3.6 V will require either two alkaline or three NiCd. Although the HC, AC and 4000B series will operate at 3 V these devices have not been optimised for this supply voltage and hence they are a compromise that will satisfy all power supply voltages. Consequently, a range of CMOS devices is now available that has been specifically designed for this lower voltage. These low-voltage devices are labelled by National Semiconductor as 74LVX or by Texas Instruments for example as 74LV, 74LVC and 74ALVC. It is also possible to obtain low-voltage BiCMOS devices which again are able to drive large capacitive loads. These not only operate with a 3 V supply but also have the capability of being driven from 5 V input signals. A typical range of devices of this type is the 74LVT series by Texas Instruments (see again Table 9.1).

**Example 9.15**

What percentage saving in dynamic power consumption will be obtained by reducing the power supply from 5 V to 3 V?

***Solution***

From Equation 9.3 the power consumption will reduce from: $C_L \times f \times 25$ to $C_L \times f \times 9$, i.e. a saving of 64% in power consumed.

## 9.6 OTHER TECHNOLOGY OPTIONS

### 9.6.1 Emitter coupled logic – ECL

The emitter coupled logic family has been available for the digital designer since the early TTL days. A circuit configuration for a two-input OR/NOR gate is shown in Fig. 9.21.

This family has delays of the order of 1 ns and achieves this by (a) ensuring that the transistors do not enter saturation and (b) having a smaller voltage swing. The circuit contains two inputs $A$ and $B$ and two outputs $V_{out1}$ (NOR) and $V_{out2}$ (OR). The outputs are taken from the emitters of T5 and T6. Although these emitters appear to be floating they are assumed to be driving other ECL gate inputs which, as can be seen, have a 50 kΩ resistor between its input and –5.2 V. The output circuit is thus acting as a *voltage follower.*[8] Note also that the power supplies are 0 V and –5.2 V which are not compatible with TTL, CMOS or BiCMOS. In ECL technology a logic 0 is defined as having a larger negative voltage than a logic 1.

This ECL logic gate functions as follows. Firstly, the circuit inside the dotted box is a voltage reference circuit providing a reference voltage at the base of T2. If

[8]B. Hart, *Introduction to Analogue Electronics*, in this series.

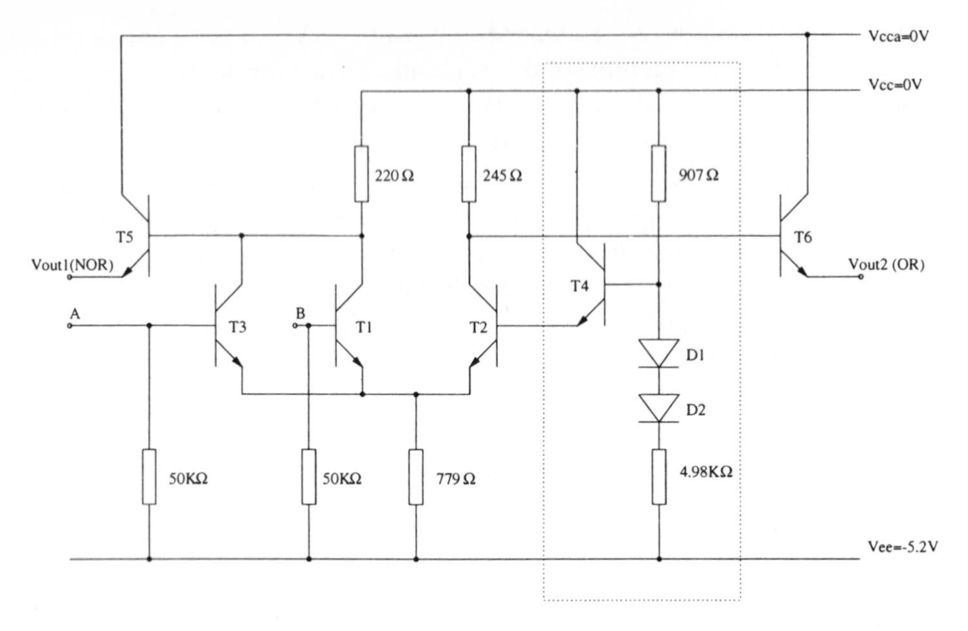


Fig. 9.21 ECL NOR/OR gate

both inputs *A* and *B* are taken low (i.e. a large negative voltage) then transistors T1 and T3 are turned off and current flows through T2 and hence the output $V_{out1}$ (NOR) is pulled towards 0 V (i.e. a logic 1) and $V_{out2}$ (OR) moves towards –5.2 V (i.e. a logic 0). When one or both of the inputs are held high then the current passes through the transistor which has a high on its base and T2 turns off. Hence $V_{out1}$ (NOR) is now pulled towards –5.2 V and $V_{out2}$ (OR) pulled towards 0 V. The circuit thus functions as an OR/NOR gate.

Two families have been marketed in ECL logic. These are the 10K series and the 100K series, the 100K series being the most recent. The voltage levels for this family are (see Table 9.1):

$$V_{OH}=-0.9\,V;\ V_{OL}=-1.7\,V;\ V_{IH}=-1.2\,V;\ \text{and}\ V_{IL}=-1.4\,V$$

As can be seen these logic levels are not TTL or CMOS level compatible.

The 100K series differs from the 10K series by having a more temperature stable characteristic and a faster speed of operation, with a delay of 0.75 ns and a power consumption of 20 mW. These devices tend to be used in specialist high-speed logic requirements such as digital telephone exchanges or high-speed super-computers.

## 9.6.2 Gallium arsenide – GaAs

All the technologies we have looked at so far have been implemented in silicon. Alternative semiconductors, such as gallium arsenide (GaAs) and germanium (Ge) exist. The use of GaAs in digital applications such as telecommunications has been marketed for some time. GaAs has an electron mobility approximately five times that of silicon and hence can operate at much higher frequencies. Consequently a lot of research effort has been spent on exploiting this speed advantage. However, the processing that is needed with GaAs is much more complicated than silicon and hence these devices are used only for the specialist high-speed digital market. Typical operating delays of 100 ps with power consumptions of 1 mW per gate are available. Since these devices are to complement ECL they use ECL logic levels with similar power supply requirements.

## 9.7 GATE CHARACTERISTICS

We should now be more familiar with the various technology options so let us investigate how the gate characteristics vary from technology to technology. A logic gate is characterised in terms of various parameters. Some of the more important parameters are: transfer characteristics; noise margin; output drive (fan out and fan in); propagation delay; power dissipation; and power delay product.

### 9.7.1 Transfer characteristics

A transfer characteristic plots the output voltage versus the input voltage for a logic gate. A transfer characteristic for a non-inverting logic gate is shown in Fig. 9.22.

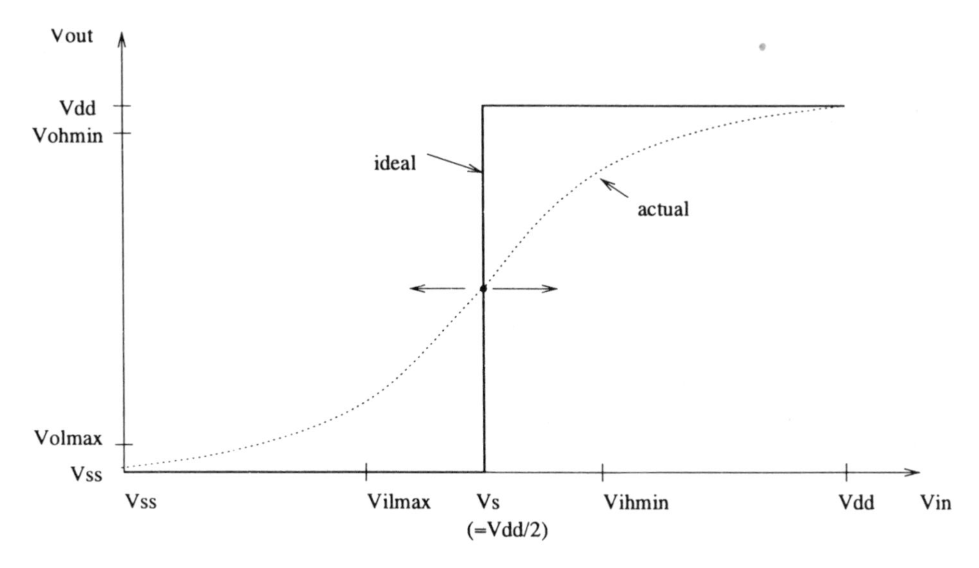


Fig. 9.22 Logic gate transfer function

The ideal characteristic (shown by the solid line) illustrates that below a switching voltage ($V_s$) the output will equal the most negative voltage (usually ground or $V_{ss}$). With the input above $V_s$ the output will equal the most positive voltage (referred to as $V_{cc}$ for bipolar or $V_{dd}$ for CMOS). The switching voltage $V_s$ is usually at half the supply voltage and the change from one logic state to the other occurs very sharply.

In reality the transfer characteristic is as shown by the dotted line. In this case the switching voltage may not be at half the supply, the switching region is gradual and the output voltage may not reach the supply rails. For TTL $V_s$ is 1.4 V whilst for CMOS $V_s = V_{dd}/2$ (unless of course the CMOS is *TTL input compatible* when it too will be 1.4 V - see Section 9.10.1).

Various significant voltages are defined for a logic gate. For the gate input, the two logic level voltages are

- $V_{ILmax}$: the maximum value of input voltage that can be recognised as a logic '0';
- $V_{IHmin}$: the minimum value of input voltage that can be recognised as a logic '1';

whilst for the gate output the two logic levels are

- $V_{OLmax}$: the maximum value of output voltage for a legal logic '0';
- $V_{OHmin}$: the minimum value of output voltage for a legal logic '1'.

Input voltages between $V_{ILmax}$ and $V_{IHmin}$ will result in an indeterminate value of output voltage and hence are not allowed. In addition the output of a gate is not allowed to have values between $V_{OLmax}$ and $V_{OHmin}$.

### 9.7.2 Noise margin

Noise in a digital system is mainly caused by switching transients which cause perturbations in the power supply or generate crosstalk between adjacent wires on the chip or circuit board. These disturbances are propagated to the output or the input and can either add or subtract to the existing signals and hence change the voltage appearing at the output. If the noise is large enough it can change a legal logic level into an illegal value. The magnitude of the voltage required to reach this illegal state is called the *noise margin* and is specified for both logic high and logic low conditions. It indicates the maximum noise voltage that can appear on an output before the output level is deemed illegal.

Consider two non-inverting gates in series as shown in Fig. 9.23 and the associated voltage levels. The maximum high and low noise voltages that can be allowed on the output of the first buffer are thus

$$NM_H = V_{OHmin} - V_{IHmin} \text{ and } NM_L = V_{ILmax} - V_{OLmax}$$

**Example 9.16**

Two inverters from a 74LS04 hex inverter IC (i.e. six inverters in one package) are connected in series such that one inverter is driving the other. From Table 9.1 determine the high and low noise margins.

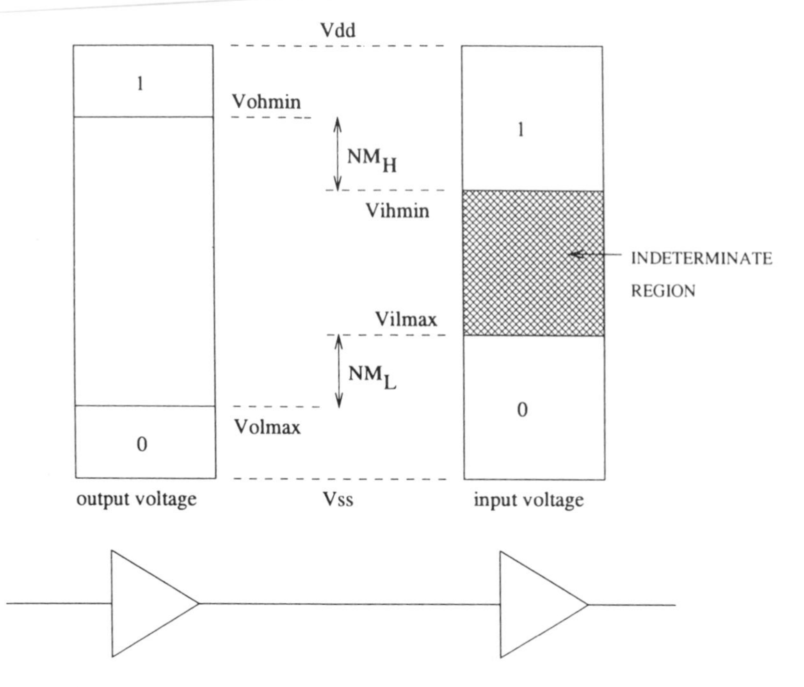


Fig. 9.23 Logic levels and noise margins

***Solution***

$$NM_{\mathrm{H}} = V_{\mathrm{OHmin}} - V_{\mathrm{IHmin}} = 2.7 - 2.0 = 0.7\ \mathrm{V}$$

$$NM_{\mathrm{L}} = V_{\mathrm{ILmax}} - V_{\mathrm{OLmax}} = 0.8 - 0.5 = 0.3\ \mathrm{V}$$

**Example 9.17**

Repeat 9.16 for a 74HCT04 (CMOS version of 74LS04).

***Solution***

$$NM_{\mathrm{H}} = V_{\mathrm{OHmin}} - V_{\mathrm{IHmin}} = 4.3 - 2.0 = 2.3\ \mathrm{V}$$

$$NM_{\mathrm{L}} = V_{\mathrm{ILmax}} - V_{\mathrm{OLmax}} = 0.8 - 0.33 = 0.47\ \mathrm{V}$$

Hence an improved noise margin is obtained with CMOS. It should be noted, however, that since the CMOS output is driving another CMOS device then the current drawn from the output is small. Hence the output voltage levels for a CMOS device will be much closer to the supply than indicated in Table 9.1 resulting in an even larger noise margin.

### 9.7.3 Output drive (fan out/fan in)

One of the requirements of a logic gate is that sufficient output current drive is available to drive other inputs. However, as the output current increases the voltage dropped across the 'on' output transistor will increase. It is essential that this voltage does not rise above the point at which the voltage levels become illegal (i.e. below $V_{OHmin}$ or above $V_{OLmax}$). The number of inputs which a gate output can drive before the output becomes invalid is called the *fan-out* or *output drive* capability. This *fan-out* is expressed as

$$\text{fanout}_{high} = I_{OHmax}/I_{IHmax} \text{ and } \text{fanout}_{low} = I_{OLmax}/I_{ILmax}$$

The *fan-in* on the other hand is the load that an input places on an output. This is sometimes expressed as the input capacitance, the input current or sometimes as the number of inputs to a gate.

**Example 9.18**

How many standard 74 series gates can a 74ACT series drive?

***Solution***

From Table 9.1 the output current drive of a 74ACT device is 24 mA.The worst case fan-out calculations will be for the logic low case since this requires the largest input current. Hence

$$\text{fanout}_{low} = I_{OLmax}/I_{ILmax} = 24/1.6 = 15$$

i.e. the 74ACT series will drive 15 standard 74 series devices. If we compare this with a 74LS series device driving a standard 74 series the output drive current is only 8 mA ($I_{OLmax}$) and thus its fan-out is only 5.

**Example 9.19**

Repeat Example 9.18 for a 74ACT driving other 74ACT devices.

***Solution***

Since the input current to a 74ACT device is negligible (due to the fact that it uses MOS transistors), then the load it places on the output is minimal and the fan out is very large (much greater than 15). However, the inputs do have an input capacitance and this will affect the propagation delay, as discussed in the next section.

### 9.7.4 Propagation delay

As we have already seen the *propagation delay* is defined as the time it takes for a signal at the input to pass to the output. It is usually defined between the 50% points as illustrated in Fig. 9.24. Two propagation delays are quoted for a logic gate depending upon whether the *output* is going low to high ($\tau_{PLH}$) or high to low

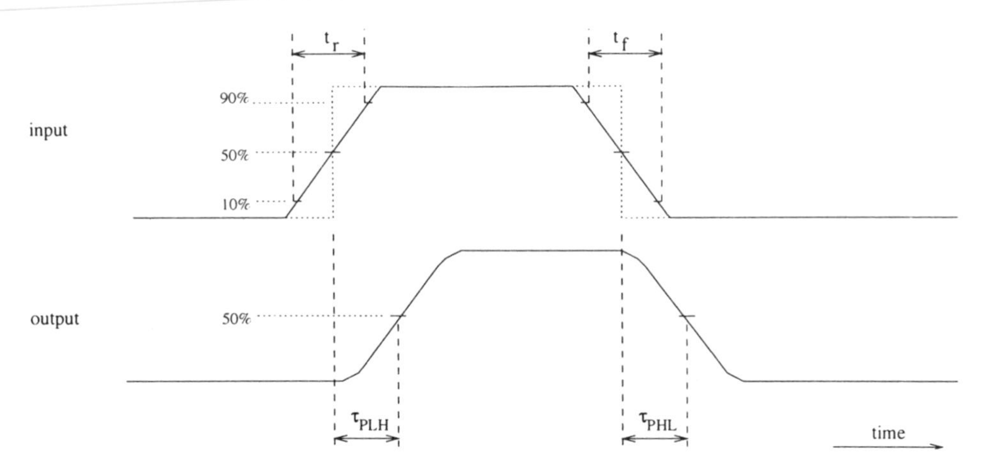


Fig. 9.24 Delay times for logic gates

($\tau_{PHL}$). Notice that the ideal input (shown dotted) has an immediate change from 0 to 1 and from 1 to 0. However, in reality this response is not as sharp as this and hence the input has a *rise* time and a *fall* time. The rise and fall times are defined as being between the 10% and the 90% points. Notice that the definition of the propagation delay time is unaffected by the value of the rise and fall time.

As we have seen from Equation 9.4 for CMOS circuits, the propagation delay depends upon the capacitance being driven and the effective value of $K$ for the output transistor – a similar relationship holds for bipolar technology. A plot of relative propagation delay versus external load capacitance is given in Fig. 9.25 for the three main families of CMOS, TTL (or BiCMOS) and ECL. At low capacitances the CMOS family has a smaller delay than TTL or BiCMOS. This is because the external load capacitance on the $x$-axis does not include the internal capacitance of the logic gate. For MOS devices this internal capacitance is smaller

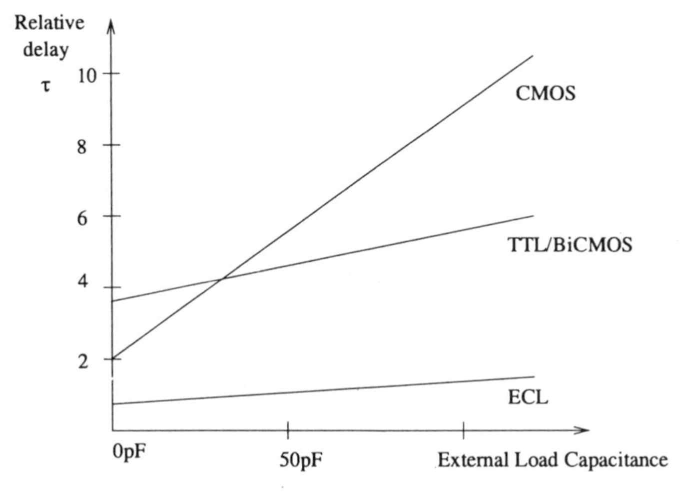


Fig. 9.25 Comparison of propagation delay versus load capacitance for different technologies

than bipolar devices partly because the processing of MOS devices has advanced considerably over the years and partly because a bipolar transistor takes up more space on the chip and hence has a higher capacitance. However, as the external load capacitance increases it soon dominates over the small internal load capacitance. The rate of increase of propagation delay with capacitance depends upon the effective $K$ of the output transistor. The effective $K$ is larger for a bipolar transistor than for a MOS transistor and hence with large load capacitance the delays for CMOS are larger than for BiCMOS or TTL. Consequently if you are driving a large capacitance, i.e. several other inputs (or large fan-in), then it is preferable to use a bipolar output. If, however, the output is driving a low capacitance, i.e. less than 30 pF, then use CMOS outputs. Drawn on the same axis is the ECL delay versus capacitance. As expected these devices are faster since they do not enter saturation.

**Example 9.20**

Compare the delay of a TTL device driving 15 CMOS devices with that of a CMOS device driving the same load. Assume that the TTL output can drive at 20 ps/pF whilst the CMOS device has a drive of 67 ps/pF and that each CMOS input has a capacitance of 10 pF.

***Solution***

$$\text{Total load from 15 CMOS devices} = 15 \times 10\,\text{pF} = 150\,\text{pF}$$

$$\text{TTL driving: loading delay} = 150\,\text{pF} \times 20\,\text{ps/pF} = 3\,\text{ns}$$

$$\text{CMOS driving: loading delay} = 150\,\text{pF} \times 67\,\text{ps/pF} = 10.05\,\text{ns}$$

### 9.7.5 Power dissipation

The power dissipation of logic gates is characterised under two modes. These are *static* and *dynamic*. Under static conditions the input is held at either logic '1' or '0'. The *static* power consumption is thus

$$P_{\text{static}} = V_{\text{dd}} \times I_{\text{supply}}$$

Under *dynamic* conditions the inputs are changing state and hence the transistors between the supplies will either be both on or require energy to charge and discharge output capacitances. Hence the dynamic power dissipation will depend upon the number of times the transistors switch per second, i.e. the signal frequency. If the rise and fall times of the input signal are small then the dynamic power dissipation is due solely to the energy required to charge and discharge the load capacitances. As seen in Equation 9.3 for CMOS, this is equal to

$$P_{\text{dynamic}} = C_{\text{L}} \times V_{\text{dd}}^2 \times f$$

where $C_L$ is the total capacitance seen at the output and $f$ is the signal frequency. This equation also applies to bipolar technology. The total power dissipated is therefore the sum of the static and dynamic power dissipations.

Consider the comparison of power consumptions between TTL and CMOS. With TTL devices the static power dissipation is quite large. Fig. 9.3 shows that for the TTL family, with a low at the input, a current must flow out of the input (typically 1.6 mA). Now, with the inputs high the second stage will be on and drawing current from the supply. However, for CMOS devices one transistor is always off between the supplies and hence the static current drawn is only due to the off transistor. This is called the *leakage* current, and is very small. Hence the total power dissipation for CMOS is due mainly to dynamic effects and is very small at low frequencies. This is the reason why CMOS is such a popular choice.

However, Equation 9.3 shows that as the frequency increases the power dissipation of the CMOS devices will increase. The same component is present in TTL devices but since the static power consumption is high in the first place it does not show itself until relatively high frequencies are reached. A typical plot of power dissipation versus operating frequency is shown in Fig. 9.26 for a 74LS00 device and a 74HC00 device (quad two-input NAND gate). Notice that it is not until frequencies above 5 MHz that the CMOS device has similar power consumption to the TTL device. Below this the power dissipation of the CMOS device is very low.

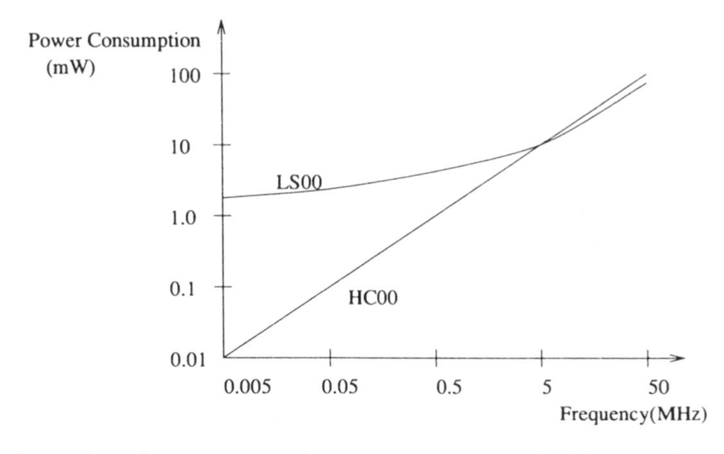


Fig. 9.26 Comparison of power consumption versus frequency for CMOS and 74LS series

## 9.8 OPEN COLLECTOR AND THREE-STATE LOGIC

In cases where data has to pass off-chip to a single interconnect that is used by other output devices (called a bus) then the traditional totem-pole output of bipolar, or the standard complementary pair of CMOS, cannot be used in its present form. This is illustrated in Fig. 9.27 where two outputs drive the same line. If the output of gate 1 is high and that of gate 2 is low then a condition called *bus*

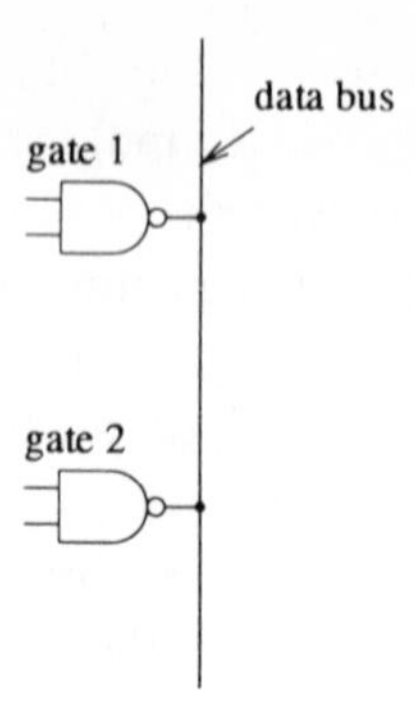


Fig. 9.27 Bus contention with logic gates

*contention* occurs. Current passes from gate 1 to gate 2 and an unknown voltage is presented to the line and in some cases may result in damage to either or both of the output stages. Two remedies exist to this problem: use open collector (or open drain) outputs; or use three-state output circuits.

### 9.8.1 Open collector/drain

The open collector output (or open drain for MOS devices) is quite simply the same output as a TTL totem pole (or CMOS output buffer) but with the top half of the output circuit removed to just leave the lower transistor with its collector open – hence its name. An example of two open collector gate outputs driving a

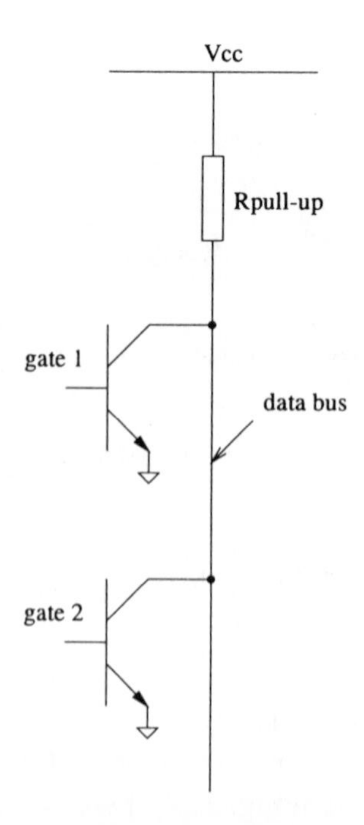


Fig. 9.28 Open collector outputs driving a common data bus

common data bus is shown in Fig. 9.28. The advantage of this type of output is that they can be wired together and bus contention is no longer a problem. If both output transistors are turned off then the output is pulled up to a logic '1' via the single pull-up resistor. The case of bus contention when one output transistor is on and another is off is avoided because the off transistor presents an open circuit to the common line and does not interfere with the logic level. Such a wiring arrangement is often called a *wired-or* connection.

### 9.8.2 Three-state logic

Two disadvantages exist with the wired-or connection. The first is that power is consumed with the pull-up resistor via the on transistor and secondly that the switching speed is reduced due to the arrangement having no active pull-up. The low-to-high switching time is now dictated by the time constant of the pull-up resistor with the external load capacitor. An alternative to the open collector or open drain output is to use a *three-state circuit.*

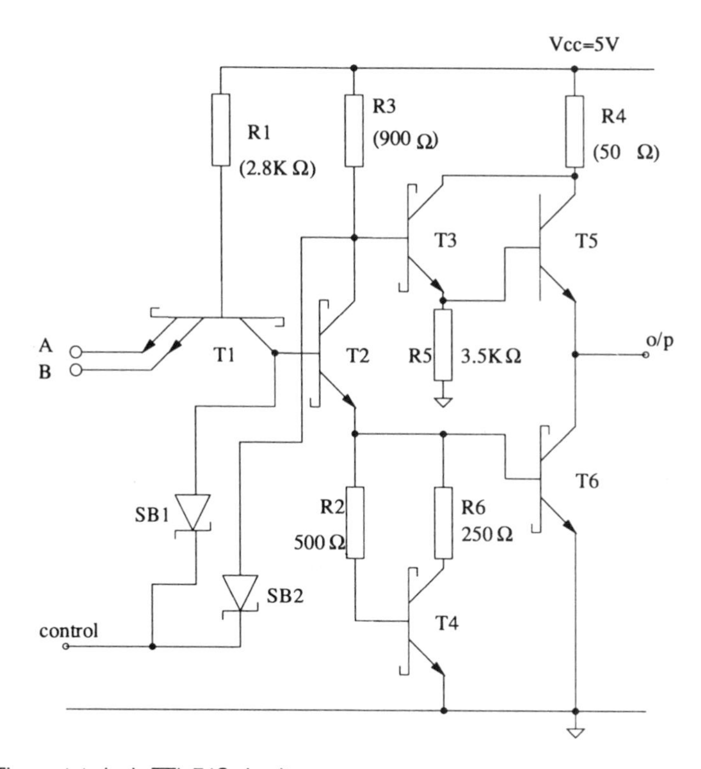


Fig. 9.29 Three-state logic TTL 74S circuit

The term three-state logic is a most misleading term in digital logic, especially since we have been using the binary system which only has two values! Three-state logic is a term given to the ability of an output stage to: drive a logic 1; drive a logic 0; and a third state where the output presents a high impedance to the common data bus and hence does not drive the output to any voltage. This third state has many other names, these being *high impedance*, *high Z* and *tri-state* to name but a few. They all produce the same function of having this third mode where the output does not drive any voltage on to the common data bus and presents a high impedance to the line as though it were not connected.

A circuit diagram for a three-state logic TTL 74S series circuit is shown in Fig. 9.29. The circuit is exactly the same as the 74S series device illustrated in Fig. 9.6 except that an extra control input is added via two Schottky barrier diodes. When the control is high both diodes SB1 and SB2 are reverse biased and the circuit operates as normal with the output either driving a logic 1 or a logic 0. When the control is held low the diodes are forward biased and the bases of T2 and T3 are held low. The two output driving transistors T5 and T6 are therefore turned off and hence present a high impedance to the output and the common data bus.

**Example 9.21**

A CMOS version of a three-state logic output buffer is shown in Fig. 9.30. Explain how the circuit operates.

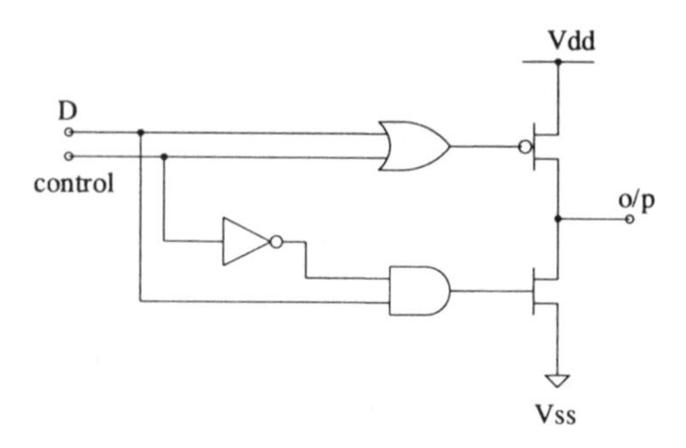


Fig. 9.30 A CMOS three-state logic circuit

***Solution***

With the control line high, a logic '1' appears at the gate of the PMOS and a logic '0' at the gate of the NMOS. Hence both the PMOS and NMOS transistors are turned off and the output presents a high impedance. However, when the control line is low the data at D can pass through to the output and the gate operates normally.

## 9.9 COMPARISONS OF LOGIC FAMILIES

So how does a designer decide which logic family to choose? The answer depends upon whether one is looking for high speed, low power, special power supply voltage, what level of noise immunity, and/or cost. In other words there is no single answer and it depends upon the requirements. As we have seen, Table 9.1 summarises the main features for each logic family.

Some of the more important observations from this table are:

- the 74F series is the fastest of the commonly available TTL series with a moderate power consumption;
- the CMOS 74ACT series is an excellent low-power, high-speed replacement for TTL devices;
- the CMOS 74AC is low power, with a good speed performance and a low voltage operation if required;
- the ECL/GaAs range offer the fastest families but with non-standard power supplies and high cost;
- the 4000B CMOS series has a wide power supply range and a good noise immunity but is relatively slow;
- the CMOS 74HC/HCT series is an earlier low-cost option to the AC/ACT series but with reduced speed of operation;
- dedicated 3 V series (LV, etc.) offer good speed and low power for battery applications;
- BiCMOS provides low power, good speed and excellent drive capability but at an increased cost compared to CMOS.

As can be seen, with the exception of cost, the combination of delay and power is the main issue facing a designer in making a logic family selection. A useful *figure of merit* is therefore the power-delay product (PDP) which has units of Joules and is the energy dissipated per logic gate. Fig. 9.31 shows a plot of power versus delay for the various technology families with constant PDP lines drawn in. As can be seen the BiCMOS family has just the edge on this figure of merit but CMOS is an excellent low cost alternative that is used in most new designs today.

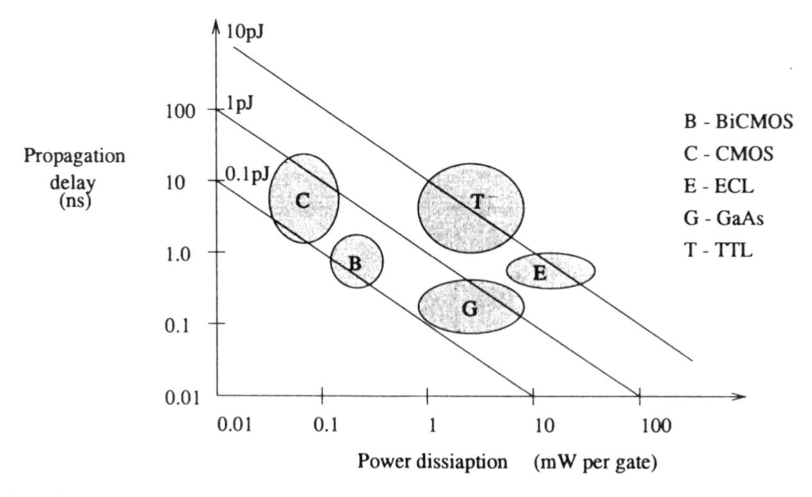


Fig. 9.31 Power delay product for different logic families

# 9.10 MISCELLANEOUS PRACTICAL PROBLEMS

The use of these various logic technologies can present many practical problems. This section will briefly discuss some of these issues.

## 9.10.1 Interfacing

It may be necessary on occasions to mix technologies on a single PCB, for example TTL and CMOS devices. The main concerns are:

1. Do the logic output levels from one device fall inside the legal input levels for the next device?
2. Can the output transistors provide sufficient current for driving the next stage without producing illegal logic levels?

Table 9.1 (on p. 192) will again help answer these questions.

**CMOS to TTL**

Figure 9.32 shows a CMOS output stage driving a typical TTL input. No special interfacing circuitry is required as long as the CMOS output can source and sink $I_{IL}$ and $I_{IH}$ for the TTL input. If we look at Table 9.1 we can see that all the CMOS outputs can source and sink at least 4 mA with output voltage levels of 4.2–4.3 V ($V_{OHmin}$) and 0.33–0.44 V ($V_{OLmax}$). This current drive is sufficient for the TTL inputs and provides adequate TTL voltage levels.

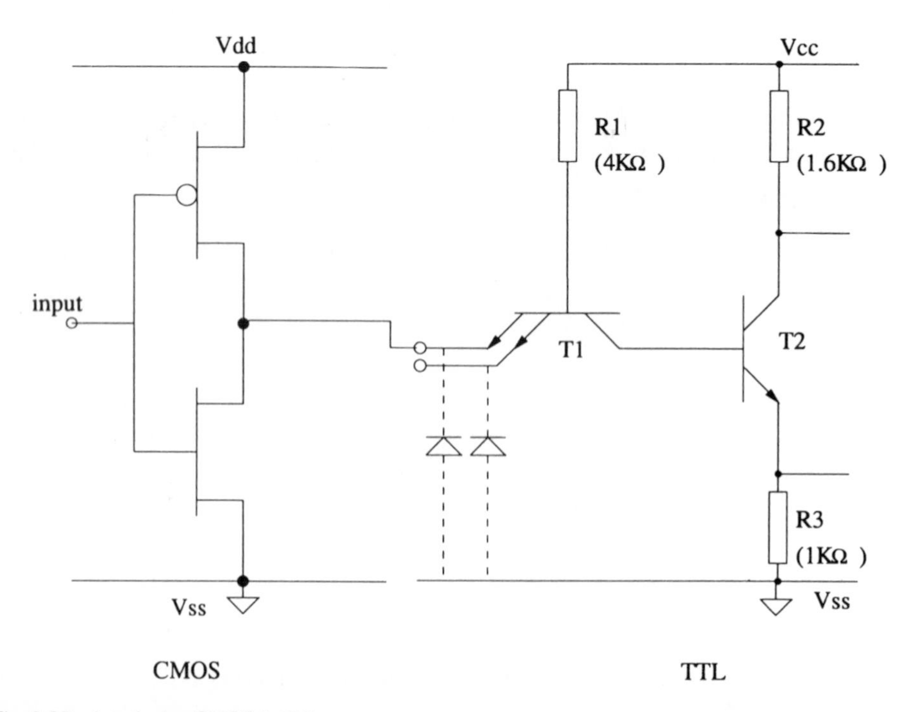


Fig. 9.32 Interfacing CMOS to TTL

## TTL to CMOS

Fig. 9.33 shows a TTL output driving a CMOS input. It is necessary to be very careful here. If we connect a 74ALS to a 74AC then when the TTL output goes high the minimum output voltage level could be as low as 2.7 V. This is less than the minimum legal high input voltage for the 74AC device which is 3.5 V and the CMOS output will be indeterminate. Two methods are used for TTL driving CMOS. The first is to use TTL input compatible CMOS devices. These are the devices which have a T in their code, i.e. 74HCT, 74ACT, etc. In this case an input inverter is included which has its *W*/*L* ratios adjusted such that the switching point is halfway between the TTL input voltage levels, i.e. at 1.4 V for 74X or 1.6 V for other TTL devices. The second method is to use the non-TTL compatible CMOS devices (i.e. 74HC, 74AC, etc.) and add a pull-up resistor at the input to the CMOS device. This is the resistor, $R_p$, in Fig. 9.33.

### Example 9.22

From Table 9.1 what is the maximum value of pull-up resistor, $R_p$, that can be used for a 74ALS device (TTL) driving a 74AC (CMOS) device? What would the input time constant be if the 74AC input capacitance is 10 pF?

***Solution***

From Fig. 9.33 when the output of the TTL circuit is driven high, due to T3

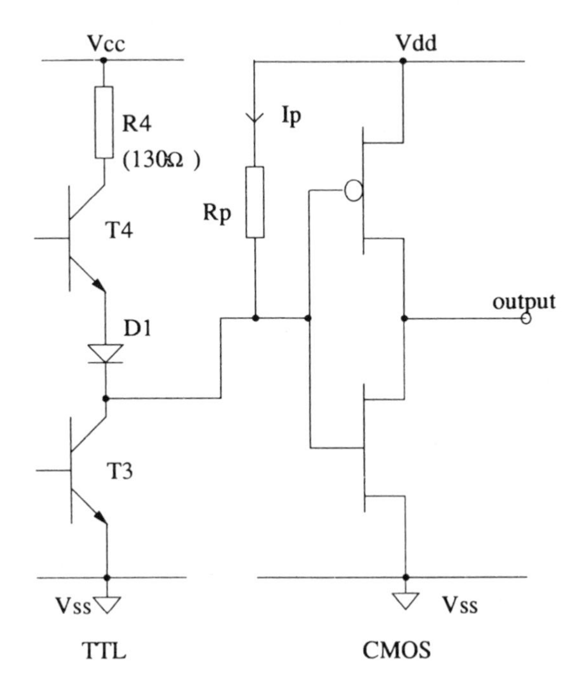


Fig. 9.33 Interfacing TTL to CMOS

turning 'off' and T4 turning 'on', the resistor $R_p$ pulls the output high and thus turns off T4. Current, $I_p$, flowing through $R_p$ is therefore due to the input current for the CMOS device (approximately 1 μA) plus the current through the off transistor T3 (approximately 10 μA). Hence

$$V_{cc} = I_p \times R_p + V_{IHmin}$$

and so $R_p = (V_{cc} - V_{IHmin})/I_p = (5-3.5)/11 \times 10^{-6} = 136.4\,\text{k}\Omega$.

The input time constant is $R_p \times C_{in} = 136.4 \times 10^3 \times 10 \times 10^{-12} = 1364$ ns! This delay is very large and can seriously affect power consumption of the following CMOS stage. This is simply because the input voltage will be around 2.5 V for a relatively long time. During this time both MOS transistors will be on between $V_{dd}$ and $V_{ss}$ and hence a large current will flow. The solution is either to use a much lower value of $R_p$ (typically 2.2 kΩ) or as stated before to use the CMOS TTL input voltage compatible series (i.e. HCT, ACT, etc).

### 9.10.2 Unused inputs

To understand what to do with unused inputs take another look at the input of each of the three technologies that we have studied so far, i.e. Fig. 9.3 for TTL, Fig. 9.11 for CMOS and Fig. 9.21 for ECL.

From Fig. 9.3, unused inputs on TTL gates can be left unconnected and will float and give the appearance of a logic 1 at the input. However, it is best that these inputs are tied to either ground directly or to the supply via a 2.2 kΩ resistor since the input may pick up noise and oscillate between a logic 1 and logic 0.

CMOS inputs (Fig. 9.11) are simply the two complementary MOS transistors. Two problems occur when the input(s) of a CMOS gate is left floating. The first problem is due to the very high input impedance of MOS devices. When an input is left unconnected the input terminal can float to high voltages due to a build up of electrostatic charge that cannot leak away. The gate oxide of a MOS device is extremely thin and hence very high fields can be generated at the input sufficient to destroy this oxide. To avoid this problem all CMOS inputs are designed with electrostatic protection (and a CMOS inverter for pulse sharpening). The resulting CMOS input circuit is shown in Fig. 9.34. To avoid undue stress on these protection circuits the inputs nevertheless should be tied to either ground or supply. The second reason for tying the inputs to ground or supply is because if they are left to float they can obtain voltages that can turn both of the MOS transistors on and thus result in a large power consumption. Hence all unused inputs on a CMOS chip *must* be tied high or low.

ECL circuits (Fig. 9.21) have their inputs already tied to -5.2 V by the 50 kΩ resistor and hence they can be left unconnected.

A final note about inputs. Any input signal that has a slow rise or fall time of the order of 50 ns or greater must not be applied to a gate until it has been sent through a pulse sharpening circuit. This is particularly true for clock signals. These pulse sharpening circuits can simply be a two-stage CMOS inverter as in the

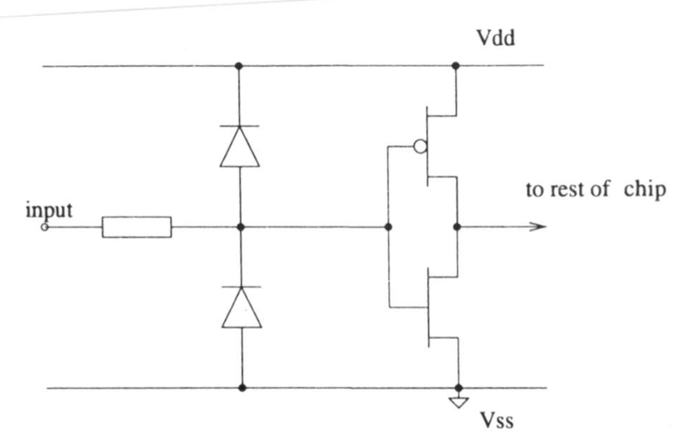


Fig. 9.34 A standard CMOS input circuit

double buffered outputs of the CMOS logic families. However, a Schmitt trigger device is the more common pulse sharpener device in use and is available in both the TTL (74LS14) and CMOS (74HC14) families.

### 9.10.3 Decoupling

When the output of a logic gate switches from one state to another then a large power supply current will flow for a very short time. These fast changing current spikes and the inductance of the power supply wiring feeding the chip will cause voltage transients on the power supply which are passed to the next stage hence appearing at its output. This is then passed on to the input of the following stage and may well produce an illegal state. The solution is to stop the voltage spike from passing down the power supply line by adding bypass or *decoupling* capacitors as close as possible to the source of the problem. This usually means adding decoupling capacitors to every chip. Typically two capacitors are placed in parallel with the supply: a 4.7 μF tantalum (good for low frequencies) and a 10 nF disk ceramic (for high frequencies). The liberal use of decoupling capacitors cannot be overemphasised.

## 9.11 SELF-ASSESSMENT

9.1 What is the voltage across a saturated bipolar transistor?

9.2 What base–emitter voltage is needed to turn on a bipolar transistor?

9.3 What is $h_{fe}$?

9.4 What do the following acronyms stand for: DTL; TTL; NMOS; CMOS; ECL; BiCMOS?

9.5 A diode is forward biased if the cathode is more negative than the anode. True or False?

9.6 Place in chronological date order the following:
74ALS; 74; 74LS; 74F.

9.7 Group into CMOS and TTL the following devices:
74ALS; 74HC; 74; 74LS; 74AC; 74ACT; 74F; 74HCT; 74AHC and 4000B.

9.8 What is the difference between 74ACT and 74AC?

9.9 For a TTL device which is the larger: $I_{ILmax}$ or $I_{IHmax}$?

9.10 Repeat Question 9.9 for a CMOS device.

9.11 When would you use the technologies CMOS and TTL?

9.12 Why is a Schottky clamped bipolar transistor faster than an unclamped device?

9.13 State the MOS transistor $I/V$ equations.

9.14 Write down the equation for $K$ for a MOS transistor.

9.15 Explain why an NMOS device is smaller than an *electrically identical* PMOS device.

9.16 A two-input NOR gate has its PMOS transistors in series. True or False?

9.17 Write down the expression for dynamic power consumption in a CMOS device.

9.18 Repeat Question 9.17 for propagation delay.

9.19 In CMOS combinational logic what is the relationship between the NMOS transistors and the PMOS transistors?

9.20 What is a CMOS transmission gate?

9.21 Place in *decreasing* speed order the following: 74LS; 74HC; 74ALS; 74AC; F100K.

9.22 Place in *increasing* power consumption order the devices in Question 9.21.

9.23 Define PDP.

9.24 Place in *increasing* PDP order the following: CMOS; TTL; ECL; BiCMOS.

9.25 A 74LS device is connected directly to a 74HC device – is this acceptable?

## 9.12 PROBLEMS

9.1 For the circuit shown in Fig. 9.1 determine the value of $R_c$ needed such that the transistor is saturated when the input is 5 V. Assume that $R_b = 100\,\text{k}\Omega$ and $h_{fe} = 100$.

9.2 A two-input DTL NAND circuit is shown in Fig. 9.2(a). If R1 = 4.9 kΩ then determine the approximate power consumption when both inputs are low and both inputs held high. Assume a low = 0.2 V, a high > 3 V and $V_{cc} = 5$ V.

9.3 How many 74S gates will a 74ALS gate drive?

9.4 From Table 9.1 determine the high and low noise margins for the following gate combinations: (a) 74 driving 74; (b) 74ALS driving 74S; (c) 74HCT driving 74AS; (d) F100K driving F100K. Comment on the accuracy of these values.

9.5 For the NMOS resistive load inverter of Fig. 9.9(a) calculate the *W*/*L* ratio of the MOS transistor to obtain an output voltage of 0.25 V when the input is 5 V. Assume that $R_L = 2\ \text{k}\Omega$, $V_T = 1$ V and $\mu C_{ox} = 32\ \mu\text{A}\ V^{-2}$.

9.6 A CMOS inverter is powered from a 5 V supply. What supply current will flow when the input voltage is 2.5 V. Assume that $V_{TN} = 0.8$ V, $K_P = 200\ \mu\text{A V}^{-2}$, $V_{TP} = -0.8$ V, $K_P = 200\ \mu\text{A V}^{-2}$ and that the gate is unloaded.

9.7 The CMOS inverter in Problem 9.6 is to drive ten similar gates each having an input capacitance of 0.3 pF. Calculate the propagation delay for this inverter. Assume that the inherent capacitance is zero. Without reprocessing this chip how could you *reduce* its delay?

9.8 The circuit shown in Fig. 9.13(a) is a two-input NOR gate. Calculate $\tau_{PLH}$ and $\tau_{PHL}$ when both inputs are tied together. Assume that $K_N = K_P = 200\ \mu\text{A V}^{-2}$ for each transistor, the output is driving a 50 pF load and a 5 V supply is used. Hint: for transistors in series the effective *K* halves; for transistors in parallel the effective *K* doubles.

9.9 Draw the minimum CMOS transistor circuit configuration that will implement the function: $f = \overline{A \cdot B \cdot (D + C)}$.

9.10 A 74HC series logic gate is to drive an LED such that when the output is high the LED will be illuminated. What value current limiting resistor would be required to switch on the LED. Assume that the forward current of the LED is 4 mA and that the voltage across the LED when on is 1.8 V.

9.11 A CMOS inverting output stage is to be designed such that it will drive a 74 series TTL load. Calculate the corresponding values of *K* required for both the NMOS and PMOS transistors. Comment on the relative transistor sizes. Assume that a CMOS legal logic '1' is no less than 4.2 V and a legal logic '0' is no more than 0.4 V. Also assume that $V_{TN} = 1$ V and $V_{TP} = -1$ V.

# 10 Semiconductor memories

## 10.1 INTRODUCTION

In general semiconductor memories are used when fast access of data is required. Semiconductor memories are historically divided into two types. These are Read Only Memory (ROM) and Random Access Memory (RAM). The main members of each family are shown in Fig. 10.1.

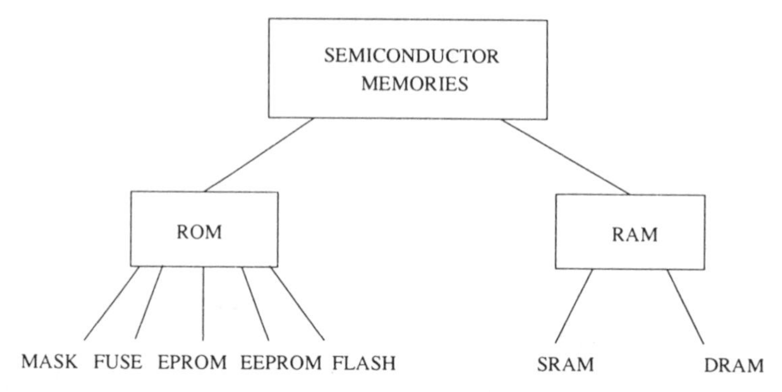


Fig. 10.1 Semiconductor memory types

**Read only memory overview**

ROM devices are used for storage of data that does not require modification, hence the name 'read only memory'. This definition however, has become less clear over the years and now includes devices whose data are occasionally modified. The original true ROM types are mask programmable ROM and fuse programmable ROM (or PROM). The mask programmed ROM devices are programmed at the factory during manufacture whilst the fuse programmed ROM devices are programmed by blowing small fuses and hence are sometimes called One Time Programmable ROM or OTPROM. Both mask ROM and OTPROM devices are true read only memory devices which are written only once. Other ROM devices that are, paradoxically, written more than once are: Erasable PROM (called EPROM) – these devices are programmed electrically but are fully erased with ultraviolet light; Electrically Erasable ROM (referred to as EEPROM or $E^2$PROM) – these devices can be both programmed *and* erased electrically; FLASH memory – these devices use the same technology as EPROM but not only are they electrically programmed, they can be erased electrically in a very short time. The great advantage of *all* these ROM devices is that they are non-volatile. This means that when the power is removed the stored data is not lost.

**Random access memory overview**

The RAM device family is divided into two types. These are Static RAM (SRAM) and Dynamic RAM (DRAM). The SRAM device retains its data as long as the supply is maintained. The storage element used is the transmission gate latch introduced in Chapter 9 (see Fig. 9.18(a)). On the other hand, DRAM devices retain their information as charge on MOS transistor gates. This charge can leak away and so must be periodically refreshed by the user. In both cases these devices are volatile, i.e. when the power is removed the data is lost. However, newer devices are available which muddy the water, such as non-volatile SRAM (NOVRAM) which have small batteries located within their packages. If the external supply is removed the data is retained by the on-board battery. Another relatively new device is the Pseudo Static RAM (PSRAM). This is a DRAM device with on-board refresh circuitry that partially relieves the user from refreshing the DRAM and hence from the outside it has a similar appearance to that of an SRAM device.

The term 'random access memory' is given to this family for historical reasons as opposed to the magnetic storage media devices, such as tape drives, which are sequential. In RAM devices any data location can be read and written in approximately equal access times hence the name 'random access memory'.

**Semiconductor memory architecture**

Most semiconductor memories are organised in the general form as shown in Fig. 10.2. The chip consists of an array of cells with each cell holding a single bit of data as either a logic '1' or a logic '0'. The memory cells are arranged in rows and columns. Each row is individually accessed by a row decoder. As seen in Chapter 4, a decoder is a device which has $n$ inputs and $2^n$ outputs such that only one output line goes high when an *n-input* data string is applied. When a row goes high the data for that row is presented at the bottom of the array on the column lines. A second decoder, called a column decoder, is used in conjunction with the row decoder thus allowing a single cell or bit to be individually accessed. Hence by applying a row and a column address the data of a single bit can be either read (R) or written (W) via a single input/output pin under the control of a R/W pin.

In some cases the data is organised as more than one bit, e.g. an eight-bit data word called a *byte* and the chip therefore has eight data inputs and outputs. In this case eight column decoders could be used to access in parallel eight different data bits from a single row. This structure is true for both RAMs and ROMs, although mask programmable ROMs of course would only have a DATA OUT pin(s) and no R/W pin.

This chapter will focus on the structure of a single memory cell and we shall assume that the row lines (sometimes called WORD) and column lines (sometimes called BIT) are decoded such that only one row and column line is high at

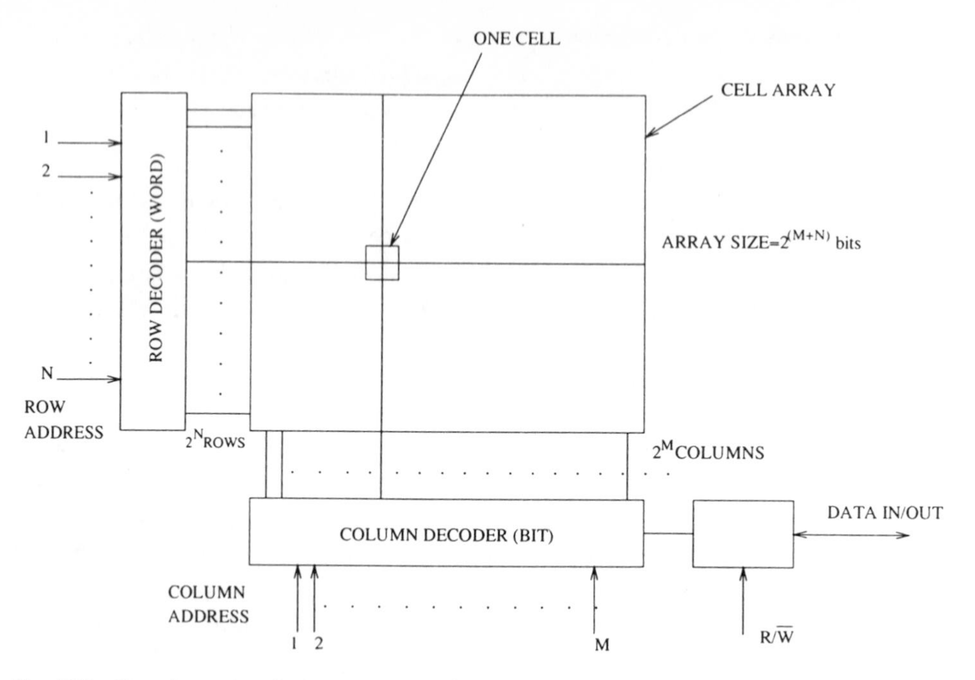


Fig. 10.2 Generic semiconductor memory architecture

any one time to select that particular cell. However, in order to visualise the memory layout, other cells may have to be included. In addition to looking at each cell we shall discuss the relative timing requirements for the input and output signals in order to read and write data.

## 10.2 READ ONLY MEMORY – ROM

ROMs are read only memory devices, or nowadays more strictly RMM (Read Mostly Memory). As stated, an important property of all ROMs is that they are non-volatile, i.e. when the power is interrupted the data is retained. Most semiconductor memories are implemented with MOS technology due to its high packing density and low power consumption compared to bipolar.

### 10.2.1 Mask programmed ROMs

Mask programmable ROMs are programmed during manufacture and hence cannot be changed once programmed. These devices are thus used for program storage where the data stored is not required to change.

A section of a programmed NMOS ROM array holding eight bits is shown in Fig. 10.3. This type of array is called a NOR type memory array. Data is stored by the presence or non-presence of an NMOS drain connection to the COLUMN lines. When a ROW is high then the NMOS transistors connected to that ROW will be turned on and the output on the column line will be '0'. Those locations

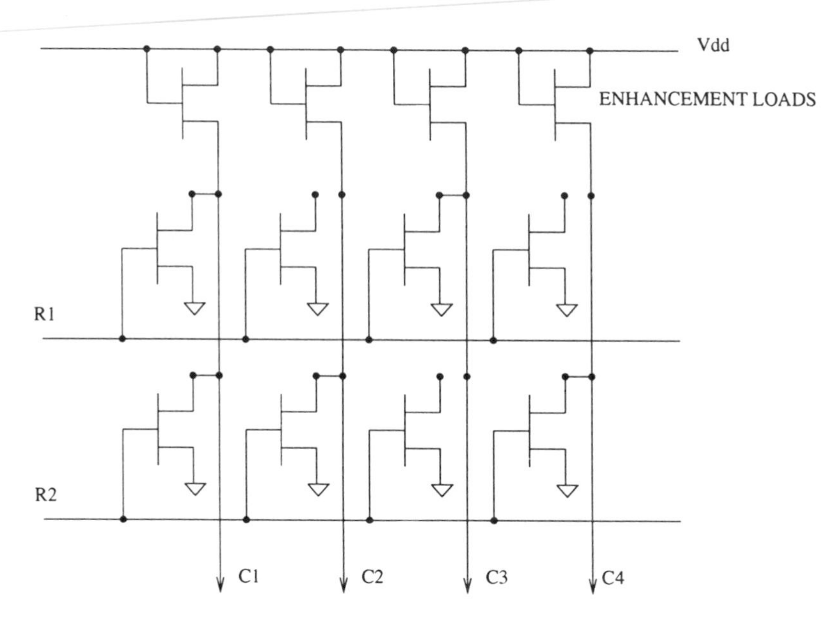


Fig. 10.3 Mask programmed ROM

with no NMOS drain connections on the column line are pulled up to '1' by the enhancement mode NMOS devices acting as a load.

Alternative ways of programming ROM devices during processing exist. For example omitting the source connection at locations where a logic '1' is desired is just as effective. Alternatively the threshold voltage of those transistors which are not to turn on can be increased by using a selective implantation of P-type dopants into the MOS substrate. These processing steps are performed by use of a processing tool called a *photo or electron beam mask* (for further reading on semiconductor processing the reader is referred to an alternative text in this series[1]). Since these processing steps are performed near the end of the process then a large inventory of uncommitted ROM wafers with most of the processing completed are kept on the shelf. All that is required is for the customer to specify the data to be stored and hence the mask step can be implemented.

It should be noted that this mask step is very expensive and usually mask programmed ROMs are used only when large volumes are required – greater than 10 000 pieces. The delivery time for such devices is approximately 3–6 weeks and hence mistakes incur a heavy financial and time penalty.

The inputs and outputs for a typical mask programmed ROM are as follows:

- Inputs: address lines; chip enable ($\overline{CE}$); output enable ($\overline{OE}$);
- Outputs: data out;
- Power: $V_{dd}$ and $V_{ss}$.

The address lines and data output lines require no explanation. The chip enable pin is quite simply used to 'wake up' the chip from its low power, stand-by mode.

[1] L. Ibbotson, *Introduction to Solid State Devices*, in this series.

When 'chip enable' is low the device operates normally but when it is high the device is 'off' and in 'stand by'. Since a low turns the device on then this pin is said to be *active low* and hence it is labelled as $\overline{CE}$. This pin can be used as an extra address line for cascading two or more ROM devices when larger memory capacities are required. This is illustrated in Fig. 10.4 where an extra address line is used with an inverter to select either ROM1 or ROM2 via $\overline{CE}$. The output enable pin is used for shared data buses, with microprocessors for example, where the data output pins can be made *tri-state* by holding $\overline{OE}$ high. When $\overline{OE}$ (and $\overline{CE}$) is low then data is presented at the output pins. Since it is active low it is therefore labelled $\overline{OE}$.

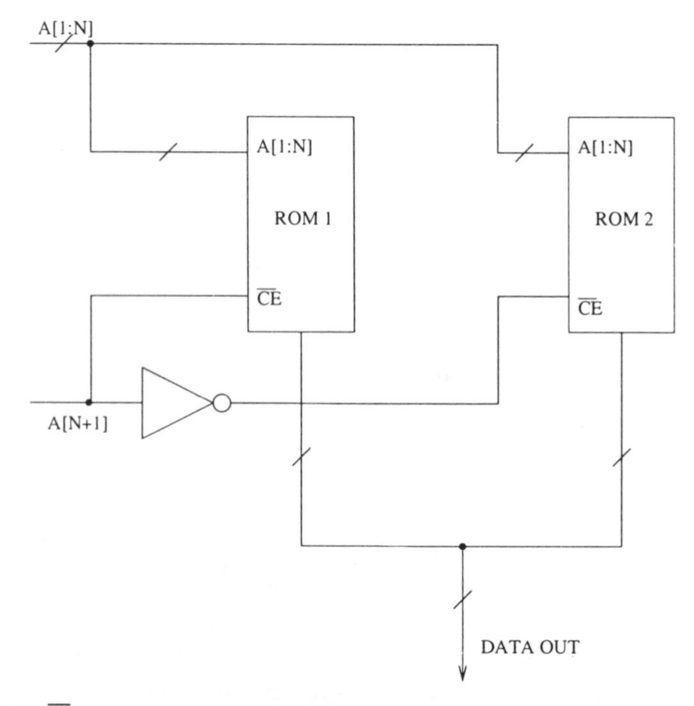


Fig. 10.4 Using $\overline{CE}$ to increase memory storage

The timing diagram for a read operation of a mask programmable ROM is shown in Fig. 10.5. Although these diagrams appear daunting at first they are in fact essential to the system designer so that the device can be correctly interfaced to other devices. This timing diagram shows that we must first set up a valid address and then bring chip enable ($\overline{CE}$) and output enable ($\overline{OE}$) low in order to read data. Three signal types 'a', 'b', and 'c' are labelled in Fig. 10.5. Type 'a' indicates that the signal can be either '1' or '0'; type 'b' is the tri-state or high impedance condition where the outputs are floating (see Chapter 9); and type 'c' refers to an unknown state that occurs whilst the system is changing states. We can see that once the address is set up (i.e. valid) and $\overline{CE}$ and $\overline{OE}$ are low then valid data can be read out of the ROM but only after a time $t_{acc}$ called the address access time. This access time is an important figure of merit for all types of memory

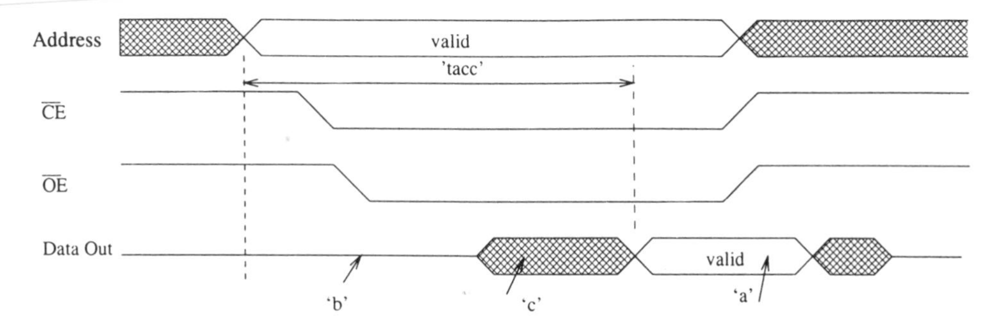


Fig. 10.5 . ROM read timing diagram

devices. For a ROM device it is typically 100 ns whilst for a hard disk on a computer it can be as large as 100 ms!

In semiconductor memories this access time is limited by the resistance and capacitance of the row and column lines which act as an RC delay line. The row and column lines are usually made from metal since this has a low resistance but unfortunately it will also have wiring capacitance as well as the gate capacitance of each of the storage transistors. Hence the larger the array the longer the row and column lines and so the larger the value of the RC component. Large memory arrays, using the same technology, therefore have longer access times.

Like all of the semiconductor memory market, single chip mask ROM packing densities have grown over the years from 256 Kbit in 1986 to 16 Mbit now.

## Example 10.1

For the mask programmable ROM layout shown in Fig. 10.3 determine the data appearing on the column lines when both rows 1 and 2 are accessed.

### *Solution*

To access a row the row decoder output (not shown) for that row must be high. When this happens all other row lines will be low. Hence to access row 1 a high must be present on the decoder output. All transistors with the drain connected to the column line will produce a 0 and those not connected will produce a 1 on the column line. Hence:

Row 1 selected: then data out $= C_1C_2C_3C_4 = 0101$

Row 2 selected: then data out $= C_1C_2C_3C_4 = 0010$

## Example 10.2

Assume that the memory array shown in Fig. 10.3 are two rows of a larger array of size 64 by 4 bits. What would the row address be to access these two rows if they are the *last two* rows in the array?

***Solution***

This chip would have 64 rows with each row four bits in length. To address all 64 rows we would need a decoder which has six address inputs and 64 outputs each connected to a single row. The address 000000 will produce a '1' on the first row with a '0' on all the other rows. Hence in order to address the last two rows we will need row addresses of 111110 and 111111, respectively.

## 10.2.2 PROMs

A PROM, or programmable read only memory device, is programmed by blowing small fusible links which are made of nichrome or polycrystalline silicon. Since fuses are blown the result is irreversible and hence the devices are sometimes called one-time programmable ROMs (or OTPROMs). The early PROM devices were mainly of bipolar form which have a higher effective $K$ than MOSFETs and thus can generate the 15–20 mA needed to blow a fuse without the use of large $W/L$ ratio MOS transistors.

A schematic for an eight-bit bipolar fuse PROM is shown in Fig. 10.6. before the fuses are blown. To blow a fuse the row is selected and the corresponding column line held low in order to program a zero. The power supply $V_{pp}$ is then held at typically 12.5 V for approximately 50μs which is sufficient to generate enough power to blow the fuse. In this case the devices are supplied to the customer with a 1 in every location and the user is able to program a 0 where

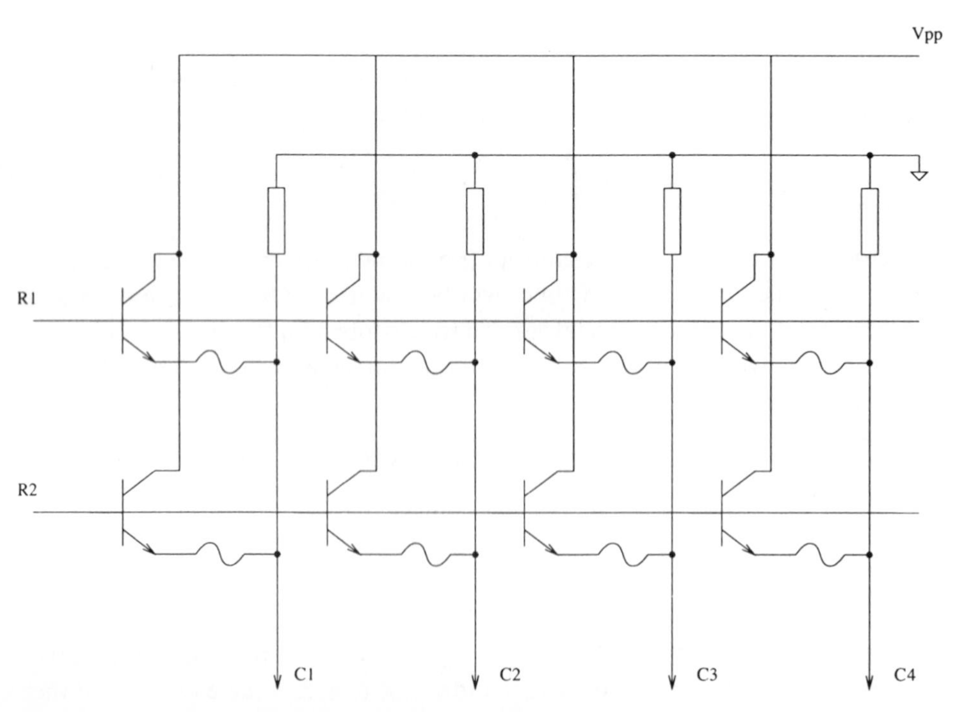


Fig. 10.6 A bipolar fuse programmable ROM – OTPROM

required by blowing the small fuses. During normal operation the power supply voltage is 5 V and so there is no danger of accidentally blowing the fuses.

The increased packing density of MOS transistors over bipolar has resulted in PROMs now being predominantly fabricated using MOS techniques. These MOS devices are in some cases quite simply MOS EPROM devices (see next section) with no transparent window hence stopping erasure by ultraviolet light. In both cases the pin-out for bipolar or MOS OTPROMs are:

- Inputs: address lines; data in; chip enable ($\overline{CE}$); output enable ($\overline{OE}$);
- Outputs: data out;
- Power: $V_{dd}$, $V_{pp}$ and $V_{ss}$ or $V_{cc}$, $V_{pp}$ and $V_{ee}$ (for bipolar).

The difference in pin-outs between mask and one-time programmable ROMs is the addition of the higher power supply voltage pin ($V_{pp}$) and that the data pins are bidirectional, i.e. having both input and output capability for programming purposes. The read timing diagram is the same as for mask programmable ROMs with similar address access times. However, since these devices need to be programmed by the user then a write program timing sequence is supplied in the data sheet. A simplified write or fuse program timing diagram is shown in Fig. 10.7 which actually consists of two stages: data programming and then data verification. As can be seen the address and data are set up first and $V_{pp}$ is pulled up to 12.5 V and $V_{cc}$ to 6.25 V. The actual programming operation occurs when $\overline{CE}$ is held low for typically 50–100 µs ($t_{pw}$). To verify that the correct data was programmed into the PROM the data is read back out again by pulling $\overline{OE}$ low for a short time (approximately 200 ns). This process is repeated for every address value and hence the total time taken to program a PROM depends upon the total number of addresses and can take as long as 10 minutes in some cases. This timing diagram is fairly complex but fortunately PROM programmers are readily available that will operate automatically from PCs.

Single-chip OTPROM packing densities are currently at 4 Mbit.

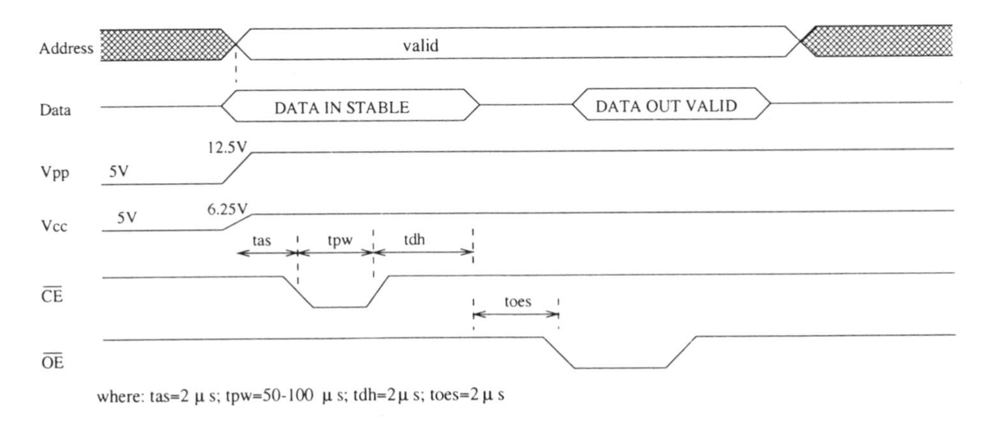


Fig. 10.7 Program timing diagram for fuse programmable ROMs

**Example 10.3**

A PROM of size 4Mbit, organised as 0.5 Mbit by 8 bit, is to be programmed using the $\overline{CE}$ pulse technique described above. If the $\overline{CE}$ minimum low pulse width is 100 μs then what is the maximum programming time to program all bits? Assume that all other timing parameters can be neglected.

***Solution***

A total of 0.5 Mbytes require to be programmed. Since each byte is programmed in 100 μs then the total approximate programming time, ignoring $t_{as}$, $t_{dh}$ and $t_{oes}$, etc., is

$$T_{total} = 0.5 \times 10^6 \times 100 \times 10^{-6} = 50 \text{ seconds}$$

## 10.2.3 EPROM

The problem with the OTPROM devices is that for program development they are inefficient since only one address change will require a completely new device. Hence OTPROMs are only used when the program is settled and contains no known bugs. However, during system development several iterations are usually required, hence devices which can be reprogrammed are more useful. The erasable programmable ROM, more commonly known as an EPROM, is not only electrically programmed but can be fully erased by exposure to ultraviolet light. Hence the device can be reused over and over again until the design is completed. The code on the EPROM can then be transferred to either an OTPROM or a mask programmable ROM.

The cross-section of a single EPROM cell is shown in Fig. 10.8. It consists of two gates:

(i) FG1 – Floating gate not connected in the array and insulated from the channel by an oxide layer of standard thickness of 0.05 μm.
(ii) G2 – Polysilicon gate used as the normal memory transistor on the row or word line.

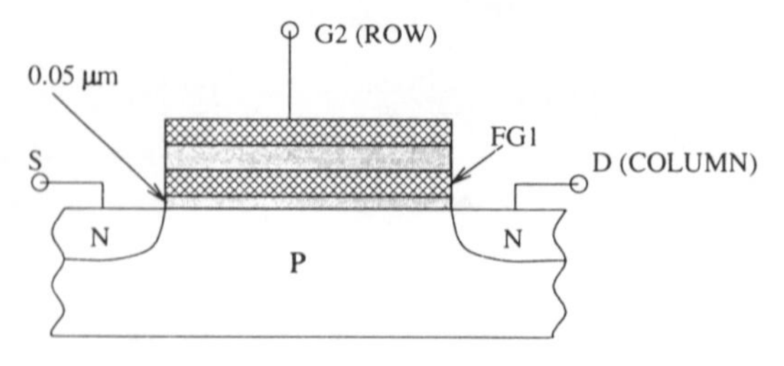


Fig. 10.8 Cross-section of a single EPROM cell

This device fits into the same NOR array as in Fig. 10.3 with inverters on the column line outputs and some additional circuitry. The floating gate (FG1) is used to store charge which thus modifies the threshold voltage ($V_t$) of G2.

Assume initially that FG1 has zero charge and thus when G2 (or ROW) is held high then the transistor conducts as normal and the column line is 1 (due to the column inverters). Hence unprogrammed EPROM devices have all bits at logic 1.

To write a '0' into the EPROM cell, the row and column lines are both held at a sufficiently high voltage (typically 12.5 V) to cause the drain to enter a condition called avalanche breakdown and a large current flows. The high field from D to S accelerates these electrons to high velocities and some of these electrons (called hot electrons) have sufficient energy to jump to FG1 where they are trapped. (It should be noted that these hot electrons pass over a potential barrier which is larger when looking back from FG to substrate. These hot electrons do not tunnel through the oxide. For a more detailed explanation of the MOS device the reader is referred to Hart[2]). Now since FG1 is totally insulated then on removing this voltage a negative charge is left on FG1 and under normal conditions it will not leak away for typically 5–10 years.

This negative charge on FG1 will attract holes to the silicon/silicon dioxide interface and so raises the threshold voltage seen by G2. Thus the row voltage at G2 will not turn this device on and so the drain will be pulled up to logic 1 and the column output will be programmed as a logic 0. This avalanche hot electron MOS technique is known as the floating gate avalanche MOS or FAMOS process.

The only way to remove the charge from FG1 and hence erase the device is to make the silicon dioxide conductive by using light of energy greater than its energy gap. This is usually light in the ultraviolet region and the whole process takes approximately 20 minutes to erase. Since these devices are erased optically by the user they must have a transparent window in the package which adds slightly to the cost of the EPROM compared to OTPROMs. It should be noted that daylight also contains the correct wavelength for erasing EPROMs. Consequently once an EPROM has been programmed then the quartz window must be covered with an opaque label. (The MOS OTPROMs discussed in the previous section are nowadays actually made by using EPROM technology but with no transparent window present for erasing. Hence the device can only be programmed once.)

The input and output pins for an EPROM device are exactly the same as for the OTPROM including power supplies (i.e. $V_{dd}$, $V_{pp}$ and $V_{ss}$). Consequently the read and write timing diagrams are also identical and it is not surprising that EPROM devices are programmed by the same programmer as that which programmed an OTPROM. However, since an EPROM device is reprogrammable it is necessary to erase the device completely before every write operation. This is because the write operation can only program a logic '0' and will not reprogram a logic '0' to a logic '1'. Finally the number of write/erase cycles is an important factor – this is usually called *endurance* in memory terminology. The use of hot electrons gradually damages the gate oxide and hence the number of write/erase cycles is typically only 100 for an EPROM device.

[2]B. Hart, *Introduction to Analogue Electronics*, in this series.

In 1986 EPROM packing densities were typically 256 Kbit. Currently this figure stands at 4 Mbit.

**Example 10.4**

How much current is required to charge the floating gate of a 1 μm EPROM cell to 5 V from 0 V over a write time of 50 μs? Assume that the capacitance of the floating gate is $8 \times 10^{-4}$ pF/μm$^2$.

***Solution***

The total capacitance of the floating gate is

$$C_{fg} = 8 \times 10^{-4} \times 1 \times 1 = 0.8\,\text{fF}$$

Hence the charge current, derived from $I = C\,\mathrm{d}V/\mathrm{d}t$, is

$$I_{charge} = 0.8 \times 10^{-15} \times 5 \div (50 \times 10^{-6}) = 80\,\text{pA}$$

i.e. only a small fraction of the avalanche current (which is of the order of 1 mA) is required to charge the floating gate.

### 10.2.4 E²PROM

Although the EPROM is an extremely mature technology having been available for more than 15 years its main disadvantage, apart from requiring an ultraviolet light source to erase, is that this erase is not selective and all cells are erased at once. The electrically erasable PROM (EEPROM or E²PROM), on the other hand, is not only programmed electrically but the cells can be erased electrically. This allows the devices to be programmed whilst still in the system. Hence these devices can be used not only for programs and program upgrades but also for storing data that occasionally require updating whilst in use (for example telephone numbers on mobile telephones).

One variant of the E²PROM device employs an MNOS (Metal Nitride Oxide Semiconductor) transistor as the memory element. This MOS transistor, shown in Fig. 10.9, consists of an insulator which is composed of two layers – a silicon nitride layer of thickness 0.05 μm and a very thin silicon dioxide layer of thickness

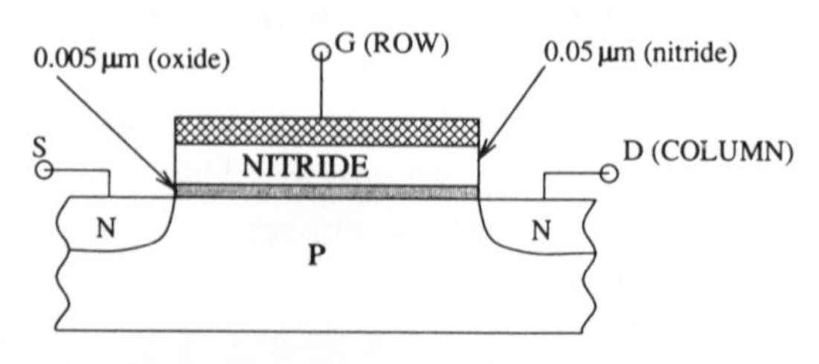


Fig. 10.9 Cross-section of a single E²PROM cell

0.005 μm. The principle of operation relies on the fact that the interface between the oxide and nitride is capable of trapping electrons. Assuming that this interface has zero charge initially then as in the EPROM a '1' on the *row* line will result in the column line reading '1' (remember that an inverter is present on the column line).

To write a '0', 12.5 V is applied to the gate with respect to the source and substrate for approximately 10 ms. The high field generated across the very thin oxide layer allows electrons to tunnel through which are caught by the electron traps at the oxide/nitride interface. The presence of this negative charge will increase the threshold voltage seen by the *row* line. When *row* goes high next time then the MNOS transistor will not conduct and the column output will indicate a logic '0'. Just as in the EPROM the charge can stay at this interface for many years.

The advantage of this tunnelling action is that it is reversible. Hence to erase, all that is required is to apply −12.5 V on the gate which repels electrons from the oxide/nitride interface. No distinct erase operation is needed; only a write cycle is required which loads in either a logic '1' or a logic '0'.

This type of device has high reliability. For example, in the event of a pinhole or a defect in the oxide then only the charge at the oxide/nitride interface above the pinhole or defect will leak away. The rest of the charge at the interface is trapped in the non-conductive layers and so the cell retains its state. The device thus has a very high endurance in that at least $10^4$ write/erase operations can be carried out before the device shows signs of degradation.

An alternative to the MNOS $E^2$PROM cell is to use a variation of the EPROM floating gate cell. In this case the floating gate is extended over the drain and here the oxide is thinned down to 0.01 μm from 0.05 μm. The cross-section for this structure is shown in Fig. 10.10. The advantage of this structure is that the thin oxide region is limited to a small area and hence reliability problems caused by defects and pinholes in the oxide are greatly reduced due to the small area occupied by this thin oxide. The device is programmed by holding the gate at a high positive voltage with respect to the drain. Electrons tunnel through to the floating gate and thus become trapped. To erase the cell the process is reversible and thus a high positive voltage is applied to the drain with respect to the gate and electrons are withdrawn from the floating gate. Note that since in a memory array all the

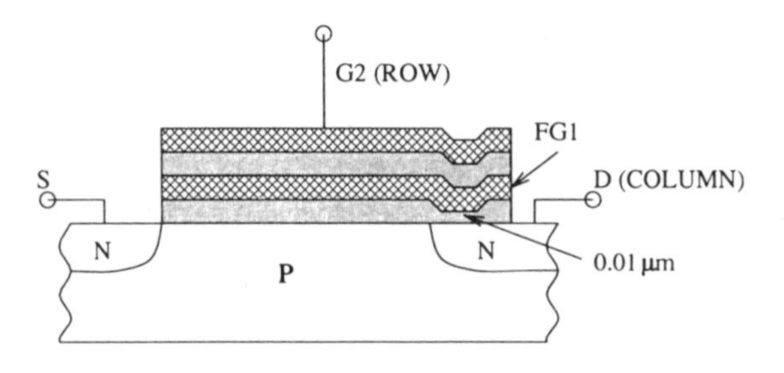


Fig. 10.10 Cross-section of a single floating gate $E^2$PROM cell

drains in one column are connected together, then the application of a large positive voltage to the drain during erase will also be passed to other cells on that column. Hence with the floating gate E²PROM device each memory element must have its own select transistor to individually access each transistor which reduces the packing density.

Whichever cell is used the pin-out for an E²PROM has typically the following pins:

- Inputs: address; data in; chip enable ($\overline{CE}$); output enable ($\overline{OE}$); write enable ($\overline{WE}$);
- Outputs: data out; RDY/$\overline{BUSY}$;
- Power: $V_{dd}$; $V_{ss}$.

The first thing to note about these pins is that no high voltage supply ($V_{pp}$) is necessary. This is generated automatically on chip for both a logic '0' and logic '1'. Since this device can be both written and read then a write enable pin ($\overline{WE}$) is needed to indicate to the array that data is to be written. Hence the read timing diagram for the E²PROM is the same as for the mask programmed ROM but with $\overline{WE}$ held high. Due to the E²PROM cell having a slightly larger capacitance on the row lines its access time is slightly larger than an EPROM at typically 150 ns.

A simplified write timing diagram for an E²PROM is shown in Figure. 10.11. Since the data is usually byte wide then eight bits at a time are written in parallel – this being true for most memories. A pin to indicate that the device is busy writing is provided, called RDY/$\overline{BUSY}$. When $\overline{WE}$ goes from low to high $\overline{BUSY}$ goes low indicating that the array is busy writing. Only when $\overline{BUSY}$ goes high impedance has the data been correctly written. The complete cycle takes approximately 10 ms. Another way of indicating successful data writing is by using the *data polling* method. In this case the data is latched into the chip and an inverted data value appears at the output. When the data has been correctly written this data changes to non-inverted.

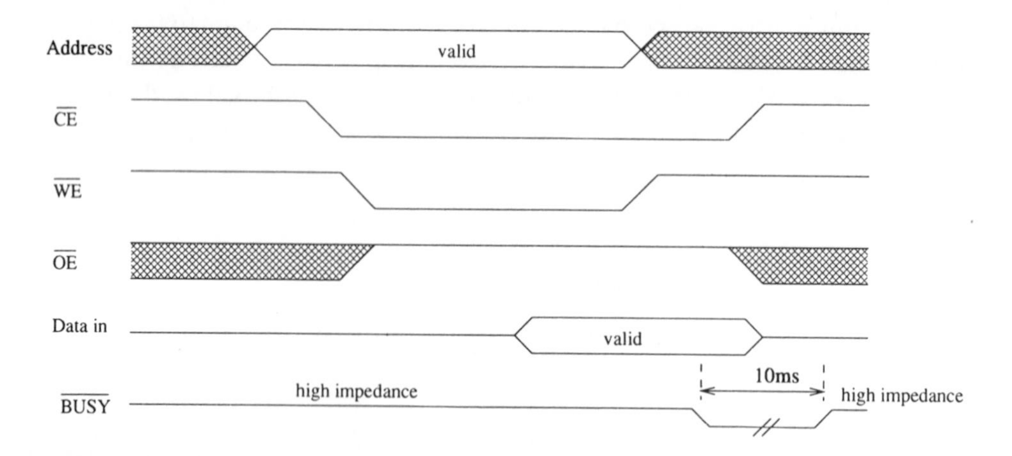


Fig. 10.11 Write timing diagram for an E²PROM device

The current state of the art of $E^2$PROM using MNOS is approximately 1 Mbit compared to 64 Kbit in 1986. However, because the floating gate version has to have an extra transistor per cell these versions are lower capacity being typically 256 Kbit.

**Example 10.5**

How many pins would an $E^2$PROM chip have if it has a capacity of 256 Kbit organised as 32 Kbits by 8 bit?

***Solution***

This chip has 32 Kbytes of storage space. To address this storage space we need 15 address lines. Hence the total pin-out would be: 15 Address lines; eight data bits; $\overline{CE}$, $\overline{OE}$, $\overline{WE}$, $\overline{BUSY}$, $V_{dd}$ and $V_{ss}$, i.e. 29 pins in total.

## 10.2.5 FLASH $E^2$PROM

Although the MNOS $E^2$PROM device has a relatively high capacity the fabrication of the nitride layer and the very thin oxide layer (0.005 μm) is more expensive to manufacture than the floating gate EPROM transistor. The alternative $E^2$PROM utilising a thinned down oxide transistor over the drain also requires extra processing steps, has an extra select transistor and it too is expensive per bit. To take advantage of the higher packing density of EPROM arrays and its lower manufacturing costs recent processing improvements have allowed a complete thinning down of the oxide under the whole length of the floating gate. The process of charge storage could now be reversed by simply reversing the applied potential. Hence the need for ultraviolet light to erase the device is no longer necessary. This type of memory element is called a Flash $E^2$PROM cell and is shown in Fig. 10.12.

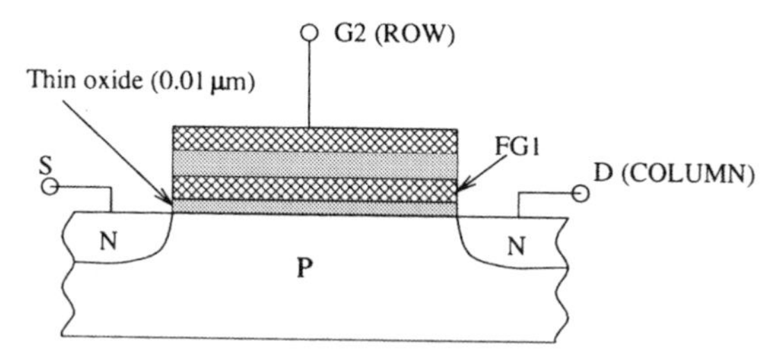


Fig. 10.12 Cross-section for a single flash $E^2$PROM cell

A single flash cell is basically the same as the EPROM cell (i.e. having a floating gate) but the oxide under the floating gate is reduced in thickness from 0.05 μm to 0.01 μm. As a reminder, to program the floating gate EPROM a large positive voltage is applied to the drain and gate with respect to the source and the device enters breakdown. This creates hot electrons of which some pass to the floating gate therefore programming the cell. Now, to program the flash device 5 V is applied to the drain and a large positive voltage (12 V) is applied to the gate with respect to the source. Since for the flash device the oxide under the floating gate is much thinner then electrons are able to tunnel through this gate oxide using a mechanism called Fowler–Nordheim tunnelling. Unlike the EPROM this process is reversible and to erase the stored charge all that is required is to apply +12 V to the source with the gate grounded (drain floating) thus returning the electrons on FG1 to the source. Since the drain is never held at a high positive voltage then the use of a select transistor to isolate other cells as in the alternative $E^2PROM$ is unnecessary. The problem with this structure is that a much larger area of thin oxide exists and hence the device is much more susceptible to damage. In addition if a cell is erased that is already erased then the MOS device will over time acquire a negative threshold voltage and will not program properly. Hence in order to ensure that this problem does not occur *all* cells must be preprogrammed before erasure can start. In this way all cells will be erased by the same amount and the problem of over erasure will not occur. Since all cells are erased at the same time the device is called *flash*. Note that before erasure a preprogram time must be allowed for of approximately the product of the number of bytes and the data write time. For 1 Mbyte arrays the total pre-write and erase time for the complete chip is typically 2–3 seconds. It should also be noted that before *programming* can commence the memory array must be erased (i.e. all cells at logic '1') since the action of programming only writes a logic '0'. The diagrams in Fig. 10.13(a), (b) and (c) illustrate these write, erase and read modes on a small 2 × 2 array.

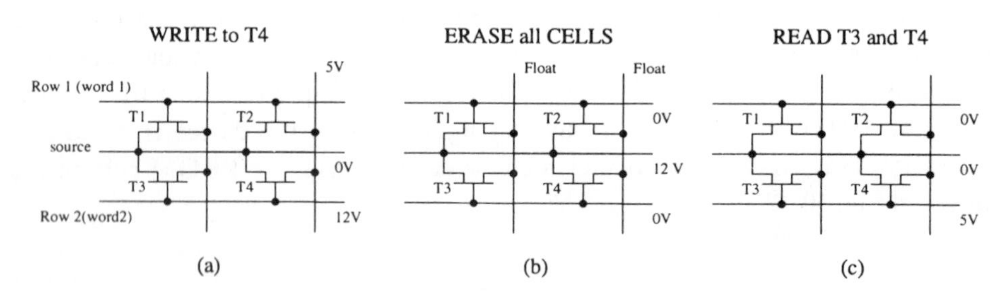


Fig. 10.13 Write, erase and read programming for a flash $E^2PROM$ array

The input and output pins for a flash $E^2PROM$ are as follows:

- Inputs: address; data in; chip enable ($\overline{CE}$); output enable ($\overline{OE}$); write enable ($\overline{WE}$)
- Outputs: RDY/$\overline{BUSY}$; data out;
- Power: $V_{dd}$; $V_{ss}$; $V_{pp}$.

However, on some of the more recent flash devices the $V_{pp}$ pin is removed as in the $E^2PROM$ and is generated on chip.

The timing diagrams for a flash read are the same as in $E^2PROM$ but the timing diagrams for write and erase are much more complicated than $E^2PROM$ and require 2–3 write cycles to load in not only data but also commands indicating whether the operation is to be a write, an erase, a data verify, etc. The timing diagrams for these devices will not be included here so as not to detract your interest from this exciting new product! For these details you are advised to consult the manufacturers' data sheets although it is expected that as with the trend of all memory products the complexity of writing and erasing will ease as more and more circuitry is included within the chip. The flash memory device is a relatively new entry into the memory market and is starting to provide strong competition to hard disk drives especially in mobile computers where the lack of moving parts is a great boost for reliability.

**Example 10.6**

A 1 Mbit flash memory, organised as 128 K by 8 bits, has a byte write time of 10 μs, a flash erase time of 10 ms and a verify time of 6 μs. Calculate the total length of time for a complete memory write and then a complete erase.

***Solution***

To write a byte of data takes 10 μs but each byte should be verified and so the total write time per byte is 16 μs. Hence the total write time for the array is: $128 \times 10^3 \times 16 \times 10^{-6} = 2.048$ seconds.

To completely erase we must first prewrite all bytes with a logic 0, then erase and finally perform an erase verify. The total erase time is thus:

$$
\begin{aligned}
T_{\text{erase}} &= 2.048 + 10\,\text{ms} + 128 \times 10^3 \times 6\,\mu\text{s} \\
&= 2.048 + 0.01 + 0.768 \\
&= 2.826 \text{ seconds}
\end{aligned}
$$

## 10.3 RANDOM ACCESS MEMORY – RAM

All the programmable ROM devices we have looked at have read times of the order of 100 ns but byte write times from 10 μs to 10 ms. This is acceptable when holding programs or storing data that change fairly infrequently but in cases where fast write times are needed such as in computers during mathematical calculations then these devices are inappropriate. For such applications RAM devices are more suitable. RAM or random access memories (more appropriately called read write memories) can be both written to or read from in a very fast time of typically 100 ns. They are classed as either static or dynamic.

Static RAMs retain their data indefinitely unless the power to the circuit is interrupted. Dynamic RAMs require that the data stored in each cell be refreshed periodically to retain the stored information. Again the data is lost if the power is interrupted.

### 10.3.1 Static RAMs

The basic schematic structure of a single static RAM cell is shown in Fig. 10.14. This cell consists of two inverters connected back to back in what is called a flip-flop arrangement. Two switches, S1 and S2, are controlled by the *row* line so that data may be read from or written to this cell. The output of one inverter reinforces the output of the other and hence the state of the circuit is locked. By closing switches S1 and S2 the cell can be accessed and data can be read. Alternatively the cell can again be accessed and by forcing *column* (and $\overline{column}$) to the required voltage, data can be written accordingly.

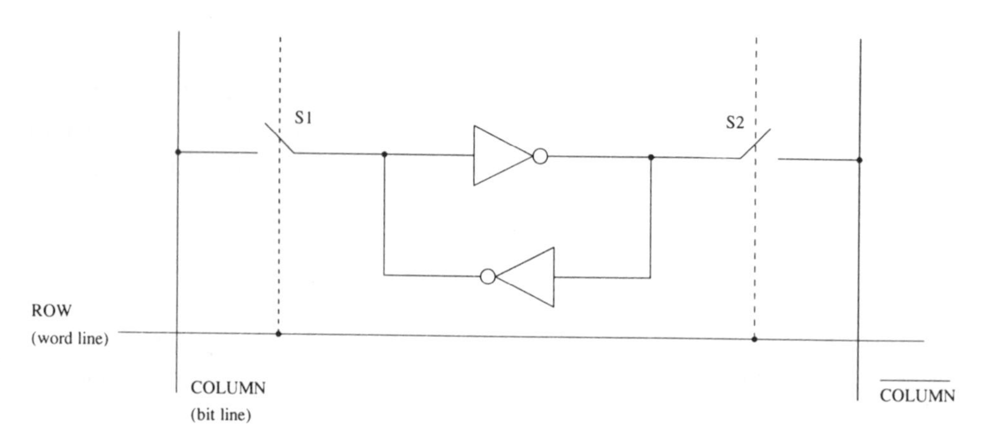


Fig. 10.14 Schematic for a single SRAM cell

SRAMs can be implemented with MOS or bipolar transistors. However, as mentioned in Chapter 9 the bipolar device is fast but consumes more space than the MOS cell and hence is only used in specialist applications. As shown in Chapter 9 BiCMOS is becoming popular when driving large capacitances which is especially important on the row lines for memory devices. Nevertheless, the MOS route is the most economical and hence is still the most popular technology for SRAMs.

The MOS transistor circuit diagram for an SRAM cell is shown in Fig. 10.15. Transistors T5 and T6 act as access transistors to the cell and implement the switches S1 and S2 from Fig. 10.14. Transistor pairs T1/T3 and T2/T4 are NMOS inverters (as described in Chapter 9) and are arranged in cross-coupled form as in Fig. 10.14. Access to the cell is achieved by a high on the *row* line which allows the state of the cell to be read out from the *column* lines. If T3 is off and T4 on then *column* will be high and $\overline{column}$ low and the cell is said to be in the logic '1' state. To write a '0' in the cell the *row* line is held high and *column* is forced low and $\overline{column}$ high. Thus T4 turns off and T3 turns on. When *row* is held low again the access transistors T5 and T6 turn off, but due to the cross-coupled nature of the two inverters the data is retained in the cell. However, if the power is removed the data will be lost and hence these devices are called

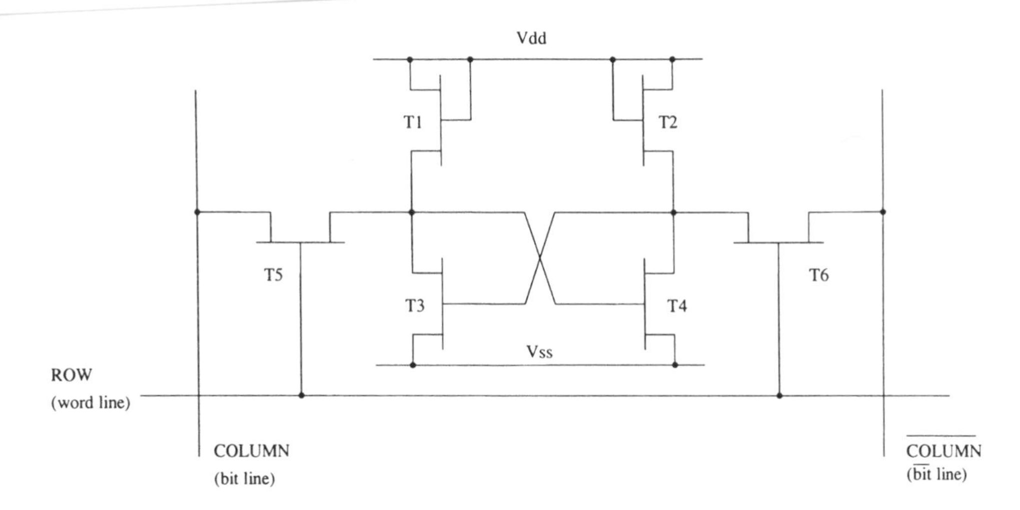


Fig. 10.15 MOS implementation of a single SRAM cell

*volatile*. It is possible to obtain pin compatible, directly replaceable, non-volatile SRAM (called NOVSRAM) which contains small batteries integrated into the package which cut-in when the main power is lost. However, although these batteries last for two years it is usually not possible to replace them since they are totally encapsulated.

It should be noted that using NMOS transistors as the load can result in unnecessary power consumption since a current will always flow between $V_{dd}$ and $V_{ss}$ when one of the driver transistors (T3 or T4) is on. Having small *W*/*L* ratios for the loads will keep this consumption low although two other techniques are used, namely: CMOS logic used for the inverters, or the load is replaced with a high-value polysilicon resistor. A cross-section of one-half of the flip-flop arrangement using this latter technique is shown in Fig. 10.16. This cross-section utilises two layers of polysilicon, one for the gate and the other for the load resistor. The resistor is positioned on top of the transistor thus saving space.

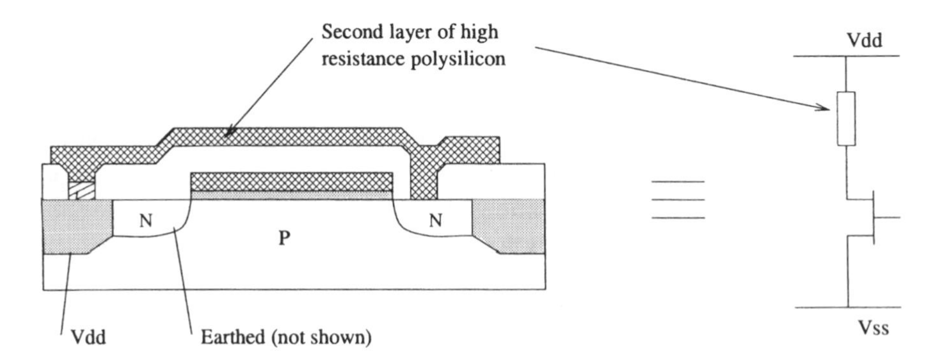


Fig. 10.16 Cross-section for one-half of an MOS SRAM cell

Typical input and output pins for an SRAM are:

- Input: address; data in; $\overline{CE}$; $\overline{OE}$; $\overline{WE}$;
- Output: data out;
- Power: $V_{dd}$; $V_{ss}$.

The $\overline{CE}$ pin is often referred to as the Chip Select pin or $\overline{CS}$. The SRAM read timing diagram is the same as in Fig. 10.5 for the mask ROM but with $\overline{WE}$ held high. The write timing diagram is the same as in Fig. 10.11 for an $E^2$PROM, but without the $\overline{BUSY}$ pin. Data is written into the SRAM as soon as $\overline{WE}$ goes high from its low state.

Low-power stand-by modes are possible with most SRAM devices, reducing the power consumption when not in use by a factor of 1000. This is achieved quite simply by holding $\overline{CE}$ high. In some devices the power can be reduced further by lowering $V_{dd}$ to 2 V without corrupting the stored data. Remember though that removing the power altogether will result in a complete loss of data.

**Example 10.7**

A single cell of an SRAM uses 20 MΩ load resistors. If the chip operates at 5 V and contains 64 Kbits then what is its power consumption? Assume that the voltage across an MOS transistor when on is 0 V.

***Solution***

When one of the inverter outputs is low then current flows through the load resistor from $V_{dd}$ to $V_{ss}$. The power consumed for one cell is thus

$$V_{dd}^2/20\times 10^6 = 1.25\ \mu W$$

Hence for a 64Kbit SRAM the total power consumed is 80 mW.

Notice that if the power supply is reduced to 2 V then the total power consumed reduces to 12.8 mW. If CMOS is used this static power consumption is reduced but with the added disadvantage of a lower packing density due to more transistors per bit.

### 10.3.2 Dynamic RAMs

Each SRAM cell either uses four transistors and two resistors or six transistors for CMOS logic. Hence each cell consumes a relatively large area of silicon compared to other semiconductor memory options. Using 1 μm technology a single SRAM cell in CMOS covers approximately 10 μm × 10 μm and thus a large multibit chip quickly becomes too large to manufacture.

The dynamic RAM or DRAM uses fewer transistors per cell than the SRAM. Hence more bits can be achieved per mm$^2$ albeit at the expense of more complex peripheral circuitry to refresh the data. All DRAMs use MOS technology and data is stored as charge on a capacitor. The highest density DRAM is obtained with the one transistor cell. There are many variations but the most common form is shown in Fig. 10.17. It consists of a single access transistor and a storage capacitor, $C_s$.

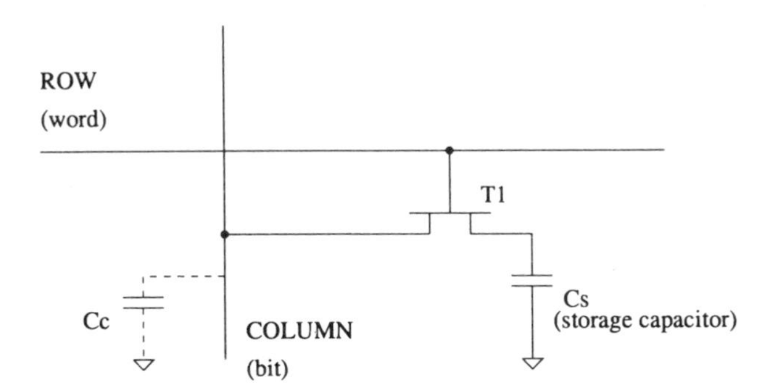


Fig. 10.17 A one-transistor dynamic RAM cell

To read the cell, *row* is held high and hence T1 is turned ON. The voltage stored on $C_s$ is transferred to the *column* line and sensed by a sense amplifier (not shown).

To write to the cell the *row* is again held high turning on T1. If a logic '1' is to be stored then the *column* line is held high and $C_s$ charges up to a logic '1'. When *row* is held low T1 is then turned off thus holding the charge on $C_s$. The charge stored on $C_s$ can, however, leak away through T1 due to its small leakage current when the transistor is off. Consequently the data must be periodically refreshed which thus requires extra circuitry on the chip. In fact the operation of reading the cell also results in the data being lost. This is because the storage capacitor is designed to be deliberately small for compactness. However, the *column* line feeds many cells and hence the capacitance of this line is very high (shown dotted as $C_c$). Each time the transistor T1 is turned on the charge on $C_s$ is distributed between these two capacitors and hence its voltage will drop (see Example 10.9 below). Consequently each time the cell is read it must be refreshed. This is usually performed automatically after every read with on-chip refresh circuitry. However, cells that lose their charge through leakage currents via the off transistor must be periodically refreshed by the user.

Typical input and output pins for a DRAM are:

- Inputs: address; data in; $\overline{RAS}$; $\overline{CAS}$; $\overline{OE}$; $\overline{WE}$;
- Outputs: data out;
- Power: $V_{dd}$; $V_{ss}$.

As can be seen two new pins, $\overline{CAS}$ and $\overline{RAS}$, are present with the DRAM and are called 'column address select' and 'row address select', respectively. Since DRAM devices have a large capacity then the number of address lines would be large. In order to reduce the number of address pins these are multiplexed into row and column addresses via the two select lines $\overline{RAS}$ and $\overline{CAS}$. Hence a 1 M by

16 bit DRAM would require 20 address lines but with address multiplexing this can be reduced to 10 address lines plus $\overline{CAS}$ and $\overline{RAS}$. This results in a smaller and hence lower cost chip. The problem with this for the user is that the timing diagrams are more complicated. For example a read timing diagram is shown in Fig. 10.18. Although not indicated, $\overline{WE}$ is held high and $\overline{OE}$ is held low throughout. Notice that the row address is latched into the chip on the falling edge of $\overline{RAS}$ and then the column address is latched on the falling edge of $\overline{CAS}$. After a short time, $t_{cac}$, called 'access time from $\overline{CAS}$' the data becomes valid on *Data out*. It remains valid until $\overline{CAS}$ goes high and $t_{off}$ seconds later *Data out* goes high impedance. Typical values of $t_{cac}$ and $t_{off}$ are 20 ns.

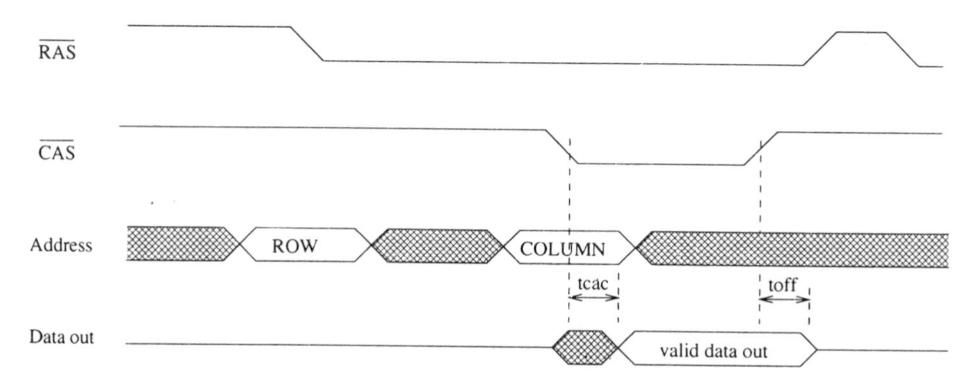


Fig. 10.18 DRAM read timing diagram

To write data into a DRAM two methods are available which provide the user with design flexibility. These are early write and late write. Both these modes are illustrated in Fig. 10.19. In both cases $\overline{RAS}$, $\overline{CAS}$ and the address lines are set up

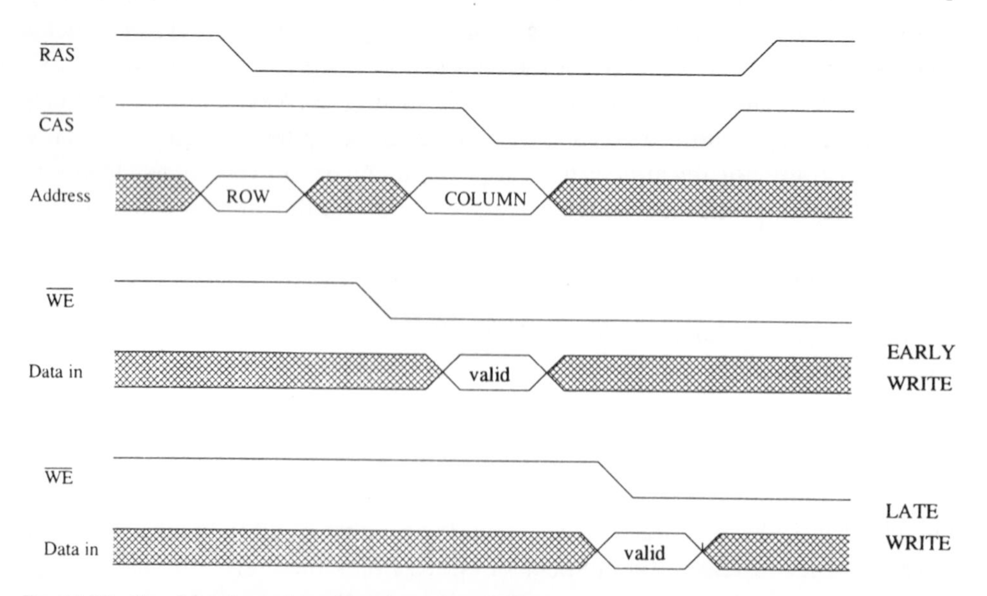


Fig. 10.19 Two techniques for writing data into a DRAM

as before, however, in early write the $\overline{WE}$ line is taken low before $\overline{CAS}$. When $\overline{CAS}$ falls the data is written into the array. For late write the $\overline{WE}$ is taken low after $\overline{CAS}$ at which point data is written into the array.

As stated the charge on the storage capacitor can leak away within a few milliseconds, hence the whole chip must be periodically refreshed. Several methods exist for the user to refresh the data. For example, every time a read operation is performed on an address then the data at that address is refreshed. Another method is called '$\overline{RAS}$ only refresh' and the timing diagram for this mode is shown in Fig. 10.20. It consists of holding $\overline{CAS}$ high and each time $\overline{RAS}$ goes low the data on the *whole* row that is being addressed will be refreshed. For other refresh modes the reader is referred to the manufacturer's data book.

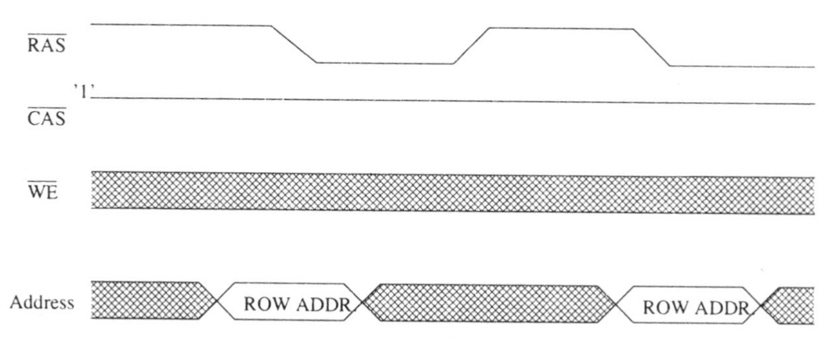


Fig. 10.20 DRAM RAS-only refresh

As can be seen the timing diagrams for DRAMs are more complex than SRAMs and hence some semiconductor manufacturers have added extra circuitry on board to relieve the system designer from complicated refreshing and address multiplexing. These devices are called Pseudo Static RAMs or PSRAM and are pin compatible with SRAMs (plus a $\overline{BUSY}$ pin) with only minor timing limitations required to allow for the chip to refresh itself.

Typical DRAM sizes are currently at 16 Mbit. The yield of such highly packed, very large, integrated circuits, needless to say, is low. However, for the cost to be kept competitive the yield must be kept high. The yield of DRAMS is dramatically increased by incorporating redundant (or spare) rows and columns of bits which can be exchanged for faulty ones.

## Example 10.8

How often must a DRAM cell be refreshed if $C_s = 0.01$ pF, the leakage current of the off transistor is 10 pA and the voltage across the capacitor is 2 V for a logic '1' and 0 V for a logic '0'.

### *Solution*

When $C_s$ is fully charged then 2 V appears across it. This voltage gradually falls as charge leaks away through the off transistor and must be refreshed before it changes to a logic '0'. The switching point is half-way between the two logic levels i.e. at 1 V. Hence using the equation for the discharge of a capacitor we can work out the time it will take for the voltage to fall from 2 V to 1 V, i.e. $I = C\delta V/\delta t$.

Rearranging to find $\delta t$:

$$\delta t = C\delta V/I = 0.01 \times 10^{-12} \times 1/10\,\text{pA} = 1\,\text{ms}$$

i.e. the data must be refreshed every 1 ms.

### Example 10.9

Consider the DRAM cell shown in Fig. 10.17. The capacitance of the column line ($C_c$) is twenty times that of the storage capacitor ($C_s$) and the voltage across each capacitor is $V_c$ and $V_s$ respectively. If a read signal is applied then what will be the change in voltage on the column line?

#### *Solution*

Before a read occurs the charges on the two capacitances are $Q_c = C_c V_c$ and $Q_s = C_s V_s$. When a read occurs the charge is divided amongst these two capacitors which are now connected in parallel. Hence the new charge on the column line is the sum of these two charges:

$$Q_{newc} = C_c V_c + C_s V_s$$

These two capacitors can be treated in parallel and hence

$$Q_{newc} = (C_c + C_s) V_{newc}$$

Equating these two expressions and substituting $20C_s = C_c$ we obtain

$$20C_s V_c + C_s V_s = (20C_s + C_s) V_{newc}$$

Thus $V_{newc} = (20V_c + V_s)/21$.

The change in voltage on the column line is $\delta V_c$ and hence $V_{newc} = V_c + \delta V_c$. Substituting for $V_{newc}$ into the above equation:

$$V_c + \delta V_c = (20V_c + V_s)/21$$

Rearranging:

$$\delta V_c = (20V_c + V_s)/21 - V_c = (V_s - V_c)/21$$

Hence if $V_s$ is 2 V and $V_c$ is 1 V then the column line voltage will only change by 47.6 mV. This is for a logic '1' stored and a similar value change (but negative) for a logic '0' stored. Consequently the sense amplifiers on the column line must be able to detect this small change in voltage. Needless to say one of the key components of a DRAM is a sensitive, noise-free, high-quality, sense amplifier.

## 10.4 MEMORY MODULES

In order to increase the memory capacity and to take advantage of the highly reliable semiconductor memories several semiconductor memory chips are incorporated onto a small PCB or substrate called 'memory modules'. Two main types of memory modules are in use today: SIMM and PCMCIA. The SIMM or single in-line memory module devices are those used in most PC based computers and consist of several DRAM chips on one board having either 30 or 72 pins arranged in a single line. The PCMCIA memory modules are actually memory cards which look like a typical bankers card. PCMCIA cards (standing for personal computer memory card international association) have 68 pins arranged in two socketed rows of 34 pins each. Most laptops and some PCs nowadays contain a PCMCIA slot. The card consists of an array of memory chips which could either be: flash, battery backed SRAM, $E^2$PROM or ROM. Three types exist: Type 1 (3.3 mm in thickness); Type 2 (5 mm in thickness); and Type 3 (10.5 mm in thickness). However, only Types 1 and 2 are used for memory whilst Types 2 and 3 are used for hard disks and fax cards. PCMCIA cards that use SRAM have a small lithium disk battery such that data is retained when the card is removed from the computer.

## 10.5 SELECTING THE APPROPRIATE MEMORY

To conclude this chapter let us try and briefly summarise these memory options as an aid to providing a selection guide. The first requirement is usually capacity – what is the maximum number of bits that can be obtained with a single chip? As you can imagine this figure is continually increasing as semiconductor processing advances. The second consideration is the write time (the read times for all types of semiconductor memory are all very similar of the order of 10 ns–100 ns). Fig. 10.21(a) and (b) shows histograms of the current capacity and the *write* time of each of the memory options discussed in this chapter. As expected these capacities have steadily increased over the years due to improvements in semiconductor processing. It can be seen that the highest capacity ROM is the mask programmable device but this is programmed once at the factory for a high cost. The DRAM is the largest read/write chip available with a very high speed write time. However, it is volatile and requires careful timing considerations. The SRAM has a medium capacity, is easy to interface to other devices, but is expensive per bit and is also volatile. The flash is a new market which is catching up with DRAMs and has the advantage of being both non-volatile and low cost but currently also requires careful timing. It is expected that these devices will become easier to use as this technology advances. EPROMs are very low cost and are ideal for program development but are totally inappropriate where in-circuit modifications are required. This area is covered by $E^2$PROM devices but write times of 10 ms/byte make its use limited. The 10 μs byte write times for flash make this the product to watch for the future.

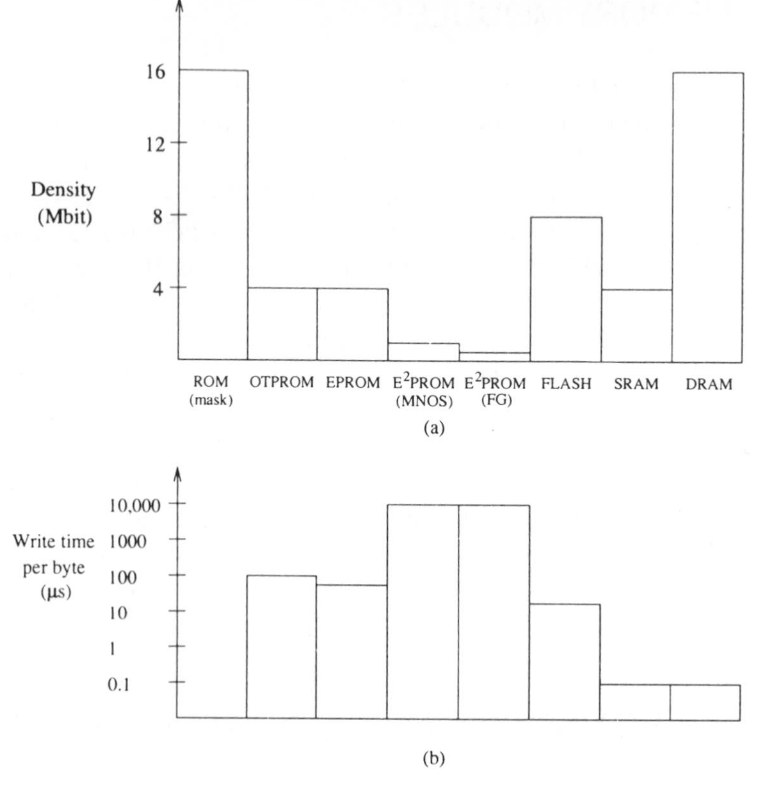


Fig. 10.21 Capacity and write time comparisons for semiconductor memories

## 10.6 SELF-ASSESSMENT

10.1 Define the terms ROM, RAM, SRAM, DRAM, EPROM, RWM, PSRAM, NOVRAM, OTPROM, $E^2$PROM.

10.2 The power is removed from the devices in Question 10.1. Which devices will retain their data?

10.3 What does non-volatile mean?

10.4 Define access time.

10.5 Which of the memory types uses the most transistors per bit and which uses the least?

10.6 Which has the longest write time: a hard disk or an SRAM?

10.7 Place in increasing write time order the memory options described in this chapter.

10.8 Which type of memory device is used as the main semiconductor memory in PC computers?

10.9 What is a SIMM?

10.10 What is a PCMCIA card?

10.11 What voltage levels should the pins $\overline{CE}$, $\overline{OE}$, $\overline{WE}$ be in order to read data from a DRAM?

10.12 Why do you have to refresh a DRAM memory device?

10.13 What power supply voltages are needed for EPROM, flash, SRAM, $E^2$PROM?

10.14 What happens when $\overline{CE}$ is held high for an SRAM device?

10.15 Which of these devices must be erased before they can be written: EPROM; $E^2$PROM; SRAM; flash?

## 10.7 PROBLEMS

10.1 A semiconductor memory chip is specified as 2 K by 8:
(a) How many words (exactly) can be stored on this chip?
(b) What is the word size?
(c) How many total bits can this chip store?

10.2 (a) How many pins would a 4 M by 8 bit mask ROM have?
(b) Repeat (a) for a multiplexed address DRAM.

10.3 Illustrate with a sketch how to combine 256 K by 4 bit SRAM chips (having a single $\overline{CE}$ pin and bidirectional I/O pins) to produce a:
(a) 256 K by 8 memory;
(b) 512 K by 4 memory (use only 19 address lines);
(c) 1024 K by 4 memory (use only 20 address lines).

10.4 A DRAM has a column line capacitance of 3.8 pF. What value of $C_s$ is required such that at least 250 mV change on the column line occurs when reading a logic zero (0 V) or a logic one (5 V)? Assume that the column line is precharged to 2.5 V before reading.

10.5 A flash memory device organised as 256 K by 8 bits has a total write time of 4.2 seconds and a total erase time of 5.8 seconds. If the flash erase time is 9.5 ms then what are the values of byte write time and byte verify time?

10.6 A 5 V NMOS SRAM cell is to have a stand-by power consumption of 0.1 mW. What value of load resistor is required? Assume the voltage across the 'on' MOS transistor is 0 V.

10.7 A 5V, 512 K by 8 bit CMOS SRAM is to be used in a three hour battery powered application. If the leakage current of an off transistor is 0.5 nA then what is the total static power dissipation and the total ampere hours required. Comment on the validity of your answer.

# 11 Selecting a design route

## 11.1 INTRODUCTION

The preceding chapters have described the various techniques used to design combinational and sequential circuits. We have also discussed the advantages and disadvantages of each of the technology options, i.e. bipolar, CMOS, ECL, etc. This final chapter describes the various design routes which can be used to implement a design. The decision regarding which of these design routes to use depends upon the following issues:

- When should the first prototype be ready?
- How many units are needed?
- What are the power requirements?
- What is the budget for the product?
- What are the physical size limitations?
- How complex is the design (gate count, if known)?
- What is the maximum frequency for the design?
- What loads will the system be driving?
- What other components are needed to complete your design?
- What experience have you or your group had to date in the design of digital systems?

These are the questions that must be asked before starting any design. The aim of this chapter is to provide background to the various design routes that are available. Armed with this knowledge, the answers (where possible) to the above questions should allow the reader to decide which route to select or recommend.

### 11.1.1 Brief overview of design routes

The various design options are illustrated in Fig. 11.1. As can be seen the choice is either to use *standard products* or to enter the world of *application specific integrated circuits* (ASICs). The 'standard product' route is to choose one, or a mixture, of the logic families discussed in Chapter 9 such as 74HCT, 74LS, 4000 series, etc. On the other hand, an ASIC is simply an IC customised by the designer for a specific application. Various ASIC options exist which can be subdivided into either *field programmable* or *mask programmable* devices. Field program-

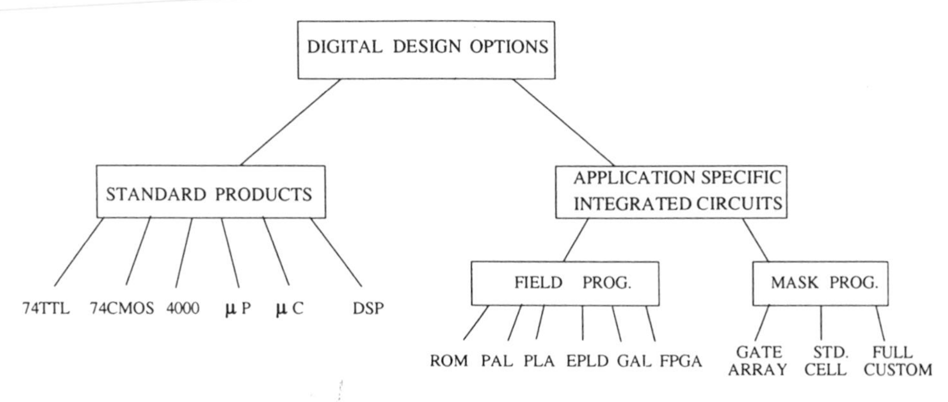


Fig. 11.1 Design options

mable devices (i.e. ROM, PAL, PLA, GAL, EPLD and FPGA) are all programmed in the laboratory. However, mask programmable devices must be sent to a manufacturer for at least one mask layer to be implemented. These mask programmable devices may be exclusively digital or analogue, or alternatively what is known as a *mixed* ASIC which will contain both.

The mask programmable devices can be further subdivided into *full custom*, *standard cell* and *gate array*. With full custom design the designer has the option of designing the whole chip, down to the transistor level, exactly as required. Standard cell design again presents the designer with a clean slice of silicon but provides standard cells (e.g. gates, flip-flops, counters, op-amps, etc.) in a software library. These can be automatically positioned and connected on the chip as required (known as *'place and route'*). Both of these levels of design complexity are used for digital and analogue design, and are characterised by long development times and high prototyping costs. The third and lowest level in terms of complexity is the gate array. With the gate array the designer is presented with a 'sea' of universal logic gates and is required only to indicate how these gates are to be connected which thus defines the circuit function. This approach offers a less complex, and hence cheaper, design route than standard cell and full custom.

Until the late 1980s the cheapest route to a digital ASIC was via the use of a mask programmable gate array. These devices are still widely used but since the late 1980s have had to face strong competition from field programmable gate arrays (FPGAs) where the interconnection and functionality are dictated by electrically programmable links and hence appear in the field programmable devices section.

With regard to the above ten questions, the overriding issue is usually when the first prototype should be ready. ASICs require computer aided design (CAD) tools of differing complexities. Designs that use such tools provide elegant solutions but can be very time consuming especially if your team have no experience in this field. However, designs that use 'standard products' are quick to realise but can be bulky and expensive when high volumes are required.

With the exception of microcontrollers/processors and DSPs this chapter will describe the design options in Fig. 11.1 in more detail. It should be noted, however, that as you move from left to right across this diagram, each option becomes more complex to implement resulting in a longer design time and greater expenditure.

## 11.2 DISCRETE IMPLEMENTATION

As has been seen in Chapter 9 the 74 series offers a whole range of devices at various levels of integration. These levels of integration are defined as:

- SSI – Small-scale integration (less than 100 transistors per chip);
- MSI – Medium-scale integration (100–1000 transistors per chip);
- LSI – Large-scale integration (1000–10 000 transistors per chip);
- VLSI – Very large-scale integration (greater than 10 000 transistors per chip).

The VLSI devices are mainly microcontrollers and microprocessors which are outside the scope of this book.

Designs using these standard parts are quick to realise and relatively easy to debug. However, they are bulky and expensive when high volumes are required. The various functions available allow all sorts of digital systems to be implemented with minimal overheads and tooling. For expediency these designs can be *ad hoc* and incorporate poor digital design techniques. We shall look at some of these pitfalls and suggest alternative safe design practices.

One such standard product is the 74HCT139 which consists of two 2-to-4 decoders in a single IC package. A logic diagram for this IC is shown in Fig. 11.2. A decoder was introduced in Chapter 4 and, as seen in Chapter 10, it can be used in memories for addressing purposes where only one output goes high for each address applied. Such a device has many other uses. However, as we saw in Chapter 4 one must be careful with this type of circuit since any of the decoder outputs can produce spurious signals called static hazards. These static hazards are called 'spikes' and 'glitches'.

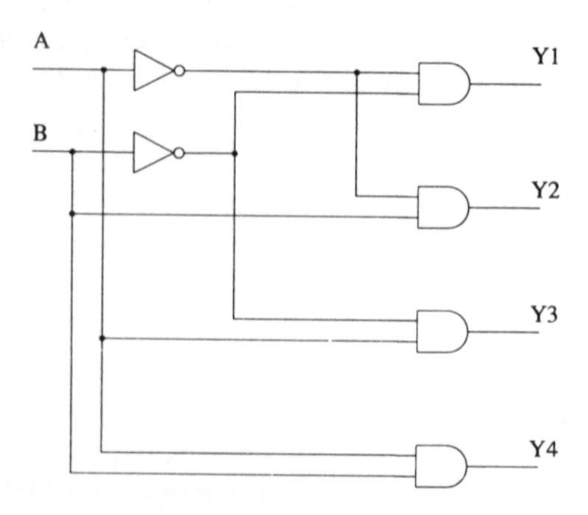


Fig. 11.2 74HCT139: two-to-four decoder

### 11.2.1 Spikes and glitches

Consider the case of output Y3 in Fig. 11.2. A timing diagram is shown in Fig. 11.3 for this output for various combinations of $A$ and $B$. At first $AB$=00 and so Y3=0. Next, $AB$=10 and Y3 goes high. With $AB$ returning to 00 the output goes low again. All seems satisfactory so far but if $AB$=11, then due to the propagation delay of the inverter the output will go high for a short time equal to the inverter propagation delay. As we shall see, although this spike is only a few nanoseconds in duration it is sufficiently long to create havoc when driving clock lines and may inadvertedly clock a flip-flop. This phenomenon is not limited to decoders. All combinational circuits will produce these *spikes* or *glitches* as they are known.

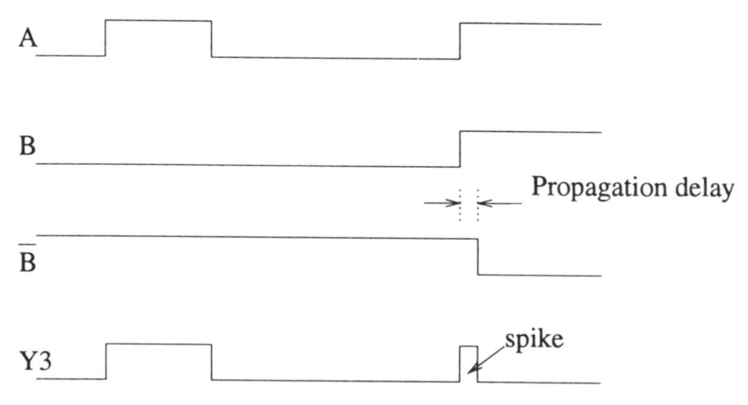


Fig. 11.3 Spike generation on output Y3 of the 2-to-4 decoder

To appreciate the problem when driving clock lines consider a circuit counting the number of times a four-bit counter produces the state 1001. A possible design using 74 series logic is shown in Fig. 11.4(a). This consists of a 4-to-16 decoder (74HC154)[1] being used to detect the state 1001 from a four-bit counter (74HC161). (For clarity the four-bit counter output connected to the four inputs of the decoder is represented as a *data bus* having more than one line. The number of signals in the line is indicated alongside the bus.) The 10th output line of the decoder is used to clock a 12 bit counter (74HC4040). However, although this will detect the state 1001 at the required time it will also detect it at other times due to the differing propagation delays in the 4-to-16 decoder. These spikes and glitches will trigger the larger counter and result in a false count.There are two solutions to this: an elegant one and one that some undergraduates fall mercy to! The latter method, illustrated in Fig. 11.4(b), is to use an $RC$ network (connected as an integrator or a low-pass filter) and a Schmitt trigger which together remove the spike or glitch. The values of $R$ and $C$ are chosen so as to filter out this fast transient – usually $RC$ is set to be 10 times the glitch or spike pulse width. Due to this long

[1]This decoder has outputs which are active low; however, for this application we shall assume that the outputs are active high.

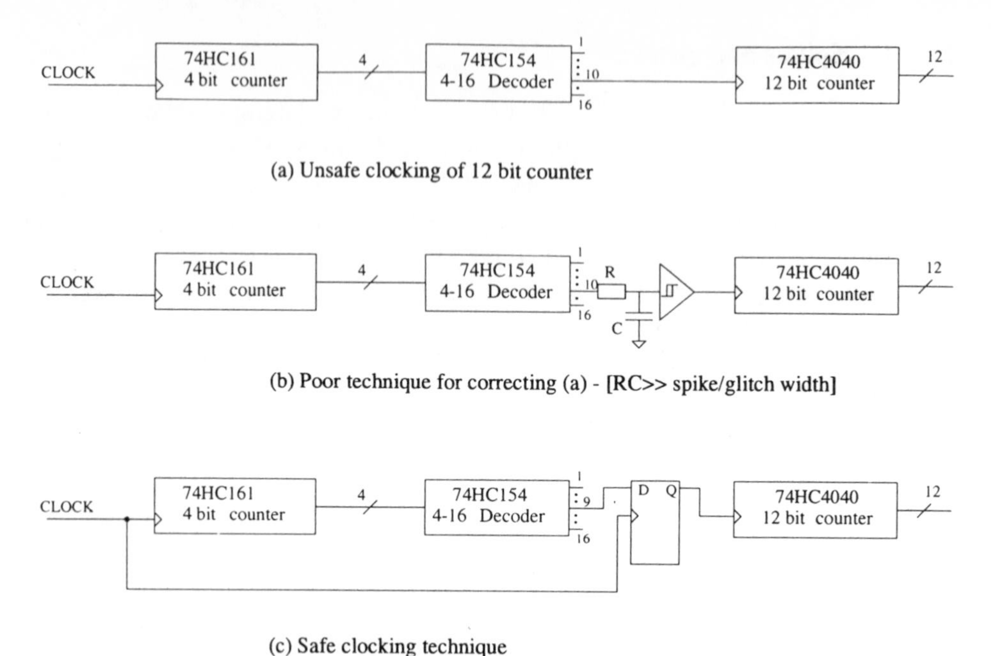


(a) Unsafe clocking of 12 bit counter

(b) Poor technique for correcting (a) - [RC>> spike/glitch width]

(c) Safe clocking technique

Fig. 11.4 Using a decoder as a state detector

time constant the signal presented at the input to the Schmitt is now only a fraction of the magnitude of the original spike. To remove this signal completely it is passed through a Schmitt. This device has a voltage transfer characteristic which has two switching points. When the input is rising (from 0 V) the Schmitt switches at typically $0.66V_{dd}$. However, when the input is falling (from $V_{dd}$) the Schmitt now switches at $0.33V_{dd}$. Hence any signal that does not deviate by more than two-thirds of the supply will be removed. This circuit, although successful, cannot be used in any of the other design options in this chapter since large values of $R$ and $C$ are not provided on chip. In addition the provision of extra inputs and outputs for these passive components will produce an unnecessarily large chip. The elegant solution, shown in Fig. 11.4(c), is to detect the *previous* state with the decoder and present this to the D input of a D-type. The *clean* output of the flip-flop is then used to drive the 12 bit counter.

To summarise, an important rule for all digital designers is that clock inputs must not be driven from *any* combinational circuit, even a single two-input logic gate. This can be stated quite succinctly as *no gated clocks*. In fact the same is true for reset and set lines since these will also respond to spikes and glitches thus causing spurious resetting of the circuit.

## 11.2.2 Monostables

Another tempting circuit much frowned upon by the purist is the monostable. The monostable or *'one shot'* produces a pulse of variable width in response to either a *high-to-low* or a *low-to-high* transition at the input. The output pulse width is set via an external resistor and capacitor.

One application of the use of a monostable is shown in Fig. 11.5(a). Suppose that we require an eight-bit parallel in, serial out shift register (PISO) to be loaded with an eight-bit data word when a line called *interrupt* goes high. An active high *load* signal must be produced which will load the eight-bit data. This *load* signal must be returned low before the *next* rising clock edge so that serial data can continue to be clocked out. It should be noted that in this case the *interrupt* line is assumed to be synchronised with the clock. By adjusting the value of $R$ and $C$ the required parallel *load* pulse width ($kRC$, where $k$ is a constant) is set to be no longer than the clock pulse width less the *load* to *clock* set-up time. The corresponding timing diagram is shown in Fig. 11.5(b).

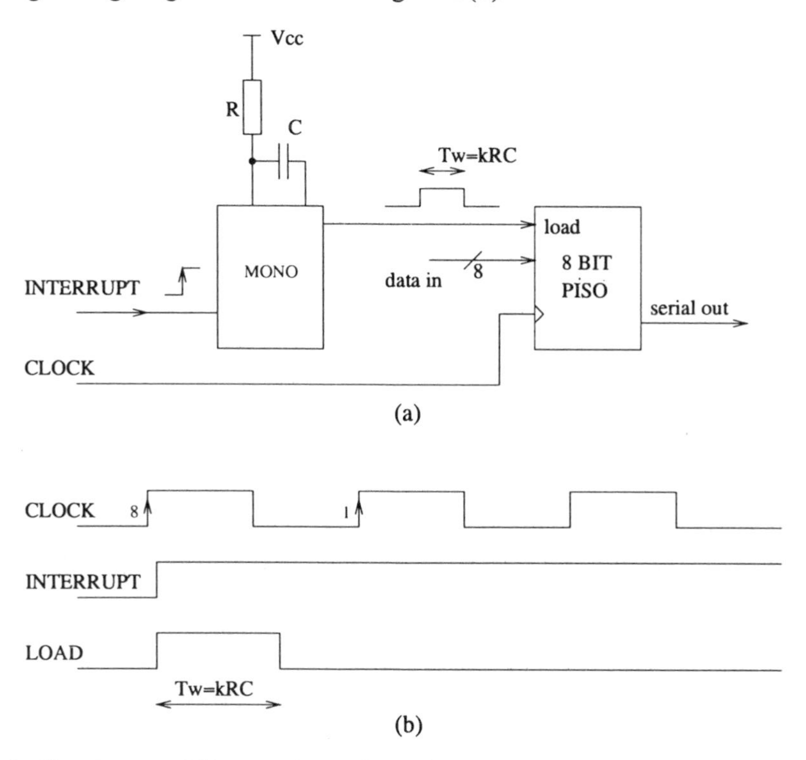


Fig. 11.5 Use of a monostable to produce a short pulse

Circuits that use monostables, however, exhibit several limitations. The first is that it is necessary to use an external $R$ and $C$ which will require a redesign when migrating to an ASIC. Other problems related to the analogue nature of the device are: the pulse width varies with temperature, $V_{cc}$ and from device to device; poor noise margin (see Chapter 9) thus generating spurious pulses; oscillatory signal edges are generated for narrow pulse widths (less than approximately 30 ns); and long pulses require large capacitors which are bulky.

An alternative to the circuit in Fig. 11.5(a) is to use the circuit in Fig. 11.6(a) which uses the *reset technique* with a purely digital synchronous approach. The

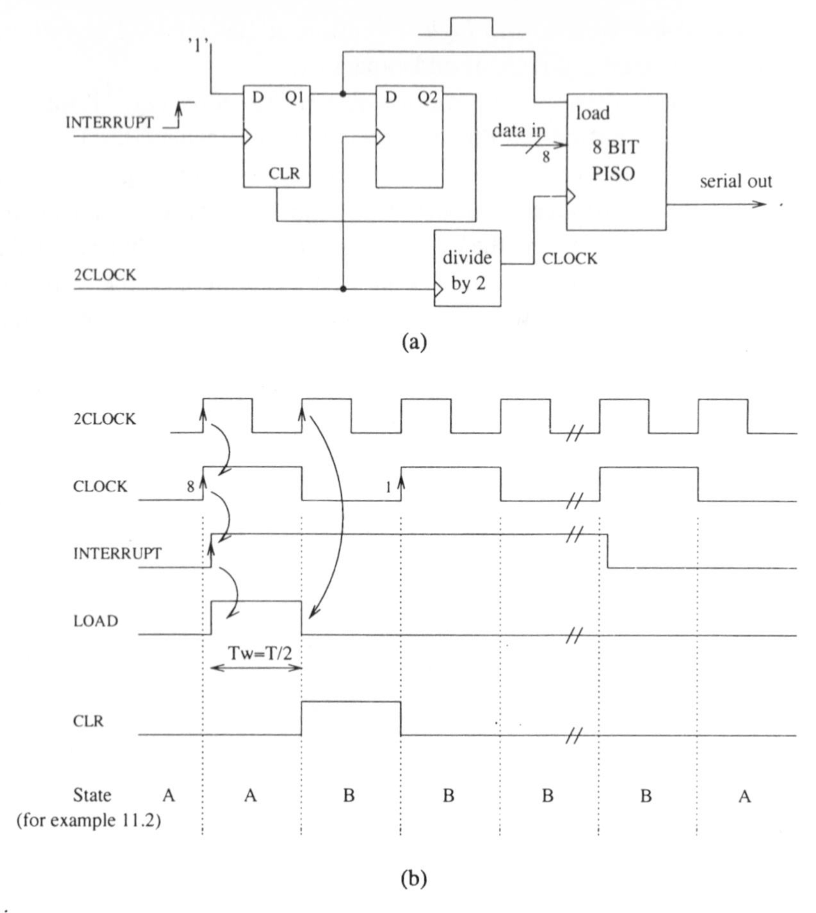


Fig. 11.6 Alternative circuit to the monostable circuit in Fig. 11.5

resulting timing diagram for this circuit is shown in Fig. 11.6(b). The circuit operates by using a clock frequency of twice the PISO register clock (*2-clock*). When *interrupt* goes from low to high, Q1 (i.e. *load*) will go high. This will load in the parallel data. At the next rising *2-clock* edge Q2 goes high (as its input, Q1, is now high) and clears or resets the *load* line. Because of the higher clock frequency used this all occurs within half a *clock* cycle. A divide-by-two counter is used to divide *2-clock* down to *clock* so that the new data loaded into the PISO can be serially shifted out on the immediately following rising edge of *clock*. *Load* will not go high again until another low to high transition on *interrupt* occurs.

The following example shows how pulses of a longer time duration can be produced.

## Example 11.1

Consider the circuit in Fig. 11.7. What pulse width is produced at the Q output of the D-type (74HC74) device? Assume that both 'CLR' and 'RESET' are active high.

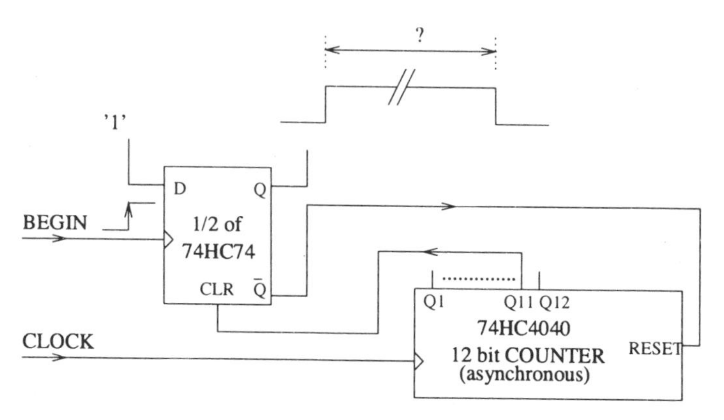


Fig. 11.7 Circuit to produce a controlled long pulse width

### *Solution*

When a *BEGIN* low-to-high transition occurs the Q output goes high which releases the counter from its reset position. The counter proceeds to count until the Q11 output goes high, at which point the D-type flip-flop is cleared and the Q output goes low again awaiting the arrival of the next *BEGIN* rising edge. The Q output is thus high for $2^{10}$ clock pulses.

Taking the clear input from any of the other outputs of the counter will produce pulses of varying width. The higher the input clock frequency the better the resolution of the pulse width.

It should be noted that if *BEGIN* is synchronised with the clock then the rising edge of the output pulse will also be synchronised (albeit delayed by one D-type flip-flop delay). However, the falling edge of the Q output pulse is delayed with respect to the clock. This is because the counter used is an asynchronous or ripple counter. The Q11 output will only go high after the clock signal has passed through 11 flip-flop delays – this could be typically 100–400 ns. This may not cause a problem but is something to be aware of. The solution is to use either a synchronous counter or detect the state before with a 10-input decoder and a D-type as described earlier.

### Example 11.2

The circuit of Fig. 11.6(a) was designed in an *ad hoc* manner with the reset technique. Using the state diagram techniques of Chapter 8 produces a circuit that will implement the same timing diagram of Fig. 11.6(b).

### *Solution*

The first task is to use the timing diagram of Fig. 11.6(b) to produce a state diagram. At the bottom of Fig. 11.6(b) are the states *A* and *B* at each rising *2-clock* edge. Remember, that the interrupt (*I*) line is generated by *2-clock* (i.e.

synchronised) and thus changes *after* the *2-clock* rising edge. Hence the state diagram, shown in Fig. 11.8(a), can be drawn. The corresponding state transition table is shown in Fig. 11.8(b) and since there are only two states then only one flip-flop is needed. Assigning $A=0$ and $B=1$ results in Fig. 11.8(c). From this we produce the K-maps for the next state $Q^+$ and present output $LOAD(L)$. These produce the functions $Q^+=I$ and $L=I\cdot\bar{Q}$. The resulting circuit diagram is shown in Fig. 11.8(e). It should be remembered that the clock input is the higher frequency *2-clock*.

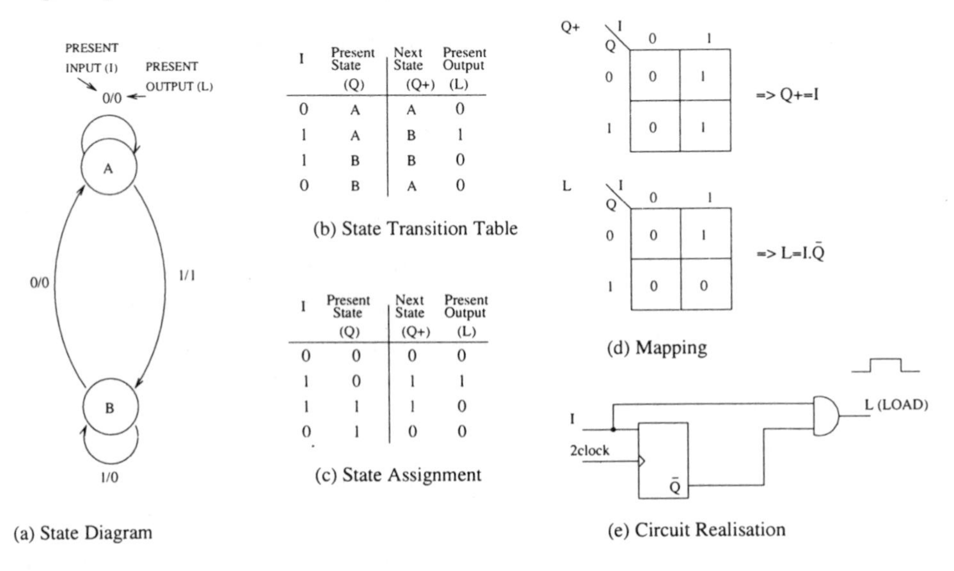


| I | Present State (Q) | Next State (Q+) | Present Output (L) |
|---|---|---|---|
| 0 | A | A | 0 |
| 1 | A | B | 1 |
| 1 | B | B | 0 |
| 0 | B | A | 0 |

(b) State Transition Table

| I | Present State (Q) | Next State (Q+) | Present Output (L) |
|---|---|---|---|
| 0 | 0 | 0 | 0 |
| 1 | 0 | 1 | 1 |
| 1 | 1 | 1 | 0 |
| 0 | 1 | 0 | 0 |

(c) State Assignment

Fig. 11.8 Using a state diagram to implement the timing diagram of Fig. 11.6(b)

### 11.2.3 CR pulse generator

The practice of using monostables has already been frowned upon and safe alternative circuit techniques have been suggested. However, monostables are tempting, quick to use and can still be found in many designs. Another design technique that is also simple and tempting to use but should be avoided is the CR pulse generator or differentiator circuit shown in Fig. 11.9(a). The circuit is the opposite of the integrator shown in Fig. 11.4. This circuit is used to 'massage' a long pulse into a shorter one and so gives the appearance of a *one-shot* reacting at

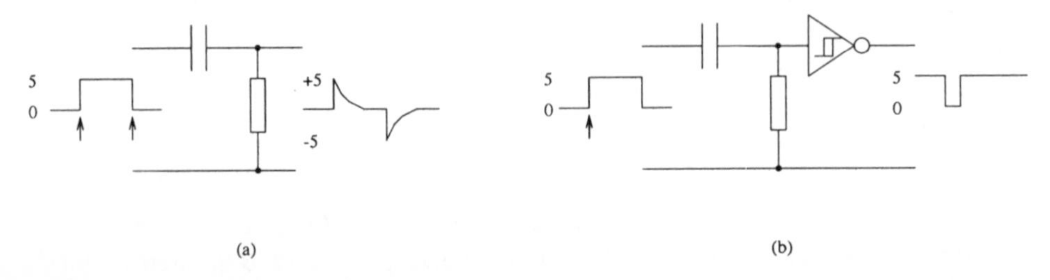


Fig. 11.9 Using a CR network to produce a narrow pulse

either rising or falling edges. If a 5 V pulse is applied to the circuit in Fig. 11.9(a) two short pulses are produced, one at the rising edge and one at the falling edge. At the rising edge when the input goes instantaneously from 0 V to 5 V the output momentarily produces 5 V. As the capacitor charges the voltage across the resistor starts to fall as the charging current falls, hence the corresponding rising edge waveform. When the input changes from 5 V to 0 V the capacitor cannot change its state instantly and so both plates of the capacitor drop by 5 V. Hence the output momentarily produces –5 V. The capacitor then discharges, resulting in the falling edge waveform.

To convert this signal into a digital form the output is fed into a Schmitt trigger and thus produces a short pulse from 5 V to 0 V whose duration is determined by the value of $R$ and $C$ and the Schmitt switching point. This pulse is only present on the rising edge of the input since the falling edge produces a negative voltage which the Schmitt does not respond to. However, this circuit should again be avoided as the migration to an ASIC would require a redesign whilst in addition the negative voltage may in time damage the Schmitt component. Consequently, it is therefore recommended that the pulse shortening techniques described earlier, which use a higher clock frequency, are employed.

## 11.3 MASK PROGRAMMABLE ASICs

The use of standard products (74 series, etc.) to implement a design becomes inefficient when large volumes are required. Hence the facility for the independent customer to design their own integrated circuits was provided by IC manufacturers. This required the designer to use either a *gate array*, *standard cell* or *full custom* approach. In each case the manufacturer uses photomasks (or electron-beam lithography) to fabricate the devices according to the customer's requirements. These devices are therefore collectively named *mask programmable ASICs*.

### 11.3.1 Safe design for mask programmable ASICs

A limitation of mask programmable ASICs is that since the layers are etched using these masks any design errors require a completely new set of masks. This is very expensive and time consuming and hence *safe* design techniques which work first time must be employed. A designer must avoid monostables and *CR*/*RC* type circuits and be aware that a manufacturing process can vary from run to run and sometimes across a wafer. Consequently, propagation delays vary quite considerably from chip to chip or even across a chip. Hence the use of gates to provide a delay (see Fig. 11.10(a)) is a poor design technique since the value of this delay cannot be guaranteed. Three *poor* ASIC circuit techniques where these delay chains are used are shown in Figs. 11.10(b)–(d) and were discussed in Section 4.3.2. Essentially the designer must use synchronised signals and a higher clock frequency to generate short predictable pulses.

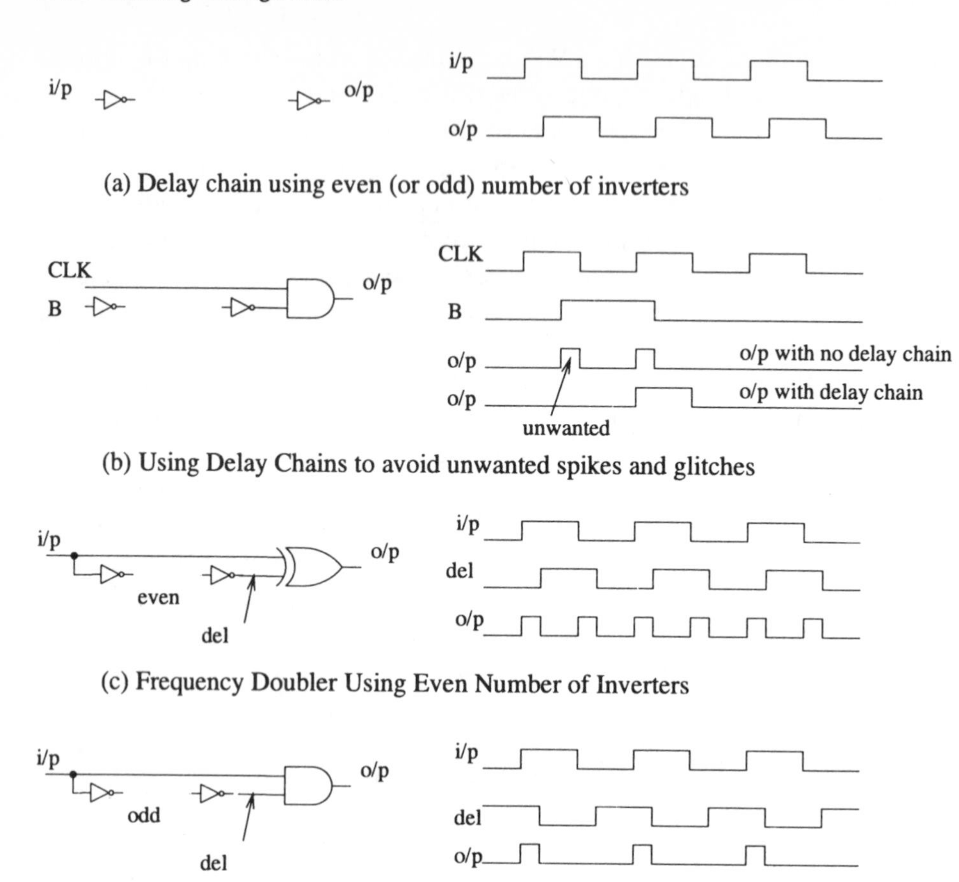


Fig. 11.10 Examples of poor ASIC circuit techniques

The use of synchronous techniques is not a panacea for all timing problems. Take for example the master clock in a synchronous system driving several different circuits. The total capacitance being driven by the master clock can be extremely large thus delaying the clock quite considerably. In order to isolate this large capacitance from the master clock, *buffers* are used leading to each circuit. These are quite simply two CMOS inverters in series. This reduces the capacitance seen directly by the master clock circuit and hence reduces the clock delay to each circuit. However, the input capacitance between the smallest and largest of these circuits may differ by an order of magnitude. Hence the clock will arrive at different times to each of these circuits and the whole system will appear asynchronous in nature (see Problem 11.10). A better buffering technique is therefore required. Two improved buffering techniques are shown in Figs 11.11(a) and (b). The first is to use an even number of inverters driving the large load. At first it just looks like our poor delay line shown in Fig. 11.10(a). However, each inverter is larger than the previous one by a factor $f$ (i.e. the $W/L$ ratios of the MOS transistors are increased by $f$ at each stage). The load capacitance gradually increases at

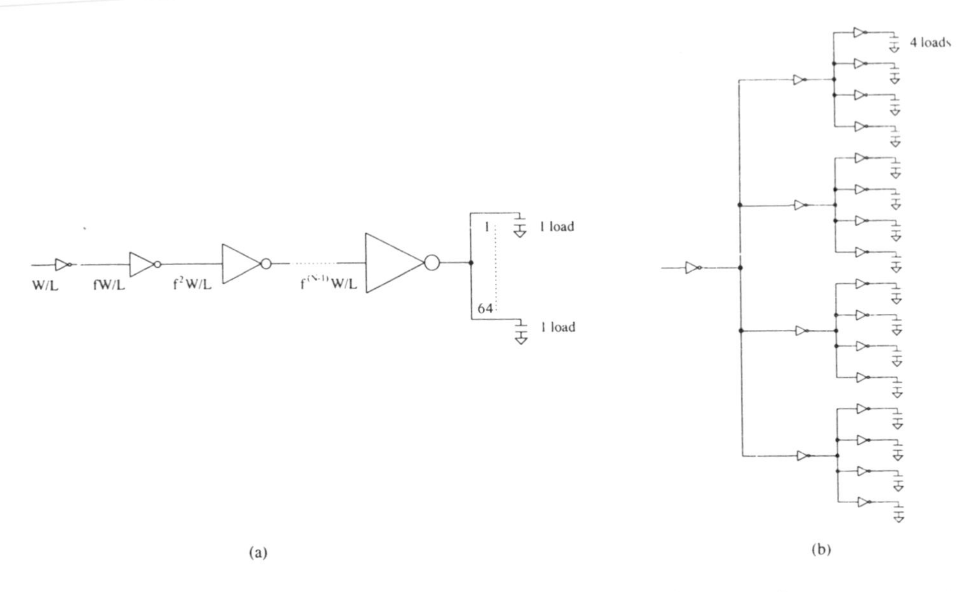


Fig. 11.11 Two techniques for buffering the ASIC clock driver from a large capacitance

each stage but the drive strength also increases. The optimum value of $f$ is in fact e or 2.718 but the number of stages required for this case would be quite large. A compromise is to use an increased value of $f$ and a reduced number of stages (see Problem 11.11). Another technique is to use *tree* buffering which consists of several small inverters arranged in a tree structure. This is illustrated in Fig. 11.11(b). In this case each inverting buffer is arranged such that it drives the same load. Hence the relative clock signal delay will be kept to a minimum.

## Example 11.3

One of the small inverters in Fig. 11.11(b) is used to drive 64 loads each of 1 pF. Determine the delay of this inverter when driv'ng this load directly and what the delay would be if the tree buffering of Fig. 11.11(b) is used. Assume that the inherent delay of a single inverter is 1 ns, its output drive capability is 20 ns/pF and has an input capacitance of 0.01 pF.

### Solution

*Unbuffered*

$$\text{Delay} = 1 + 20 \times 64 = 1281\text{ ns}$$

*Buffered*

$$\text{Delay} = (1 + 20 \times 0.01 \times 4) + (1 + 20 \times 0.01 \times 4) + (1 + 20 \times 4)$$

$$\text{Delay} = 1.8\text{ ns} + 1.8\text{ ns} + 81\text{ ns} = 84.6\text{ ns}$$

Hence a great saving in delay is achieved at the expense of more gates.

These *safe* mask programmable ASIC design techniques can therefore be summarised as follows:

1. no gated clocks or resets;
2. no monostables;
3. no *RC* or *CR* type circuits;
4. use synchronous techniques wherever possible;
5. use a high-frequency clock subdivided down for control;
6. no delay chains;
7. use clock tree buffering.

In the early days ASIC designs were breadboarded (i.e. a hardware prototype was produced) using 74 series devices in order to confirm that the design functions correctly. However, nowadays the designer has available very accurate computer simulators that can be run in conjunction with drawing packages and chip layout. Together these computer programs are called computer aided design (CAD) tools. Since a mask programmable ASIC cannot be modified once fabricated without incurring additional charges, the design cycle relies very heavily upon these CAD tools. The process of fabricating a chip and then finding a design fault is an unforgivable and costly error. We shall look at the various CAD tools employed to guarantee a 'right first time' design.

### 11.3.2 Mask programmable gate arrays

The first mask programmable ASIC that we shall look at is the mask programmable gate array. This device consists of a large array of *unconnected* blocks of transistors called *gates*. All the layers required to form these gates are prefabricated except for the metal interconnect. The IC manufacturer therefore has a 'stock-pile' of uncommitted wafers awaiting a metal mask. The user or designer only needs to specify to the manufacturer how these gates are to be connected with the metal layer (i.e. customised).

The basic building block or gate in a CMOS gate array is a versatile cell consisting of four transistors. These blocks of four transistors are repeated many times across the array. Mask programmable gate arrays are characterised in terms of the number of four transistor blocks or *gates* in the array. The gate is called a versatile cell since it contains two NMOS and two PMOS transistors which can form simple logic gates such as NOR and NAND as illustrated in Chapter 9.

Two types of arrays exist – *channelled* and *sea of gates*. These are illustrated in Figs. 11.12(a) and (b). The channelled array has a routing channel between each row of gates. These routing channels allow metal tracks on a fixed pitch to be used for interconnection across the array. Each channel can contain typically 20 wiring routes. The sea of gates on the other hand does not contain any dedicated routing channels and as a result contains more gates. The routing is implemented *across* each gate at points where no other metal exists. However, with the sea of gates the routing over long distances is more difficult and hence places a limit on the

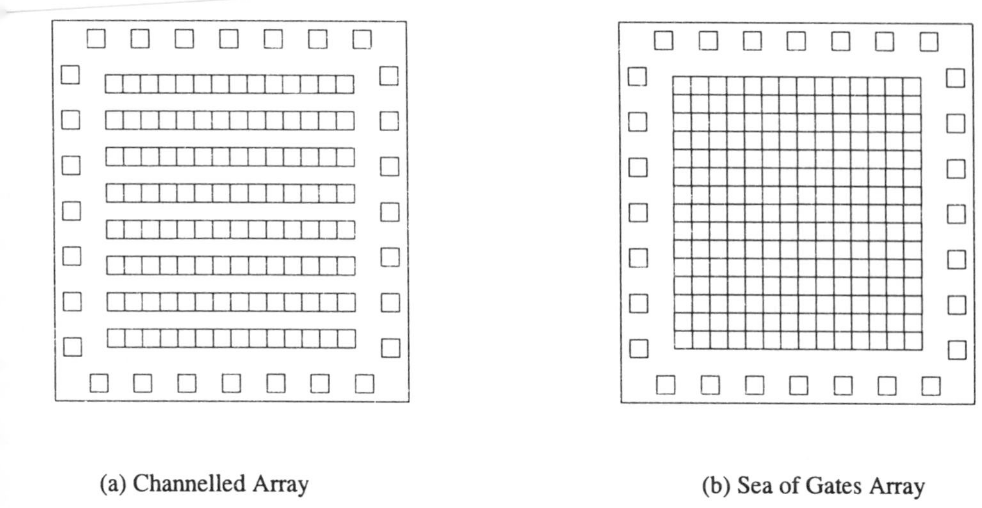


Fig. 11.12 Channelled and sea of gates mask programmable gate arrays

number of gates that can be accessed. This raises the important issue of *utilisation*. This is the percentage of gates which the designer can access. As more gates on the array are utilised the routing ability for both array types is reduced. There comes a point where there are not enough routes available to complete the design and because of this manufacturers quote a utilisation figure. As you can imagine the channelled array has a better utilisation than the sea of gates. A simple single layer metal channelled array has a utilisation of 80% whilst a double layer metal has a utilisation of 95%. Many mask programmable gate array manufacturers use three and four layer metal processes in order to fully utilise the array.

For any design it is the gate count that is the most important issue. It is therefore useful to know how many gates typical functions consume in CMOS technology. For example a two-input NOR or NAND uses one gate, whilst a D-type and a JK consume five and eight gates, respectively. Hence if a design schematic exists then a quick gate count is always useful to specify what gate array size to use. The selection of an optimum array size is crucial in gate array design since array sizes can vary from 1000 to 500 000 gates!

The cost of a mask programmable gate array depends upon:

1. number of gates required (or the number of I/Os);
2. number of parts required per year;
3. maximum frequency of operation;
4. number of metal layers.

All mask programmable gate array manufacturers charge a tooling cost for production of the metal mask(s). This charge is called a non-recurring expenditure or NRE. Quotes from three reputable ASIC suppliers for a 2000 gate design, commercialised by the authors, revealed the following prices on a small volume of 1000 parts per year:

- Firm X (2 micron) NRE of £10 000 at £4.00 unit cost;
- Firm Y (1.2 micron) NRE of £12 000 at £6.80 unit cost;
- Firm Z (3 micron) NRE of £5000 at £2.80 unit cost.

The numbers in brackets indicate the minimum feature size on the chip which is inversely proportional to the maximum operating frequency. Although the products are not fully comparable one can see that the costs of mask programmable gate arrays involves the user in large initial charges. Hence the importance of accurate CAD simulator tools prior to mask manufacture.

Because the gate array wafers *before* metallisation are customer independent, the costs up to this stage are divided amongst *all* customers. It is only the metallisation masks that are customer dependent and so these costs make up the bulk of the NRE. These NRE charges can be greatly reduced by sharing the prototyping costs even further by using a technique called a *multiproject wafer* (MPW). This is a metal mask which contains many different customer designs. The NREs are thus reduced approximately by a factor of $N$ where $N$ is the number of designers sharing that mask. Hence prototyping costs with mask programmable gate arrays are less of a financial risk when a manufacturer offers an MPW service. The typical prototyping costs for a 2000 gate design, with MPW, are now as low as £1000 for 10 devices.

Of all the mask programmable ASICs the gate array has the fastest fabrication route, since a reduced mask set is required depending upon the number of metal layers used for the interconnect. The typical time to manufacture such a device (referred as the turnaround time) is four weeks.

**Example 11.4**

How many masks are needed for a double layer metal, mask programmable gate array?

***Solution***

The answer is not two since it is necessary to insulate one metal layer from the next and provide vias (holes etched in the insulating layers deposited between the first and second layer metal) where connections are needed between layers. Hence the number is three, i.e. two metal masks and one via mask.

**Example 11.5**

A schematic for a control circuit consists of four 16 bit D-type based synchronous counters, 20 two-input NAND gates and 24 two-input NOR gates. Estimate the total number of gates required for this design.

***Solution***

Gate count for each part:

A 16 bit synchronous counter contains 16 D-type bistables plus combinational logic to generate the next state. This logic is typically comparable to the total gate count of the bistable part of the counter. Hence the total gate count for the counter will be approximately 160 gates (i.e. 16×5×2). A two-input NAND gate will require four transistors and hence one gate. Thus 20 will consume 20 gates of the array. Finally a single two-input NOR gate can be made from four transistors. Hence 24 will consume 24 gates of the array.

The total gate count required for this control circuit is 160 + 20 + 24= 204 gates.

**CAD tools for mask programmable gate arrays**

A mask programmable gate array cannot be modified once it has been fabricated without incurring a second NRE. Consequently a large reliance is placed upon the CAD tools, in particular the simulator, before releasing a design for fabrication. The generic CAD stages involved in the design of both mask and field programmable ASICs is illustrated in Fig. 11.13. For mask programmable gate arrays this design flow is discussed below:

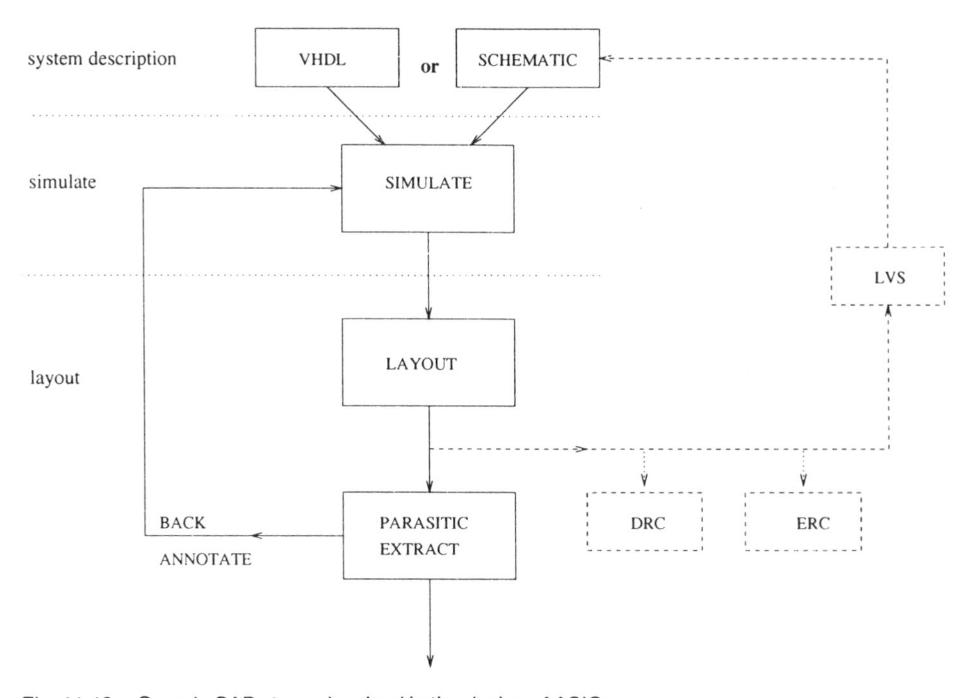


Fig. 11.13 Generic CAD stages involved in the design of ASICs

1. *System description* The most common way of entering the circuit description is via a drawing package, called *schematic capture*. The user has a library of components to call upon, varying in complexity from a two-input NAND gate through to counters/decoders, PISO/SIPOs and arithmetic logic units (as described in Problem 4.10). At no stage does the designer see the individual

transistors that make up the logic gates. For large circuits (greater than approximately 10 000 gates) the description of the circuit using schematic capture becomes rather tedious and error prone. Consequently high-level, textual, programming languages have been developed to describe the system in terms of its behaviour. The one language adopted as a standard is that recommended by the USA Department of Defense called VHDL. A brief introduction to VHDL is presented later in this chapter.

If the system is described in schematic form it is then converted into a *net-list*. This is a textual description of how the circuit is interconnected and is needed for the simulator. If the system is described in VHDL form then for the sake of brevity this can be considered as a net-list description already.

2. *Prelayout simulation* Once the system has been described the next stage is to simulate the system prior to layout. The components used in the schematic or VHDL are represented as digital (or behavioural) models. A digital simulator, called an *event-driven* simulator, is used to simulate the system by applying input vectors to the system, i.e. a stream of 1's and 0's. This simulator obtains its name since only the gates whose inputs are changing (i.e. an event is occurring) are updated. The outputs then drive other gates and hence a new event is scheduled some time later. In some cases, to simplify the simulation, all gates are assumed to have a 1 ns delay or a *unit delay* and wire delays are set at zero. This is because the chip has not been laid out and therefore no information is available yet about wire delays. This type of simulation is called in some CAD manuals *functional simulation*. It is, however, advisable to simulate with the gate propagation delays which include fan-out loading thus allowing the simulator to perform more realistic flip-flop timing checks such as: set-up and hold times; minimum clock and reset pulse widths, etc. This will identify, early in the design cycle, poor design techniques such as asynchronous events which violate set-up and hold time, or gated clocks which are revealed as spikes and glitches on clock lines.
3. *Layout* Next, the chip is *laid out* and this consists of a two-stage process of *place* and *route*. First the gates used to describe the system are placed on to the array and implemented using the versatile four transistor cell. Optimum placement algorithms are run which aim to reduce the total wire length. The cells are then connected together by using the available routing channels. The I/O positions may be left to the software to decide on the best position so as to assist the place and routing software, or may be specified by the user at the placement stage.
4. *Back annotation of routing delays* The metal used for the interconnect contains resistance and capacitance and will introduce delays. Hence these delays need to be added to the original system description, i.e. the schematic or VHDL file. This step is called *back annotation* and these extra delays are referred to as *wiring parasitics*.
5. *Postlayout simulation* The performance of the original prelayout system will now have changed, which in some cases may result in the delays increasing

from 1 ns to 100 ns. The system therefore needs to be resimulated with the parasitic delays included. This final simulation is called *postlayout simulation* and includes the timing delays of both the wiring and logic gates. The simulation is now called a full timing simulation since the true delays of the chip are included.[2] Any errors appearing in the simulation at this stage must be corrected by modifying the original schematic or VHDL file and rerunning the layout. This iterative process is characteristic of all ASIC CAD design tools.

An example of a layout induced timing error is demonstrated with a two-stage shift register in Fig. 11.14. The delay element indicated by the dotted box represents additional wire delay on the clock line. If this delay is greater than the propagation delay of the flip-flop then data is lost. This is because when a shift register shifts data it is assumed that all clocks arrive at the same time at each flip-flop. However, if a clock arrives at the first flip-flop before the second by at least one flip-flop delay then the data at the input to the second flip-flop will change before the arrival of its clock pulse. This data has been overwritten and therefore lost. To avoid this problem occurring the place and route software allows the designer to influence the layout in several ways. Firstly, the clock line can be given priority (called *seeding*) and it is routed first before all the other routes. It will therefore have the shortest and hence the fastest path. Another technique is to label groups such as shift registers so that they are not broken up during placement. All flip-flops are consequently placed close to each other and hence clock delays are reduced.

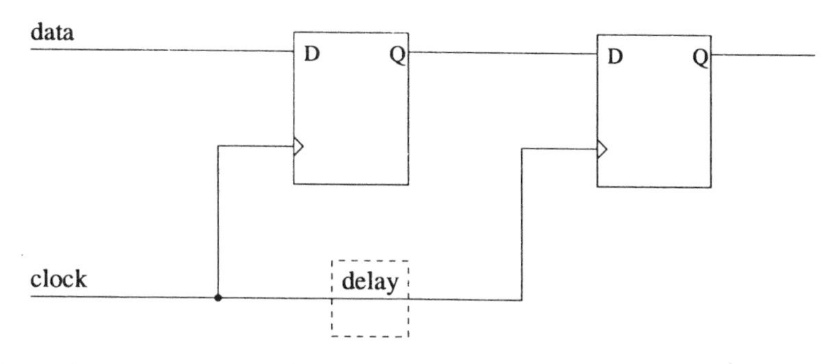


Fig. 11.14 Layout delays on clock lines can cause a shift register to malfunction

When the postlayout simulation has been successfully completed the designer has to pass an intensive *sign-off* procedure which needs to be countersigned by the project manager and an engineer at the ASIC manufacturer. The final file that is passed to the manufacturer is in a syntax which is applicable for mask manufac-

[2]In some environments a separate *static timing analyser* is available. This checks all timing delays around bistables with regard to clock and data and confirms that no set-up and hold time violations are present. This removes the time consuming process of writing a stimulus file for the timing simulator that covers all possible combinations of inputs around all bistables.

turing machines and allows the metal interconnection layer(s) to be added to the base wafers in order to customise the array.

The CAD tools described here are either supplied by the IC manufacturer or by generic CAD software houses such as *Mentor* and *Cadence*. These tools take a design from schematic through to layout. Alternative tools, such as *Viewlogic*, are used for just the prelayout stage. These so-called *front end* tools are popular PC based commodities and are used extensively in FPGA design.

### 11.3.3 Standard cell

The advantages of fast turnaround time and relatively low cost offered by gate arrays is counterbalanced by several problems. The first is that silicon is wasted because a design does not use all the available gates on the array. Also, it is not known by the manufacturer which pad on the array is to be an input or an output and so silicon is further wasted by the inclusion of both input and output circuits at every pad. As the chip price is proportional to die size then this can be uneconomical when large volumes are required. In addition, because all the transistors in a gate array are the same size then when transistors are placed in series long delays occur. This happens on the PMOS chain for NOR and the NMOS chain for NAND. Consequently the gates cannot be optimally designed and the delays $\tau_{plh}$ and $\tau_{phl}$ are asymmetrical. If the $W/L$'s of the transistors were individually adjusted for each gate type the delays would be shorter.

The standard cell approach gets around these problems. Here, the designer again has available a library of logic gates but the design starts with a clean piece of silicon. Hence only those gates selected for a design appear on the final chip and no silicon is wasted. It is also known which pads are to be input and output thus further saving silicon. The standard cell chip is therefore smaller than the gate array. This device is also faster partly because it is smaller and the routing is shorter (hence smaller wire delays) and partly because the library of logic gates is optimally designed by the manufacturer. This is achieved by adjusting the $W/L$'s of the transistors in each gate so as to achieve optimum delay.

Since the standard cell only uses those gates that are needed for a design then each chip is of different size and is unique. Hence all masks are required, which can be of the order of 8–16 masks where each mask costs £1000–£2000! The NRE costs are therefore considerably higher and the production times longer compared to a mask programmable gate array. This approach is therefore only economical when relatively large volumes are involved. However, reduced prototyping costs are again available by using multiproject wafers.

Libraries for standard cell (and gate arrays) have become quite sophisticated. Not only are the basic and complex gates provided but also counters and UARTs (serial interface) exist. Incredibly some manufacturers are even offering complete processor cores such as the Z180 by VLSI Technology, TMS320C50 by TI and the 80486 by SGS Thomson.

## Example 11.6

Compare the transistor count of a complex combinational gate that is offered in the manufacturer's library that produces the function $f=\overline{AB+CD}$ implemented in a mask programmable gate array with a standard cell approach.

### *Solution*

The gate array approach would require De Morgan's theorem to implement this function using the blocks of four transistors (i.e. using either two-input NAND or two-input NOR gates). Choosing NAND gates results in:

$$f=\overline{AB+CD}=\overline{AB}\cdot\overline{CD}$$

The function using NAND gates is shown in Fig. 11.15(a). Note that it is not possible to directly produce an AND gate with CMOS. This must be produced by using a NAND with an inverter. Thus the total number of gates required is 3.5 or 14 transistors.

Consider now the standard cell. To implement the above function the library designer uses the technique presented in Chapter 9:

1. Concentrate on the NMOS network first: those terms that are AND'd are placed in series whilst those OR'd are placed in parallel.
2. The PMOS network is just a reverse of the NMOS network.

The final circuit diagram is shown in Fig. 11.15(b). Notice that the number of transistors used is now only eight, a great saving on silicon. In addition the gate

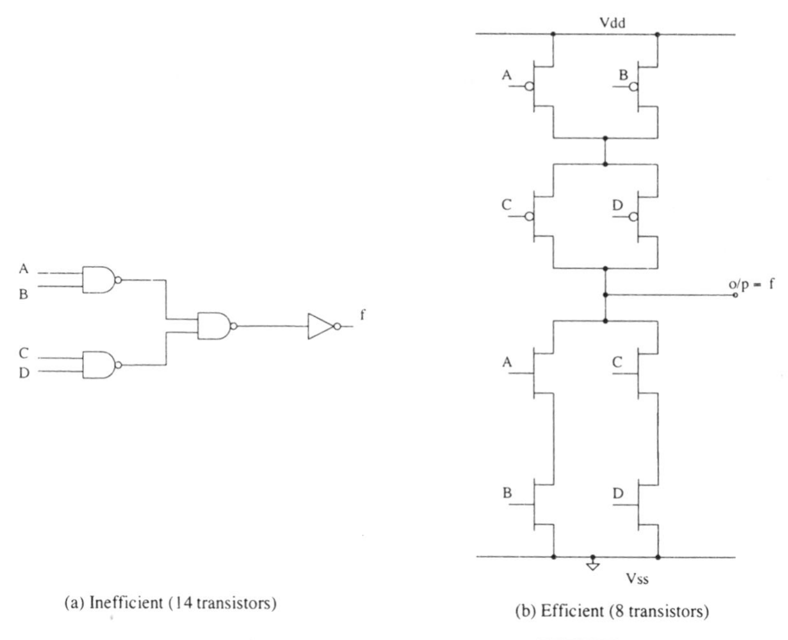


(a) Inefficient (14 transistors)

(b) Efficient (8 transistors)

Fig. 11.15 Inefficient and efficient implementation of the function $\overline{AB+CD}$

array approach uses a three-level logic whilst the standard cell uses only a single-level, giving the gate a much smaller propagation delay.

**CAD tools for standard cell**

The CAD tools for a standard cell follow those for mask programmable gate arrays with a slight exception at the layout stage. Here the designer can interconnect each cell without the restriction of a fixed number of routing channels. This results in a chip that is much easier to route but may cause errors in the layout due to incorrect connectivity caused by designer intervention. To avoid this problem the designer has available *layout verification tools* which perform various checks on the layout. These are shown dotted in Fig. 11.13 and consist of: design rule check (DRC), where the spacing of the metal interconnect is checked; electrical rule check (ERC), where the electrical correctness of the circuit is confirmed, i.e. outputs not shorted to supply, no outputs tied together etc.; and finally layout versus schematic (LVS), where a net-list is extracted from the layout and is compared with the original schematic. Since the NRE costs are high (especially for non-MPW processes) these verification tools are an essential component in standard cell design. Both Mentor and Cadence offer such tools and so are suitable for standard cell design.

### 11.3.4 Full custom

This is the traditional method of designing integrated circuits. With a standard cell and gate array the lowest level that the design is performed at is the logic gate level, i.e. NAND, NOR, D-Type, etc. No individual transistors are seen. However, full custom design involves working down at this transistor level where each transistor is handcrafted depending upon what it is driving. Thus a much longer development time occurs and consequently the development costs are larger. The production costs are also large since all masks are required and each design presents new production problems.

Full custom integrated circuits are not so common nowadays unless it is for an analogue or a high-speed digital design. A mixed approach tends to be used which combines full custom and standard cells. In this way a designer can use previously designed cells and for those parts of the circuit that require a higher performance then a full custom part can be made.

**CAD tools for full custom**

The CAD tools follow the general form described for a standard cell. However, since the design of full custom parts involves more manual human involvement then the chances of error are increased. The designer thus relies very heavily on simulation and verification tools. In addition since cells are designed from individually handcrafted transistors then they must be simulated with an analogue circuit simulator such as SPICE before being released as a digital part. Needless to say, the choice of a design route that incorporates full custom design is one that should not be taken lightly.

## 11.4 FIELD PROGRAMMABLE LOGIC

So far we have seen two extremes in the design options available to a digital designer – namely standard products and mask programmable ASICs. Although mask programmable ASICs offer extremely high performance they carry a large risk in terms of time and expenditure. To provide the designer with the flexibility of both, the industry has gradually developed a class of logic that can be programmed with a personal computer in the laboratory. These devices are called *field programmable logic* and can be either one-time programmable (utilising small fuses) or many times programmable (using either ultraviolet erasable connections or an SRAM/MUX). Because these devices contain the extra circuitry to control interconnect and functionality this overhead results in a family which is less complex and slower than the mask programmable ASICs. However, the attraction of a much lower risk can outweigh the performance problems especially for prototyping purposes.

These field programmable logic devices are divided into two groups:

- AND-OR programmable architectures;
- field programmable gate arrays or FPGAs.

### 11.4.1 AND-OR programmable architectures

The AND-OR programmable architecture devices were the first programmable logic chips available on the market and still exist today. The reason for the interest in such structures is because *all* combinational logic circuits can be expressed in this AND-OR form.

Three types of programmable AND-OR arrays are available:

- fixed AND – programmable OR (ROM);
- programmable AND – fixed OR (PAL);
- programmable AND – programmable OR (PLA).

A block schematic of an AND-OR array is shown in Fig. 11.16. Inputs are passed to the AND array whose outputs are fed into the OR array which provide

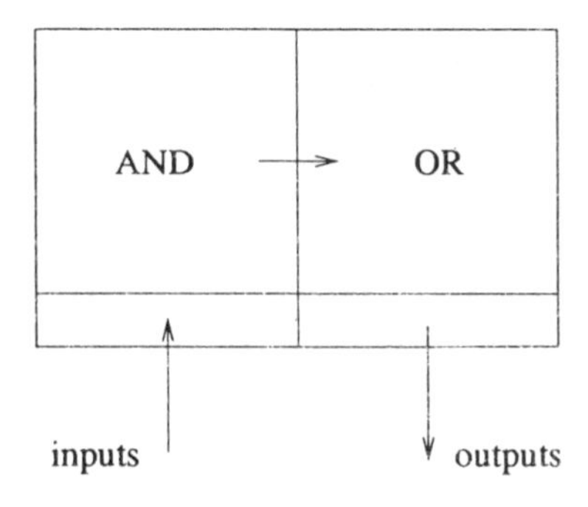


Fig. 11.16 Schematic for an AND-OR array

the outputs of the chip. Each of these AND-OR array types will now be discussed in more detail.

### 11.4.2 ROM: fixed AND-programmable OR

As was seen in Chapter 10 a ROM is a read only memory device. It consists of a decoder with $n$ inputs (or addresses) whose $2^n$ outputs drive a memory array. As seen in Fig. 11.2 a decoder can be implemented with AND gates and hence this is called the AND array. Since all possible input and output combinations exist then this is classed as a *fixed* array, i.e. an $n$ input decoder requires $2^n$ $n$-input AND gates to generate *all* product terms. As we have also seen (see Fig. 10.3), the memory array is in fact a NOR array. However, the inclusion of an inverter on each column line will turn this into an OR array. Hence if the decoder has $2^n$ outputs then the OR array must contain $m$ OR gates with each gate having *up to* $2^n$ inputs, where in this case $m$ is the number of bits in a word. Notice that we have said *'up to'* $2^n$ inputs. This is because the OR array contains the data which is programmable. The ROM architecture is thus a *fixed AND-programmable OR array*.

The complete circuit for a 4×3 bit ROM is shown in Fig. 11.17(a). Note that it consists of a fixed AND structure (i.e. a 2-to-4 decoder) and a programmable OR array (i.e. a 4-to-3 encoder).The three-bit words stored in the four addresses are programmed by simply connecting each decoder output to the appropriate input of an OR gate when a logic '1' is to be stored.

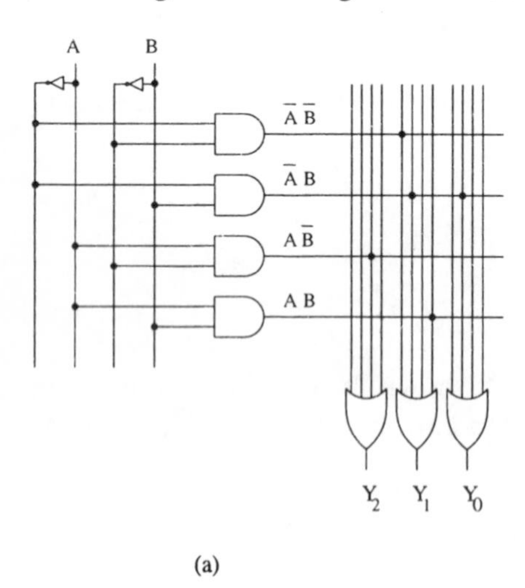


(a)

| A | B | Y2 | Y1 | Y0 |
|---|---|---|---|---|
| 0 | 0 | 0 | 1 | 0 |
| 0 | 1 | 0 | 1 | 1 |
| 1 | 0 | 1 | 0 | 0 |
| 1 | 1 | 0 | 1 | 0 |

(b)

Fig. 11.17 A 4×3 bit ROM shown storing the data in the truth table

This circuit shows the ROM storing the data in the truth table of Fig. 11.17(b). The Boolean equations, in fundamental sum of products form, are:

$$Y_2 = A\bar{B}$$
$$Y_1 = \bar{A}\bar{B} + \bar{A}B + AB$$
$$Y_0 = \bar{A}B$$

Note that rather than thinking of the ROM storing four three-bit words, an alternative view is that it is implementing a two-input, three-output truth table.

The same circuit is shown again in Fig. 11.18 but this time the $2^n$ inputs to each OR gate are shown, for simplicity, as a single input. A cross indicates a connection from the address line to the gate. The same data as in Fig. 11.17 are shown stored.

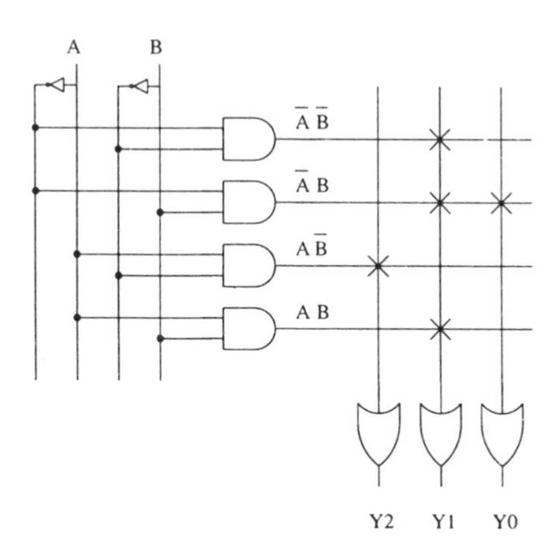


Fig. 11.18 A 4×3 bit ROM using an abbreviated notation for the OR array

As seen in Chapter 10 the physical implementation of the programmable OR array is achieved via the presence or absence of a transistor connection. This is achieved either by omitting the source or drain connections of MOS transistors or blowing fuses which are connected to the transistor terminals. Apart from using ROMs to store data or programs they can also be used to perform many digital operations, some of which are described below.

**Universal combinational logic function**

As we have seen a ROM has all fundamental product terms available for summing and can implement an $m$-output, $n$-input truth table. This is simply achieved by connecting the address lines to the $n$ input variables, and each output line programmed to give the appropriate output values. The advantages of such a ROM based design are: it is particularly applicable if $n$ is large; no minimisation is needed; it is cheap if mass produced; and only one IC is needed.

**Example 11.7**

How would the truth table shown in Fig. 11.19(a) be implemented using a ROM?

***Solution***

A ROM of at least size 16×3 would be needed. The four address lines would be connected to the input variables $A$, $B$, $C$ and $D$ with the three outputs providing $X$, $Y$ and $Z$. The required outputs (three-bit word) for each of the 16 possible

| A | B | C | D | X | Y | Z |
|---|---|---|---|---|---|---|
| 0 | 0 | 0 | 0 | 0 | 0 | 0 |
| 0 | 0 | 0 | 1 | 1 | 0 | 1 |
| 0 | 0 | 1 | 0 | 0 | 0 | 0 |
| 0 | 0 | 1 | 1 | 0 | 1 | 1 |
| 0 | 1 | 0 | 0 | 0 | 1 | 0 |
| 0 | 1 | 0 | 1 | 1 | 0 | 0 |
| 0 | 1 | 1 | 0 | 0 | 0 | 0 |
| 0 | 1 | 1 | 1 | 1 | 0 | 1 |
| 1 | 0 | 0 | 0 | 0 | 1 | 1 |
| 1 | 0 | 0 | 1 | 0 | 0 | 0 |
| 1 | 0 | 1 | 0 | 1 | 1 | 1 |
| 1 | 0 | 1 | 1 | 0 | 0 | 0 |
| 1 | 1 | 0 | 0 | 0 | 1 | 0 |
| 1 | 1 | 0 | 1 | 0 | 0 | 0 |
| 1 | 1 | 1 | 0 | 1 | 1 | 1 |
| 1 | 1 | 1 | 1 | 0 | 0 | 0 |

(a)

| | A0 | A1 | A2 | A3 | O0 | O1 | O2 |
|---|---|---|---|---|---|---|---|
| WORD 0 | 0 | 0 | 0 | 0 | 0 | 0 | 0 |
| WORD 1 | 0 | 0 | 0 | 1 | 1 | 0 | 1 |
| WORD 2 | 0 | 0 | 1 | 0 | 0 | 0 | 0 |
| WORD 3 | 0 | 0 | 1 | 1 | 0 | 1 | 1 |

(b)

Fig. 11.19 Truth table used in Example 11.7 for implementation in ROM

input combinations would be programmed into the ROM, straight from the truth table. This is shown in Fig. 11.19(b) for the first four addresses, where $A_n$ and $O_n$ are the *n*th address line and output respectively of the ROM.

Note that because all the fundamental product terms are produced by the fixed AND array of the ROM then no minimisation can take place.

**Code converter and look-up table**

A ROM can be used to convert an *n*-bit binary code (presented to the address lines) into an *m*-bit code (which appears at the outputs).The desired *m*-bit code is simply stored at the appropriate address location. Considered in this way it is a general *n*-to-*m* encoder or *code converter.*

Another ROM application similar to the code converter is the *look-up table.* Here, a ROM could be used to look up the values of, for example, a trigonometric function (e.g. sin *x*), by storing the values of the function in ROM. By addressing the appropriate location with a digitised version of *x* the value for the function stored would be output.

**Sequence generator and waveform generator**

A ROM can be used as a *sequence generator* in that if the data from an $n \times m$ ROM are output, address by address, then this will generate *n* binary data sequences. Also, if the ROM output is passed to an *m* bit digital-to-analogue converter (DAC) then an analogue representation of the stored function will be produced. Hence a ROM with a DAC can be used as a *waveform generator.*

### 11.4.3 PAL: programmable AND–fixed OR

ROM provides a fixed AND–programmable OR array in which all fundamental product terms are available, thus providing a universal combinational logic solution. However, ROM is only available in limited sizes and with a restricted number of inputs. Adding an extra input means doubling the size of the ROM. Clearly a means of retaining the flexibility of the AND-OR structure whilst also overcoming this problem would produce a useful structure.

Virtually all combinational logic functions can be minimised to some degree, therefore allowing non-fundamental product terms to be used. Therefore, a programmable AND array would allow only the necessary product terms, after minimisation, to be produced. Followed by a fixed OR array this would allow a fixed number of product terms to be summed and so a minimised sum of products expression implemented. This type of structure is called a *programmable array logic* or *PAL*.

The structure of a hypothetical PAL is shown in Fig. 11.20. This circuit has two input variables and three outputs, each of which can be composed of two product terms. The product terms are programmable via the AND array. For the connections shown the outputs are:

$$Y_2 = \bar{A}\bar{B} + AB$$
$$Y_1 = A + B$$
$$Y_0 = AB$$

(Note that $Y_0$ only has one product term so only one of the two available AND gates is used.)

Commercially available PAL part numbers are coded according to the number of inputs and outputs. For example the hypothetical PAL shown in Fig. 11.20

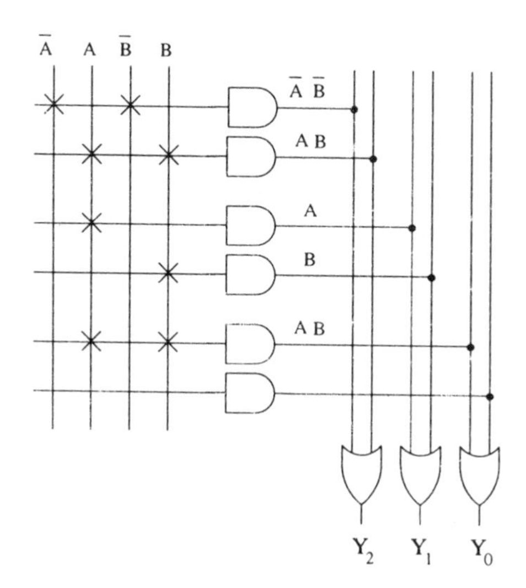


Fig. 11.20 A programmable AND-fixed OR logic structure (i.e. PAL) with two inputs, six programmable product terms and three outputs (each summing two of the six product terms)

would be coded PAL2H3, i.e. it is a PAL having two inputs and three outputs. The *H* indicates that the outputs are active high. One of the smallest PALs on the market is a PAL16L8 offered by Texas Instruments, AMD and several other manufacturers. This has 16 input terms and eight outputs. The *L* indicates that the outputs are active low. This device actually shares some of its inputs with its ouputs, i.e. it has feedback. Hence if all eight outputs are required then only eight inputs are available. The other piece of information that is required about a PAL is how many product terms each OR gate can support. This is supplied on the data sheet, and for the PAL16L8, for example, it is seven.

By adding flip-flops at the output, the designer is able to use PALs as sequential elements. The nomenclature for the device would now be PAL16R8 for example where *R* stands for registered output. The early PALs were fuse programmable. However, companies such as Altera, Intel and Texas Instruments added EPROM technology to these registered output PALs so that the devices could be programmed many times. These devices are called *erasable programmable logic devices* or EPLDs.

Very large PALs exist having gate equivalents of over 2000 gates quoted (remember a *gate* is defined as a two-input NAND gate). The inflexibility of only having the flip-flops at the outputs and not buried within the array (as in mask programmable ASICs) resulted in the GAL. The GAL (generic array logic) is an ultraviolet-erasable PAL with a programmable cell at each output, called an output logic macro cell (OLMC). Each OLMC contains a register and multiplexers to allow connections to and from adjacent OLMCs and from the AND/OR array. The GAL (trademark of Lattice Semiconductors) has a similar nomenclature to PALs. For example the GAL16V8 has 16 inputs and eight outputs using a versatile cell (i.e. *V* in the device name). However, because it uses OLMCs then it can emulate many different PAL devices in one package, having a range of inputs (up to 16) and outputs (up to eight).

### Example 11.8

How could the truth table in Fig. 11.19(a) be implemented using a (hypothetical) PAL with four inputs, three outputs and a total of 12 programmable product terms (i.e. four to each output)?

#### *Solution*

First, we use Karnaugh maps (Fig. 11.21) to minimise the functions $X$, $Y$ and $Z$. From these Karnaugh maps:

$$Z=\bar{A}\bar{B}D+\bar{A}CD+AC\bar{D}+A\bar{B}\bar{D}$$

$$Y=\bar{A}\bar{B}CD+B\bar{C}\bar{D}+A\bar{D}$$

$$X=\bar{A}\bar{C}D+\bar{A}BD+AC\bar{D}$$

The PAL, a PAL4H3, would therefore be programmed as shown in Fig. 11.22.

| Z | $\bar{A}\bar{B}$ | $\bar{A}B$ | $AB$ | $A\bar{B}$ |
|---|---|---|---|---|
| $\bar{C}\bar{D}$ | 0 | 0 | 0 | 1 |
| $\bar{C}D$ | 1 | 0 | 0 | 0 |
| $CD$ | 1 | 1 | 0 | 0 |
| $C\bar{D}$ | 0 | 0 | 1 | 1 |

| Y | $\bar{A}\bar{B}$ | $\bar{A}B$ | $AB$ | $A\bar{B}$ |
|---|---|---|---|---|
| $\bar{C}\bar{D}$ | 0 | 1 | 1 | 1 |
| $\bar{C}D$ | 0 | 0 | 0 | 0 |
| $CD$ | 1 | 0 | 0 | 0 |
| $C\bar{D}$ | 0 | 0 | 1 | 1 |

| X | $\bar{A}\bar{B}$ | $\bar{A}B$ | $AB$ | $A\bar{B}$ |
|---|---|---|---|---|
| $\bar{C}\bar{D}$ | 0 | 0 | 0 | 0 |
| $\bar{C}D$ | 1 | 1 | 0 | 0 |
| $CD$ | 0 | 1 | 0 | 0 |
| $C\bar{D}$ | 0 | 0 | 1 | 1 |

Fig. 11.21 Karnaugh maps for Example 11.8

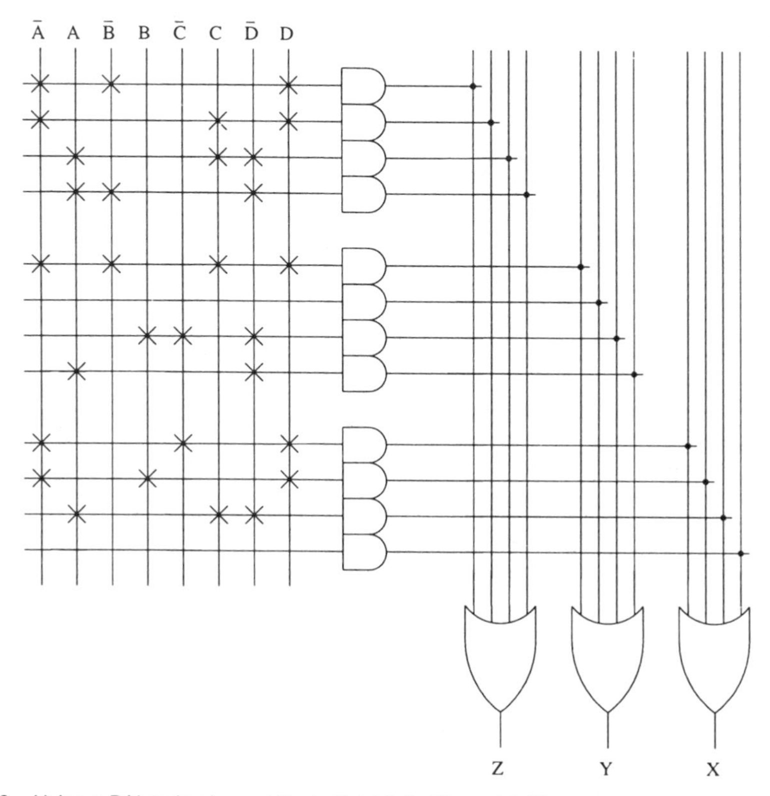


Fig. 11.22 Using a PAL to implement the truth table in Figure 11.19

## 11.4.4 PLA: programmable AND–programmable OR

The final variant of the AND-OR architectures is the programmable AND–programmable OR array or *programmable logic array* (PLA). With this the desired product terms can be programmed using the AND array and then as many of these terms summed together as required, via a programmable OR array, to give the desired function.

The structure of such an array with two inputs, three outputs and six programmable product terms available is shown in Fig. 11.23. For the connections shown the outputs are:

$$Y_0 = \bar{A}\bar{B} + AB + B$$
$$Y_1 = \bar{A}\bar{B} + \bar{A}B + A\bar{B}$$
$$Y_2 = \bar{A}\bar{B} + \bar{A} + B + A\bar{B}$$

Note that any product term can be formed by the AND gates, and that any number of these product terms can be summed by the OR gates.

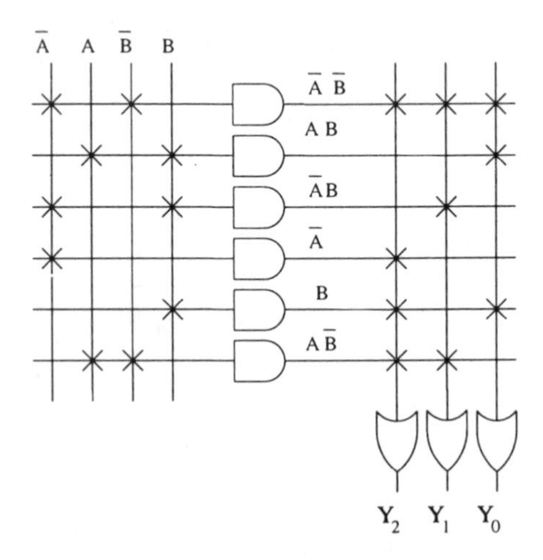


Fig. 11.23 A programmable AND–programmable OR logic array (i.e. PLA) with two inputs, six programmable product terms and three programmable outputs

## Example 11.9

How would the truth table shown in Fig. 11.19(a) be implemented using a (hypothetical) four-input, three-output PLA with eight product terms?

### *Solution*

From the minimisation performed to implement this truth table using the PAL in Example 11.8 it can be seen that the three Boolean expressions for $X$, $Y$ and $Z$ contain a total of nine different product terms ($AC\bar{D}$ is common to both $X$ and $Z$). This PLA can only produce eight which means that product terms common to the three expressions must be found, effectively de-minimising them to some degree.

This can be achieved by reconsidering the Karnaugh maps and *not* fully minimising them, but rather looking for common implicants in the three expressions:

$$Z = \bar{A}\bar{B}D + \bar{A}BCD + AC\bar{D} + A\bar{B}\bar{C}\bar{D}$$
$$Y = B\bar{C}\bar{D} + \bar{A}\bar{B}CD + AC\bar{D} + A\bar{B}\bar{C}\bar{D}$$
$$X = \bar{A}\bar{C}D + \bar{A}BCD + AC\bar{D}$$

In this form only seven different product terms are required to implement all three functions and so the given PLA can be used as shown in Fig. 11.24.

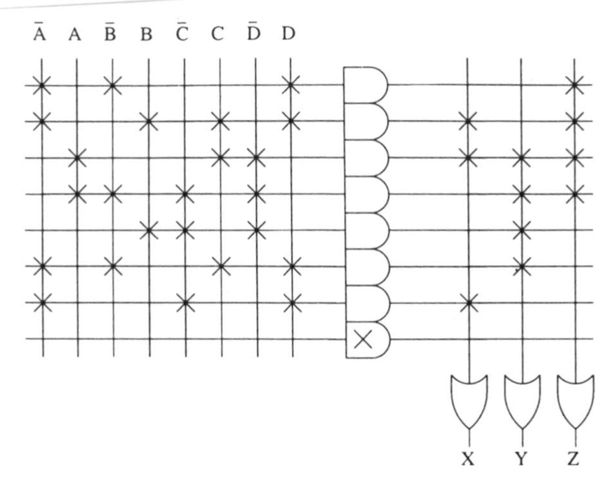


Fig. 11.24 A programmable AND–programmable OR logic array (PLA) with four inputs, eight programmable product terms and three programmable outputs

## 11.4.5 Field programmable gate arrays

The advancement in on-chip field programmable techniques combined with ever increasing packing densities has led to the introduction of *field programmable gate arrays* or FPGAs. These devices can be considered as being the same as mask programmable gate arrays except the functionality and interconnect is programmed in the laboratory at a greatly reduced financial risk. The popularity of FPGAs is indicated by the large number of companies who currently manufacture such devices. These include Actel, Altera, AMD, Atmel, Crosspoint, Lattice, Plessey, Quicklogic, Texas Instruments, and Xilinx, to name but a few. Of these, the three that are perhaps the best known are Altera, Xilinx and Actel. In order to introduce FPGAs, some of the devices provided by these three companies will therefore be discussed. Essentially they differ in terms of: *granularity*; *programming technique*; *volatility*; and *reprogrammability*. All FPGAs consist of a *versatile cell* that is repeated across the chip with its size and hence cell complexity referred to as the *granularity*. These cells are multifunctional such that they can produce many different logic gates from a single cell. The larger the cell the greater the complexity of gate each cell can produce. Those arrays that use a small simple cell, duplicated many times, are referred to as having fine granularity, whilst arrays with few, but large, complex cells are defined as coarse grain. These versatile cells have been given different names by the manufacturers, for example: modules; macrocells; and combinatorial logic blocks. The programming of the function of each cell and how each cell is interconnected is achieved via either: small fuses; on-board RAM elements that control multiplexers; or erasable programmable read only memory (EPROM) type transistors. Consequently some devices are *volatile* and lose their functionality when the power is removed whilst others retain their functionality even with no supply connected. Finally these devices can be divided into those that can be *reprogrammed* many times and those that are *one-time programmable*.

Let us now look more closely at the FPGA types, which will be divided into: EPROM type; SRAM/MUX type; and fuse type.

**EPROM type FPGAs**

The most common EPROM type FPGA device is that supplied by Altera. The range of devices available from Altera are the MAX 5000, 7000 and 9000 series (part numbers: EPM5XXX, EPM7XXX and EPM9XXX). These devices are the furthest from the true FPGAs and can be considered really as large PAL structures. They offer coarse granularity and are more an extension to Altera's own range of electrically programmable, ultraviolet-erasable logic devices (EPLD). The versatile cell of these devices is called a 'macrocell'. This cell is basically a PAL with a registered output. Between 16 and 256 macrocells are grouped together into an array inside another block called a logic array block (LAB) of which an FPGA can contain between 1 and 16. In addition to the macrocell array each LAB contains an I/O block and an expander which allows a larger number of product terms to be summed. Routing between the LABs is achieved via a programmable interconnect array (PIA) which has a fixed delay (3 ns worst case) that reduces the routing dependence of a design's timing characteristics.

Since these devices are derived from EPLD technology the programming is achieved in a similar manner to an EPROM via an Altera logic programmer card in a PC connected to a master programming unit. The MAX 7000 is similar to the 5000 series except that the logic block has two more input variables. The MAX 9000 is similar to the 7000 device except that it has two levels of PIA. One is a PIA local to each LAB whilst the other PIA connects all LABs together. Both the 7000 and 9000 series are $E^2$PROM devices and hence do not need an ultraviolet source to be erased.

It should be noted though that these devices are not true FPGAs and have a limited number of flip-flops available (one per macrocell). Hence the Altera Max 5000/7000/9000 series is more suited to combinatorially intensive circuits. For more register intensive designs Altera offer the Flex 8000 and 10K series of FPGAs which uses an SRAM memory cell based programming technique (as used by Xilinx – see next section); although currently rather expensive it will in time become an attractive economical option. The Flex 8000 series (part number: EPF8XXX) has gate counts from 2000 to 16 000 gates. The 10K series (part number: EPF10XXX) however, has gate counts from 10 000 to 100 000 gates!

**SRAM/MUX type FPGAs**

The most common FPGA that uses the SRAM/MUX programming environment is that supplied by Xilinx. The range of devices provided by Xilinx consists of the XC2000, XC3000 and XC4000. The versatile cell of these devices is the 'configurable logic block' (CLB) with each FPGA consisting of an array of these surrounded by a periphery of I/O blocks. Each CLB contains combinational logic, registers and multiplexers and so, like the Altera devices, has a relatively coarse granularity. The Xilinx devices are programmed via the contents of an on- board static RAM array which gives these FPGAs the capability of being reprogrammed (even whilst in operation). However, the volatility of SRAM memory cells requires the circuit configuration to be held in an EPROM alongside the Xilinx FPGA.

A recent addition to the Xilinx family is the XC6000 range. This family has the same reprogrammability nature as the other devices except it is possible to partially reconfigure these devices. This opens up the potential for fast in-circuit reprogramming of small parts of the device for learning applications such as neural networks.

**Fuse type FPGAs**

The most common fuse type FPGA is that supplied by Actel. These devices are divided into the Act1, Act2 and Act3 families. The Act1 FPGAs (part numbers: A10XX) contain two programmable cells: 'Actmod' and 'IOmod'. The versatile core cell is the 'Actmod' which is simply based around a 4-to-1 multiplexer for Act1. This versatile cell is shown in Fig. 11.25. Since this cell is relatively small the array is classed as *fine grain*. By tying the inputs to either a logic '0' or logic '1' this versatile cell can perform 722 different digital functions. The programmable 'IOmod' cell is used to connect the logic created from the 'Actmods' to the outside world. This cell can be configured as various types of inputs and/or outputs (bi-directional, tristate, CMOS, TTL, etc.). Unlike the Xilinx and Altera devices the Actel range are programmed using fuse technology with *desired* connections simply blown (strictly called an *antifuse*). These devices are therefore 'one time programmable' (OTP) and cannot be reprogrammed. The arrays have an architecture similar to a channelled gate array with the repeating cell (Actmod) arranged in rows with routing between each row.The routing contains horizontal and vertical metal wires with antifuses at the intersection points.

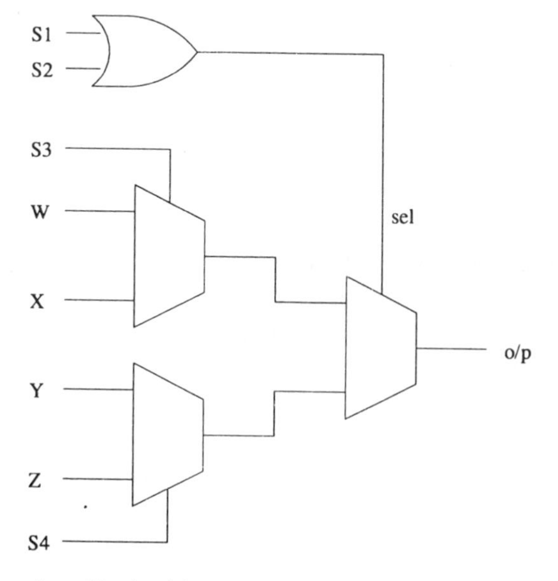


Fig. 11.25 Versatile cell used for Actel Act1 range of fused FPGAs

Other devices in the Actel range are the Act2 (part numbers:A12XX) and the Act3 (part numbers:A14XX) devices. These use two repeating cells in the array. The first is a more complex 'Actmod' cell called *Cmod* used for combinational purposes whilst the other cell is a *Cmod* with a flip-flop.

Table 11.1 shows a comparison of some of the FPGA devices offered by Altera, Xilinx and Actel.

Table 11.1 Comparison of some FPGA types

| Manufacturer | Part | Gates (k) | D-types | Cost (£) | Programming technique | Speed (MHz) | I/Os |
|---|---|---|---|---|---|---|---|
| Altera | EPM5X | 8 | 21–252 | 14–55 | EPROM | 60 | 100 |
| | EPM7X | 10 | 40–400 | 12–100 | EPROM | 70 | 288 |
| | EPM9X | 20 | 400–700 | 50–140 | EPROM | 65 | 100 |
| | EPF8X | 2.5–16 | 78–208 | 15–90 | SRAM | 100 | 208 |
| | EPF10KX | 10–100 | 148–420 | 25–550 | SRAM | 120 | 300 |
| Xilinx | XC2X | 0.6–1.5 | 64–100 | 10–15 | SRAM | 60 | 74 |
| | XC3X | 1–7.5 | 256–1320 | 10–60 | SRAM | 70 | 176 |
| | XC4X | 2–25 | 256–2560 | 15–190 | SRAM | 60 | 256 |
| Actel | A10X(Act1) | 1.2–2 | 147–273 | 12–20 | Antifuse | 37 | 69 |
| | A12X(Act2) | 2–8 | 565–998 | 17–55 | Antifuse | 41 | 140 |
| | A14X(Act3) | 1.5–10 | 264–1153 | 15–250 | Antifuse | 125 | 228 |

**Example 11.10**

Consider the versatile Actel cell shown in Fig. 11.25. What functions are produced if the following signals are applied to its inputs:

(a) $S_1S_2S_3S_4$=000A and $WXYZ$=0001
(b) $S_1S_2S_3S_4$=0BAA and $WXYZ$=1110
(c) $S_1S_2S_3S_4$=0CA1 and $WXYZ$=0B11.

Assume that the signals $A$ and $B$ are inputs having values of '0' or '1' and that for each multiplexer when the select line is low the lower input is selected.

***Solution***

(a) In this case the OR gate output is always zero and so the lower line is selected. This is derived from the lower multiplexer whose select line is controlled by the only input 'A'. With the inputs to this multiplexer at '0' for the upper line and '1' for the lower line the output thus follows that of an inverter.
(b) Here it is helpful to construct a truth table and include in this table the output of the OR gate, called *sel*.

| *A* | *B* | *Sel* | *o/p* |
|---|---|---|---|
| 0 | 0 | 0 | 0 |
| 0 | 1 | 1 | 1 |
| 1 | 0 | 0 | 1 |
| 1 | 1 | 1 | 1 |

We can thus see that the function is a two-input OR gate.

(c) Again a truth table (including *sel*) is useful to work out the function implemented:

| $A$ | $B$ | $C$ | $Sel$ | $o/p$ |
|---|---|---|---|---|
| 0 | 0 | 0 | 0 | 1 |
| 0 | 0 | 1 | 1 | 0 |
| 0 | 1 | 0 | 0 | 1 |
| 0 | 1 | 1 | 1 | 1 |
| 1 | 0 | 0 | 0 | 1 |
| 1 | 0 | 1 | 1 | 0 |
| 1 | 1 | 0 | 0 | 1 |
| 1 | 1 | 1 | 1 | 0 |

A Karnaugh map for the output is shown below which generates the function

$$o/p = \bar{C} + \bar{A}B$$

|  | $\bar{A}\bar{B}$ | $\bar{A}B$ | $AB$ | $A\bar{B}$ |
|---|---|---|---|---|
| $\bar{C}$ | 1 | 1 | 1 | 1 |
| C | 0 | 1 | 0 | 0 |

### 11.4.6 CAD tools for field programmable logic

The programming of field programmable logic devices is implemented directly via a computer. The software needed for PALs and PLAs is usually a simple matter of producing a programming file called a fuse or an EPROM bit file. This file has a standard format (called JEDEC) and contains a list of 1's and 0's. This file is automatically generated from either Boolean equations, truth tables or state diagrams using programs such as ABEL (DataIO Corp.), PALASM (AMD Inc.) and CUPL (Logical Devices Inc.). In other words the minimisation is done for you and it is not necessary to draw out any Karnaugh maps. Software programs that can directly convert a schematic representation into a JEDEC file are also available. Since these devices have only an MSI complexity level then the software tools are relatively simple to use and also inexpensive.

The FPGAs, on the other hand, have capacities of LSI and VLSI level and are much more complex. Since FPGAs are similar in nature to mask programmable gate arrays the associated CAD tools have been derived from mask programmable ASICs and follow that of Fig.11.13; that is: schematic capture (or VHDL), prelayout simulation, layout, back annotation and postlayout simulation.

It should be noted that FPGA simulation philosophy is somewhat different from mask programmable gate arrays. With mask programmable devices, 100% simulation is absolutely essential since these circuits cannot be rectified after fabrication without incurring large financial and time penalties. These penalties are virtually eliminated with FPGA technology due to the fast programming time in the laboratory and the low cost of devices. For one-time programmable devices (such as Actel) the penalty is the price of one chip whilst for erasable devices (such as Xilinx) the devices can simply be reprogrammed. Hence the pressure to simulate 100% is not as great.

For those devices that are reprogrammable this results in an inexpensive iterative procedure whereby a device is programmed and then tested in the final system. If the device fails it can be reprogrammed with the fault corrected. For OTP type FPGAs then a new device will have to be blown at each iteration; although it will incur a small charge the cost is considerably less than mask programmable arrays. It is not uncommon for FPGA designs (both reprogrammable and OTP) to experience four iterations before a working device is obtained. This is totally unthinkable for mask programmable designs where a 'right first time approach' has to be employed – hence the reliance on the simulator.

Since fuses, SRAM/MUX cells, etc., are used to control the connectivity the delays caused by these elements must be added to the wire delays for postlayout simulation. Hence it is for this reason that FPGAs operate at a lower frequency than mask programmable gate arrays. The large delays in the routing path also mean that timing characteristics are routing dependent. Hence, changing the placement positions of core cells (by altering the pin out for example) will result in a different timing performance. If the design is synchronous then this should not be a problem with the exception of the shift register problem referred to in Figure. 11.14. It should also be noted that the prelayout simulation of FPGAs on some occasions is only a unit delay (i.e. 1 ns for all gates) or functional simulation. It does not take into account fan-out, individual gate delays, set-up and hold time, minimum clock pulse widths (i.e. spike and glitch detector), etc., and does not make any estimate of the wire delay. Hence the simulation at this stage is not reflective of how the final design will perform. To obtain the true delays the FPGA must be laid out and the delays *back annotated* for a postlayout simulation. This will provide an accurate simulation and hence reveal any design errors. Unfortunately, if a mistake is found then the designer must return all the way back to the original schematic. The design must again be prelayout simulated, laid out and delays back annotated before the postlayout simulation can be repeated. This tedious iterative procedure is another reason why FPGAs are usually programmed prematurely with a limited simulation. It should be mentioned that an FPGA is sometimes used as a prototyping route prior to migrating to a mask programmable ASIC. Hence the practice of postlayout simulation using back annotated delays is an important discipline for an engineer to learn in preparation for moving to mask programmable ASICs.

When all the CAD stages are completed the FPGA net-list file is converted into a programming file to program the device. This is either a standard EPROM bit file for the Xilinx and Altera arrays or a fuse file for the Actel devices. Once a device is programmed, debug and diagnostic facilities are available. These allow the logic state of any node in the circuit to be investigated after a series of signals has been passed to the chip via the PC serial or parallel port. This feature is unique to FPGAs since each node is addressable unlike mask programmable devices.

FPGA CAD tools are usually divided into two parts. The first is the prelayout stage or front-end software, i.e. schematic and prelayout simulation. The CAD tools here are generic (suitable for any FPGA) and are provided by proprietary packages such as Mentor Graphics, Cadence, Viewlogic, Orcad, etc. However, to access the FPGAs the corresponding libraries are required for schematic symbols and models.

The second part is called the *back-end* software incorporating: layout; back annotation of routing delays; programming file generation and debug. The software for this part is usually tied to a particular type of FPGA and is supplied by the FPGA manufacturer.

For example consider a typical CAD route with Actel on a PC. The prelayout (or front end) tools supplied by Viewlogic can be used to draw the schematic using a package called Viewdraw and the prelayout functional simulation is performed with Viewsim. In both cases library files are needed for the desired FPGA. Once the design is correct it can be converted into an Actel net-list using a *net-list translator*. This new file is then passed into the CAD tools supplied by Actel (called Actel Logic System – ALS) ready for place and routing. The parasitic delays can be extracted and back annotated out of ALS back into Viewlogic so that a post-layout simulation can be performed again with Viewsim. If the simulation is not correct then the circuit schematic must be modified and the array is placed and routed again. Actel provide a static timer to check set-up and hold time and calculate the delays down all wires indicating which wire is the heaviest loaded. A useful facility is the *net criticality* assignment which allows nets to be tagged depending on how speed critical they are. This facility controls the placing and routing of the logic in order to minimise wiring delays wherever possible. The device is finally programmed by first creating a fuse file and then blowing the fuses via a piece of hardware called an *activator*. This connects to an Actel programming card inside the PC. As an example of the length of time the place and route software can take to complete the authors ran a design for a 68 pin Actel 1020 device. The layout process took approximately 10 minutes using a 486, 66 MHz PC and utilised 514 (approximately 1200 gates) of the 547 modules available (i.e. a utilisation of 94%). In addition on the same computer the fuse programming via the activator took around 1 minute to complete its program. With mask programmable ASICs, however, the programming step can take at least four weeks to complete! This is one of the great advantages that FPGAs have over mask programmable ASICs. Note, however, that as with mask programmable arrays the FPGA manufacturers only provide a limited range of array sizes. The final design thus never ever uses all of the gates available and hence silicon is wasted. Also, as the gates are used up on the array the ability for the router to access the remaining gates decreases and hence although a manufacturer may quote a maximum gate count for the array the important figure is the percentage utilisation.

Actel FPGAs also have comprehensive postprogramming test facilities available under the option 'Debug'. These consist of: the functional debug option; and the in-circuit diagnostic tool. The *functional debug* test involves sending test vectors from the PC to the activator, which houses the FPGA during programming, and simple tests can be carried out. The *in-circuit diagnostic* tool is used to check the real time operation of the device when in the final PCB. This test is 100% observable in that any node within the chip can be monitored in real time with an oscilloscope via two dedicated pins on the FPGA.

The Xilinx FPGA devices are programmed in a similar way by using two pieces of software. Again typical front-end software for these devices is Viewlogic utilising Viewdraw and Viewsim for circuit entry and functional simulation respectively. The net-list for the schematic is this time converted into a Xilinx net-list and the design can now move into the Xilinx development software supplied by Xilinx (called XACT). Although individual programs exist for place and route, parasitic extract, programming file generation, etc., Xilinx provide a simple to use compilation utility called XMAKE. This runs all of these steps in one process. Parasitic delays can again be back annotated to Viewsim for a timing simulation with parasitics included. A static timing analyser is again available so that the effects of delays can be observed on set-up and hold time without having to apply input stimuli. Bit stream configuration data, used in conjunction with a Xilinx provided cable, allow the data to be down-loaded to the chip for configuration. As with Actel both debug and diagnostic software exist such that the device can be tested and any node in the circuit monitored in real time. The bit stream data can be converted into either Intel (MCS-86), Motorola (EXORMAX) or Tektronix (TEKHEX) PROM file formats for subsequent PROM or EPROM programming. The one disadvantage of these devices as compared to the Actel devices is that when in final use the device needs to have an associated PROM or EPROM which increases the component count.

## 11.5 VHDL

As systems become more complex the use of schematic capture programs to specify the design becomes unmanageable. For designs above 10 000 gates an alternative design entry technique of *behavioural specification* is invariably employed. This is a high-level programming language that is textual in nature, describes behaviour and maps to hardware. The most commonly accepted behavioural language is that standardised by the IEEE (standard 1076) in 1987 called VHDL. VHDL is an acronym for *VHSIC Hardware Description Language* where VHSIC (Very High Scale Integrated Circuits) is the next level of integration above VLSI. This language was developed by the USA Department of Defense and is now a world-wide standard for describing general digital hardware. The language allows a system to be described at many different levels from the lowest level of logic gates (called structural) through to behavioural level. At behavioural level the design is represented in terms of programming statements which makes no use of logic gates. This behaviour can use a digital (i.e. Boolean), integer or real representation of the circuit operation. A system designer can specify the design at a high level (i.e. in integer behavioural) and then pass this source code to another member of the group to break the design down into individual smaller blocks (partitioning). A block in behavioural form requires only a few lines of code and hence is not as complex as a structural logic gate description and hence has a shorter simulation time. Since the code supports mixed levels (i.e. gate and behaviour) then the system can be represented with some blocks represented at the gate level and the rest at behavioural. Thus the complete system can be simulated in a much shorter time.

One of the biggest problems of designing an ASIC is the interpretation of the specification required by the customer. Because VHDL has a high-level description capability it can be used also as a formal specification language and establishes a common communication between contractors or within a group. Another problem of ASIC design is that you have to choose a foundry *before* a design is started thus committing you to that manufacturer. Hence, it is usual to insist on second sourcing outlets to avoid production hold-ups. However, VHDL at the high level is technology and process independent and is therefore transportable into other processes and CAD tools. It is not surprising that many companies are now insisting on a VHDL description for their system as an extra deliverable as well as the chip itself.

A simple example of a VHDL behavioural code for a 2-to-1 multiplexer is shown in Table 11.2. This source code is divided into two parts: *entity* and *architecture*. The *entity* lists the input and output pins and what form they are – bit or binary in this case – whilst the *architecture* describes the behaviour of the multiplexer. The process labelled *f1* is only run if any of the inputs *d0*, *d1* or *sel* change, i.e. it is an event driven simulator. If one of these events occurs the IF statement is processed and the output *q* is set depending upon the value of *sel*.

Table 11.2 VHDL behavioural code for a 2-to-1 multiplexer

```
ENTITY mux IS
   PORT (d0, d1, sel:IN bit; q:OUT bit);
END mux;
ARCHITECTURE behaviour OF mux IS
   BEGIN
   f1:
   PROCESS (d0,d1,sel) —sensitivity list
   BEGIN
   IF sel=0 THEN
      q<=d1;
      ELSE
      q<=d0;
   END IF;
END behaviour;
```

Notice that since this is a behavioural description then no logic gates are used in the architecture. The next stage would be to convert this design into logic gates. This can be performed in two ways: automatically or manually. With the automatic approach an additional CAD software package is required called a *synthesiser*. These are available at an extra charge and will generate the logic gates required to implement the desired behaviour. Alternatively this step can be performed manually. A typical VHDL structural description of the above multiplexer implemented with logic gates is shown in Table 11.3.

Table 11.3 VHDL structure code for a 2-to-1 multiplexer

```
ENTITY mux IS
   PORT (d0, d1, sel:IN bit; q:OUT bit);
END mux;
ARCHITECTURE structure OF mux IS
   COMPONENT and2
      PORT (in1, in2:IN bit; f:OUT bit);
   END COMPONENT;
   COMPONENT or2
      PORT (in1, in2:IN bit; f:OUT bit);
   END COMPONENT;
   COMPONENT inv
      PORT (in1, in2:IN bit; f:OUT bit);
   END COMPONENT;
   SIGNAL x, y, nsel: bit;
   FOR U1: inv USE ENTITY work.inv;
   FOR U2: and2 USE ENTITY work.and2;
   FOR U3: or2 USE ENTITY work.or2;
   BEGIN
      U1: inv PORT MAP(sel, nsel)
      U2: and2 PORT MAP(nsel, d1, y)
      U3: and2 PORT MAP(d0, sel, x)
      U4: or2 PORT MAP(x,y,q)
END structure;
```

Since this is only a trivial example then a manual synthesis is possible. It is also apparent that a behavioural code is more succinct than a structural one hence the simulation is faster. As the design becomes more complex then the use of an automatic synthesiser is essential.

## 11.6 CHOOSING A DESIGN ROUTE

So we now know all the options available to a digital circuit designer. The decision is now to choose the appropriate route. It is wise at this point to revisit the ten questions that were raised at the beginning of this chapter and to consider them in the light of the summarised information given in Table 11.4.

A standard part (called 'Std. Part' in the table) design route (i.e. 74HCT, etc.) is certainly the quickest to get started and can handle large and complex designs. However, it may well be limited when the design needs to be miniaturised or put into production. Also design techniques with standard products often tend to be

Table 11.4 Comparison of digital design routes

| | FC | SC | MPGA | FPGA | PAL/PLA | Std. Part |
|---|---|---|---|---|---|---|
| Design time (months) | 6–12 | 2–6 | 1–6 | 1–30 days | 1–14days | 1–30days |
| Fab. time (months) | 2–4 | 1–3 | 2–6 wks | 1–10 mins | 1–5mins | 14 days |
| Time to mkt. (months) | 8–16 | 3–9 | 1.5–7.5 | 30 days | 14 days | 6 wks |
| Prototype cost | hi | hi | med. | lo | V.lo | V.lo |
| Production cost | med./lo | lo | lo | hi | med./lo | hi |
| Speed | V.hi | hi | hi/med. | med./slow | med./hi | med./hi |
| Complexity | V.hi | hi | hi/med. | med./lo | lo | V.hi |
| Redesign time (months) | 3–5 | 2–4 | 3–6 wks | 5 days | 2 days | 1–14 days |
| CAD complexity | V.hi | hi/med. | med. | med./lo | lo | lo |
| Risk | V.hi | hi | med. | lo | V.lo | V.lo |

*ad hoc* and in some instances not synchronous. The design may use *RC* components, 555 timers and gated clocks. If the design is only a one-off and it functions correctly then this will be perfectly satisfactory if size and power are not an issue. However, if the design requires miniaturisation or transfer to an ASIC for power consumption reasons then the circuit will have to be redesigned for a totally synchronous approach as discussed in Chapter 8.

Of the AND-OR array devices a ROM device can be used to efficiently perform a number of digital tasks, whilst PALs and PLAs are also widely available providing much of the capability of ROM but with smaller circuits. However, it is necessary to minimise the Boolean functions before they can be implemented in a PAL or PLA. A PAL allows a fixed number of minimised product terms to be summed whilst a PLA enables any number of the product terms formed to be summed. However, although PALs and PLAs are quite adequate for gate counts of the order of 500 they are limited due to having no buried registers.

Obviously no one route will satisfy all options but FPGAs are becoming a strong prototyping contender. At present FPGA performance is still below that of mask programmable gate arrays (MPGA) which still have the edge in terms of high performance (high speed, low power consumption), high gate density or large volumes. However, for small volumes, FPGAs offer a virtually immediate turn around time and relatively low cost (particularly in terms of the non-recurring engineering (NRE) costs which can be very large for many ASIC designs). In addition since a single FPGA is relatively inexpensive the risk factor is significantly less and hence the emphasis on the simulation stage is reduced. An FPGA can thus be programmed and tested rapidly. However, if a design fault exists then the fault can be quickly corrected and the device reprogrammed. The use of an FPGA is a very powerful vehicle for testing out an idea before going to volume production. Translators are available which can convert FPGA designs into mask programmable ASICs once the design is confirmed at the prototype stage. FPGAs also support VHDL description and since VHDL is transportable it can be transferred into standard cell (SC) options if sufficient volume is expected.

The standard cell (SC) and full custom (FC) route appear at first sight to be much too expensive to consider in small volumes. However, processes such as multiproject wafers and the *Europractice* initiative have brought these routes into the reach of small companies and universities. Europractice (previously called *Eurochip*) provides low-cost access to both CAD software and foundries. A very competitive service for standard cell and full custom designs is offered. For example a typical standard cell charge for a 2000 gate design costs approximately £500 for 10 devices. It is initiatives such as these that provide training for future IC designers to move into industry and take advantage of the latest technology. For those of you who wish to explore further the Europractice route then visit the web site: *http://www.te.rl.ac.uk/europractice* or *http://www.imec.be/europractice*.

## 11.7 SELF-ASSESSMENT

11.1 Define the acronym 'ASIC'.

11.2 List the *safe* rules to follow for mask programmable ASIC design.

11.3 Explain what is meant by the term *gated clock*.

11.4 Why should gated clocks be avoided?

11.5 What problems do monostables have for digital design?

11.6 Why should delay lines be avoided in ASIC designs?

11.7 Name three ways in which programmable logic devices can be programmed.

11.8 What is the difference in architecture between a ROM, PAL and a PLA?

11.9 Define *utilisation* with respect to gate arrays.

11.10 What does the acronym *VHDL* stand for?

11.11 Why is it necessary to resimulate after a chip is laid out?

11.12 What is meant by a *functional simulation*?

11.13 Define an *event driven simulator*.

11.14 Is the following statement true or false: 'A gate array has more customer specific masks than a standard cell design'?

11.15 Name two pieces of software for producing a JEDEC file for a PAL type device.

11.16 How does a GAL differ from a PAL?

11.17 A hypothetical PAL has 10 input terms and four outputs (active high). What is its part number?

11.18 Repeat Question 11.17 for a GAL.

11.19 A VHDL description is divided into two parts. Name these parts.

11.20 With respect to VHDL what is the difference between a *structural* description and a *behavioural* description?

11.21 Define *NRE* with respect to ASICs.

## 11.8 PROBLEMS

11.1 Assuming that you are the first digital designer employed by a company, choose a design option for the following digital circuits:
(a) seven input, three output truth table (low volume);
(b) controller using 200 gates (high volume);
(c) controller using 1500 gates (small size, low volume);
(d) high-speed synchronous sequencer with 1000 miscellaneous logic gates (very high volume);
(e) controller using 15 000 gates (high volume);
(f) a single PCB for the control of a multisite temperature measurement system (no size and power restrictions but required within two weeks).

11.2 How could an eight-bit ROM with 2048 memory locations be used as a look-up table for the function $y = \sin x$?

11.3 How could the ROM used as a look-up table for $y = \sin x$ in Problem 11.2 be used to generate a 500 Hz sine wave?

11.4 A digital designer designing with a gate array requires a JK flip-flop. However, the library contains only D-type flip-flops and basic gates. Using a state diagram approach design a circuit that will implement the JK function from a D-type and these basic gates.

11.5 The system in Fig. 11.8 assumes that the interrupt input line is synchronised to the clock. If *interrupt* can now arrive at any time (i.e. asynchronously) then redesign the circuit such that the load line must go high for one clock pulse after *interrupt* goes high. Hint: redraw the timing diagram and label the states (you will now need three states).

11.6 Using the PAL in Fig. 11.22 produce a fuse map for the truth table shown in Fig. 11.26. What software is available to produce a JEDEC file directly from a truth table without having to draw the Karnaugh maps?

11.7 A 10 MHz clock is applied to the 12 bit counter in Fig. 11.7 with the 74HC74 device cleared by the *Q4* output. Given that the propagation delay of each bistable in the counter is 10 ns what is the *exact* pulse width produced at the Q output of the 74HC74 device? Assume all other delays are zero.

| A | B | C | D | X | Y | Z |
|---|---|---|---|---|---|---|
| 0 | 0 | 0 | 0 | 0 | 0 | 0 |
| 0 | 0 | 0 | 1 | 0 | 1 | 1 |
| 0 | 0 | 1 | 0 | 0 | 0 | 0 |
| 0 | 0 | 1 | 1 | 1 | 0 | 0 |
| 0 | 1 | 0 | 0 | 0 | 0 | 1 |
| 0 | 1 | 0 | 1 | 0 | 1 | 0 |
| 0 | 1 | 1 | 0 | 0 | 0 | 0 |
| 0 | 1 | 1 | 1 | 0 | 0 | 1 |
| 1 | 0 | 0 | 0 | 1 | 0 | 0 |
| 1 | 0 | 0 | 1 | 1 | 0 | 0 |
| 1 | 0 | 1 | 0 | 0 | 0 | 0 |
| 1 | 0 | 1 | 1 | 0 | 1 | 0 |
| 1 | 1 | 0 | 0 | 1 | 1 | 0 |
| 1 | 1 | 0 | 1 | 1 | 0 | 1 |
| 1 | 1 | 1 | 0 | 0 | 0 | 0 |
| 1 | 1 | 1 | 1 | 0 | 1 | 0 |

Fig. 11.26 Truth table for Problem 11.6

11.8 The inverter chain buffer circuit of Fig. 11.11(a) is used to drive a load of 24 unit loads where one unit load is equal to the input capacitance of the first inverter. If $N=4, f=3$ and the delay of the first inverter driving itself is 2 ns then what is the total delay for this network?

11.9 The basic cell used in the Actel Act1 FPGA (see Fig. 11.25) has the following input signals applied: $S_1S_2S_3S_4$=0ABC and $WXYZ$=0A01. If '$A$', '$B$' and '$C$' are inputs having values of '0' or '1', and for each multiplexer when the select line is low the lower input is selected, what function is performed by this circuit?

11.10 A master clock circuit is to drive two circuits, $A$ and $B$, having input capacitances of 5 and 10 unit loads respectively. Each circuit is buffered by a non-inverting buffer having an inherent delay of 1 ns and output drive capability of 15 ns/unit load. What is the relative delay between the clock signals arriving at each circuit?

11.11 A small CMOS inverter is to drive a 50 pF load via two inverting buffers whose $W/L$ ratios increase by a factor of $f$, as shown in Figure 11.11(a). Calculate the value of $f$ to achieve a minimum delay and the magnitude of this delay. Assume that each inverter has a zero inherent delay and the small CMOS inverter has a loading delay of 12.5 ns/pF and an input capacitance of 0.01 pF.

# 12 Answers to selected self-assessment questions and problems

## Chapter 1

### Self-assessment

**1.1:** 0 or 1 (HIGH or LOW). **1.2:** 3; NOT, AND, OR. **1.3:** six columns; 16 rows. **1.4:** As variables are AND'd together. **1.6:** A number of product terms are OR'd (summed) together. **1.7:** Can implement NOT function; and then use duality for other functions. **1.8:** Boolean algebra; truth table; circuit diagram; timing diagram. **1.9:** (a) 0; (b) no change; (c) no change; (d) 1.

### Problems

**1.1:** Use Equation 1.7, then see Ex. 1.13. **1.2:** Similar to Ex. 1.14, except use Equation 1.15 first. **1.3:** Begin by using De Morgan's theorem twice. **1.6:** $Y=\overline{A}+\overline{B\oplus C}$. **1.7:** See Ex. 1.26; 3 NANDs. **1.8:** (c) AND gate; (d) 2 NAND; 3 NOR.

## Chapter 2

### Self-assessment

**2.1:** Variables can only be 0 or 1. **2.2:** 8 units; 0 through to 7. **2.3:** $8^0$, $8^1$ and $8^2$. **2.4:** 16 units; 0 to 9, then A to F. **2.5:** 4×4. **2.6:** Adjacent codes differ only by 1 bit. **2.7:** 000, 001, 011, 010, 110, 111, 101, 100. **2.8:** $2^{12}=4096$. **2.9:** As can multiply/divide by 2 by simply shifting binary representation of number.

### Problems

**2.1:** 230. **2.2:** 317. **2.3:** $142_5$. **2.4:** $3452_6$. **2.5:** 100101100. **2.6:** 4DE. **2.7:** 85. **2.8:** 2750. **2.9:** A59. **2.10:** 0010:0100:0011. **2.11:** 4×200×30=24 000 bits=3000 bytes. **2.12:** $5\div(2^4-1)=333$mV. **2.13:** 100011 (Wire 0's to gate via inverters). **2.15:** 76+43=119; giving +19. **2.16:** 64+17=81; but is negative result; giving −19.

## Chapter 3

### Self-assessment

**3.1:** Outputs only depend upon inputs (no memory). **3.2:** Respectively variables AND'd/OR'd together. **3.3:** Sum of products/product of sums. **3.4:** As for SA question 1.8 plus a Karnaugh map. **3.5:** For $m$-input and $n$-output variables; get $(m+n)$ columns and $2^m$ rows. **3.6:** Has *all* input variables AND'd together; is linked to single row of the truth table. **3.7:** Sum of products or product of sums;

*theoretically* the fastest. **3.8:** Gives sum of products expression with fewer, non-fundamental product terms. **3.9:** Property of inverse elements (after use of distributive law). **3.10:** They place logically adjacent product terms next to each other. **3.12:** Groupings of 8, 4 and 2 minterms. **3.13:** Single cell in Karnaugh map; minimised product term; minimised product term that *must* be used for minimisation. **3.14:** A product term whose value is not defined; can be set to 0 or 1 to aid minimisation. **3.15:** SOP/POS describe where the 1's/0's are on a Karnaugh map. **3.16:** $\log_2 n$.

## Chapter 4

### Self-assessment

**4.1:** To route 1 of 8 inputs to single output (selected via three control lines). **4.2:** As all fundamental product terms are decoded by the multiplexer (mux). **4.3:** Routes single input to one of $n$ outputs (selected via $m$ control lines ($n=2^m$)). **4.4:** Just hold input to de-mux HIGH or LOW as required. **4.5:** Programmable inverter; parity generator/checker; comparator. **4.6:** Adds two bits plus carry-in bit to give sum and carry-out bits. **4.7:** Iterative array; full-adder; rippling carry makes it slow. **4.8:** Carry signals generated directly from inputs (hence name); therefore faster. **4.9:** Race conditions so that signals do *not* have expected values at all times. **4.10:** Signal goes transiently to other state; caused by race conditions for two paths; static-0 and static-1. **4.11:** Signal expected to change state actually does so twice; caused by race conditions for three or more paths. **4.12:** By including non-essential prime implicants in minimised expression.

### Problems

**4.2:** (a) 19; (b) 18 gates. **4.6:** three-input XOR gate. **4.7:** Swap final XOR for NOR gate. **4.8:** Full adder with inverted carry-out. **4.9:** Invert all inputs and add 1 using carry-in. **4.11:** Gives two's complement of input word. **4.12:** $(m+n)$ input and output columns; $2^{(m+n)}$ rows. **4.13:** Need NOT and OR gate. **4.14:** Static-1 hazard for $\bar{B}C$; use this product term for blanking gate.

## Chapter 5

### Self-assessment

**5.1:** Sequential circuits have 'memory' because their outputs depend, in part, upon past outputs. **5.2:** Combinational logic plus 'memory'. **5.3:** For $n$-outputs from 'memory', and $m$-external inputs; have: $2^n$ internal and $2^{m+n}$ possible total states. **5.4:** Memory elements in synchronous circuits are flip-flops which are clocked. Asynchronous circuits are unclocked. **5.5:** The internal inputs and outputs *must* match (as they are connected). **5.6:** Only one input can change at a time (fundamental mode operation). **5.7:** 'Cutting' the connection between internal inputs and outputs. **5.9:** (a) Horizontal; (b) vertical. **5.10:** Oscillation. **5.11:** Non-critical races do *not* affect final output; critical races do.

**Problems**

**5.1:** Detects input sequence (1,0), (1,1), as Circuit 4. **5.2:** State diagram has same form as P5.1. **5.3:** Functionally same state diagram as others; $Z=AB\bar{y}$. **5.4:** $\overline{SR}$ flip-flop. **5.5:** Circuit similar in form to 5.1, 5.2 and 5.3; two stable states for inputs (0, 0): $Z=1$ when entered from (0,1).

## Chapter 6

**Self-assessment**

**6.1:** SR, T, JK, D. **6.2:** They can be toggled so if unclocked when toggled would oscillate. **6.3:** Truth table; excitation table; Karnaugh map; next state equations. Gives necessary inputs for specific outputs. **6.5:** Gives next state of output, $Q^+$, in terms of present output, $Q$, and inputs. **6.6:** Inputs affect outputs: immediately; when clock line reaches (and remains at) a certain level; at the instant an edge occurs on the clock line. **6.7:** Goes to 1 or 0; either immediately or when next clocked. **6.8:** Series of flip-flops whose outputs/inputs are linked. Shifting performs multiplication or division. **6.9:** 12 and 6.

**Problems**

**6.1,2:** Use short pulses as inputs. **6.3:** Feedback path is broken. **6.4:** (a) $J=\bar{K}$; (b) $J=K$; (c) $D=T\oplus Q$. **6.5:** $Q=1$ when $X=1$ else $Q=\bar{C}$. **6.6:** Oscillates between state 010 and 101.

## Chapter 7

**Self-assessment**

**7.1:** The number of count states it possesses. **7.2:** Asynchronous: each flip-flop is clocked by the last (ripple counters); synchronous: all flip-flops clocked simultaneously under control of a combinational logic function of flip-flops' outputs. **7.3:** Set count state; reset count to 0. **7.4:** Either use $\bar{Q}$ outputs as clock or use positive edge triggered flip-flops. **7.5:** Decode $N$th count state and use it to reset all flip-flops; spikes in output. **7.6:** Use T-type flip-flops; always toggle first flip-flop and only toggle others when *all* preceding flip-flops have an output of 1. **7.7:** Use present outputs (via combinational logic) to determine next outputs. **7.8:** Need $M$, with $2^M \geq N$; will have $(2^M - N)$ unused states; depends upon minimisation of 'don't care' states.

**Problems**

**7.1:** Mux to route either $Q$ or $\bar{Q}$ to next FF's clock. **7.2:** Mod-8 binary ripple down counter; $3=\overline{Q_2+\bar{Q}_1+\bar{Q}_0}$. **7.3:** Mod-6 binary ripple up counter. **7.4:** Reset using $Q_2Q_0$. **7.6(a):** $D_2=Q_1Q_0$, $D_1=Q_1\oplus Q_0$, $D_0=\bar{Q}_2\bar{Q}_0$; $J_2=Q_1Q_0$ $K_2=1$, $J_1=Q_0$ $K_1=Q_0$, $J_0=\bar{Q}_2$ $K_0=1$. **7.7:** $J_2=Q_1Q_0$, $K_2=Q_1$, $J_1=Q_0$, $K_1=Q_2+Q_0$, $J_0=\bar{Q}_2+\bar{Q}_1$, $K_0=1$.

## Chapter 8

### Self-assessment

**8.1:** Possesses memory in the form of flip-flops which are clocked together. **8.2:** Autonomous (no external inputs); general synchronous sequential circuit with outputs either depending upon internal inputs only (Moore model) or upon external inputs as well (Mealy model). **8.3:** Write table for present and next states; produce Karnaugh maps for next state variables; minimise to find inputs to flip-flops. **8.4:** Those not required in the design. If entered, erroneous circuit operation may occur; so often made to lead to some specified state.

### Problems

**8.2:** Autonomous. **8.5:** States 4 and 5 cycle; lead them to another state. **8.6:** $I$ controls direction through states. **8.7:** For $I = 1$ sequence of states. A, B, C, D, A; for $I = 0$ sequence is A, B, D, C, A. **8.8:** Moore. **8.9:** Mealy; serial adder.

## Chapter 9

### Self-assessment

**9.1:** 0.2 V. **9.2:** 0.7 V. **9.3:** Transistor current gain or $I_c/I_b$ . **9.4:** Diode Transistor Logic, Transistor Transistor Logic, N-channel Metal Oxide Semiconductor, Complementary MOS, Emitter Coupled Logic, Bipolar and CMOS. **9.5:** True. **9.6:** 74, 74LS, 74F, 74ALS. **9.7:** CMOS: 74HC, 74AC, 74ACT, 74HCT, 74AHC, 4000B. TTL: 74ALS, 74LS, 74F, 74. **9.8:** Both devices are CMOS and pin compatible with TTL devices but the ACT device has input voltage levels that are TTL voltage levels. **9.9:** $I_{ILmax}$. **9.10:** Both currents are equal and very low (0.1 μA). **9.11:** CMOS: low power, high density. TTL: high speed but now being superseded by high-speed CMOS processes. **9.12:** Because the transistors are prevented from entering saturation. **9.13:** If $V_{DS} < V_{GS} - V_T$ then the device is in the *linear* region and $I_{DS} = K[(V_{GS} - V_T).V_{DS} - V_{DS}^2/2]$. If however, $V_{DS} > V_{GS} - V_T$ then the device is in the *saturation* region and $I_{DS} = [K/2][V_{GS} - V_T]^2$. **9.14:** $K = (W/L)\mu C_{ox}$. **9.15:** Since $\mu_n > \mu_p$ then in order to ensure that $K_N = K_P$ the PMOS device is larger than the NMOS. **9.16:** True. **9.17:** $P_{dynamic} = C_L.V_{dd}^2.f$. **9.18:** $\tau_p = 2C_L/(K_N.V_{dd})$. **9.19:** Series NMOS parallel PMOS, parallel NMOS series PMOS. **9.20:** One NMOS and one PMOS transistor back to back. **9.21:** F100K, 74AC/74ALS, 74HC, 74LS. **9.22:** 74AC/74HC, 74ALS, 74LS, F100K. **9.23:** Product of power and delay. **9.24:** BiCMOS, CMOS, GaAs, ECL, TTL. **9.25:** No.

### Problems

**9.1:** 1.17 kΩ. **9.2:** $P_{low} = 4.18$ mW (T1 is off), $P_{high} = 26.96$ mW (T1 is on). **9.3:** 8 (high), 4 (low). **9.4:** 0.4(high)/0.4(low), 0.7/0.3, 2.3/0.47, 0.3/0.3. This assumes that o/p's are at the minimum and maximum conditions. **9.5:** 76.6. **9.6:** 0.289 mA. **9.7:** 6 ns, increase $V_{dd}$ and/or decrease temperature. **9.8:** 200 ns, 50 ns. **9.9:** NMOS: A in series with B in series with a parallel combination of D and C. Reverse for PMOS. **9.10:** 625 Ω. **9.11:** $K_N = 1053\ \mu A\ V^{-2}$, $K_P = 13.9\ \mu A\ V^{-2}$.

## Chapter 10

### Self-assessment

**10.1:** Read only memory; random access memory; static RAM; dynamic RAM; erasable PROM; read–write memory; pseudo SRAM; non-volatile SRAM; one-time PROM; electrically erasable and programmable ROM. **10.2:** Retains data; loses data; loses; loses; retains; depends if EEPROM/FLASH (retains) or RAM (loses); loses; retains; retains; retains. **10.3:** When the power is removed the data is retained. **10.4:** Time taken to read data from memory. **10.5:** In ascending transistor count per bit: mask ROM; EPROM; FLASH; EEPROM(MNOS); OTPROM; DRAM; EEPROM(floating gate); SRAM. **10.6:** Hard disk write times are typically 100 ms whilst SRAM is 0.1 μs. **10.7:** SRAM/DRAM (0.1 μs); FLASH (10 μs); EPROM (50 μs but takes 20 minutes to erase); OTPROM(100 μs but only one write operation); EEPROM (10 ms); hard disk (100 ms). **10.8:** DRAM due to its high capacity and fast write times. **10.9:** Single in line memory module. **10.10:** A thin plastic card containing memory chips with a standard 68 pin connector. **10.11:** $\overline{CE}=0$, $\overline{OE}=0$, $\overline{WE}=1$. **10.12:** Data is stored as charge on a capacitor via an off transistor. This charge can leak away through the off transistor and hence must be periodically recharged. **10.13:** All need $V_{dd}$ and $V_{ss}$ except the EPROM and FLASH which currently require an extra high voltage pin, $V_{pp}$. **10.14:** SRAM device goes into low-power mode. **10.15:** EPROM and FLASH.

### Problems

**10.1:** (a) 2048 words, (b) 8 bits, (c) 16 384 bits. **10.2:** (a) 32 pins, (b) 24 pins. **10.3:** (a), (b) drive A19 to $\overline{CE}$ of one chip and $\overline{A19}$ to $\overline{CE}$ on the other chip, (c) $\overline{CE}_1 = A19+A20$, $\overline{CE}_2=\overline{A19}+A20$, $\overline{CE}_3=A19+\overline{A20}$, $\overline{CE}_4=\overline{A19.A20}$. **10.4:** 0.42 pF. **10.5:** $t_{\text{byte verify}}=6.21$ μs, $t_{\text{byte write}}=10.19$ μs. **10.6:** 250 kΩ. **10.7:** 20 mW, 12 mA hours – but does not include dynamic power dissipation.

## Chapter 11

### Self-assessment

**11.1:** Application Specific Integrated Circuit. **11.2:** No gated clocks/resets, monostables, *RC*/*CR* circuits and delay chains; use synchronous techniques, use a high frequency clock; use clock buffering. **11.3:** A clock signal passed through a logic gate such as an AND gate. **11.4:** These can cause spikes and glitches on clock lines and hence can cause incorrect clocking of flip-flops. **11.5:** A monostable is a device that in response to a rising (or falling edge) will produce a pulse of duration dictated by external *C* and *R*. External *R* and *C* required (not suitable for ASICs); pulse width varies with temperature, $V_{cc}$ and from device to device; poor noise margin; unclean signal for narrow pulses. **11.6:** The propagation delay on a chip (and across it) varies considerably. **11.7:** SRAM, fuse, EPROM. **11.8:** ROM: fixed AND–programmable OR; PAL: programmable OR–fixed OR; PLA: programmable OR–programmable AND. **11.9:** The percentage of gates used in a design. **11.10:** VHSIC (Very High Scale Integrated Circuits) Hardware Description

Language. **11.11:** When a chip is laid out the interconnect and other layers add $R$ and $C$ and so slow down signals – hence it must be resimulated. **11.12:** All gates have the same delay (typically 1 ns) and no set-up, hold and pulse width checks are carried out. **11.13:** Only when an input to a gate changes will the gate output be computed. **11.14:** False. **11.15:** ABEL, PALASM, CUPL. **11.16:** Generic Array Logic – a PAL with a versatile cell at the output called an OLMC (output logic macro cell). **11.17:** PAL10H4. **11.18:** GAL10V4. **11.19:** Entity and architecture. **11.20:** Structure: logic gates and blocks connected together; behaviour: use of high-level statements (i.e. IF THEN etc.) to describe function of system. **11.21:** Non-recurrent engineering charges – tooling costs.

**Problems**

**11.1:** (a) PAL, FPGA; (b) GAL, standard cell; (c) FPGA, standard products; (d) standard cell, full custom; (e) gate array, standard cell; (f) standard products, FPGA. **11.2:** ROM would need to have 11 address lines (i.e. 2048 locations) and eight data bits. **11.3:** Add DAC to output of ROM at an addressing clock frequency of 0.9 MHz. **11.4:** $D = J.\bar{Q} + Q.\bar{K}$. **11.5:** Two D-types: $D_0 = \bar{Q}_0.Q_1 + I.Q_0$; $D_1 = I.\bar{Q}_0.\bar{Q}_1$ where $D_1$ output is LOAD. **11.6:** $X = A.\bar{C} + \bar{A}.\bar{B}.C.D$; $Y = A.C.D + \bar{A}.\bar{C}.D + A.B.\bar{C}.\bar{D}$; $Z = \bar{A}.B.\bar{C}.\bar{D} + \bar{A}.\bar{B}.\bar{C}.D + A.B.\bar{C}.D + \bar{A}.B.C.D$; use ABEL or PALASM. **11.7:** 840ns. **11.8:** 19.8 ns (both the input capacitance and $K$ increases by '$f$' at each stage). **11.9:** output $= \bar{A}.\bar{C} + A.\bar{B}$. **11.10:** 75 ns. **11.11:** $f = 17.1$, minimum delay $= 6.42$ ns.

# Index

HAVERING COLLEGE OF F & H E
193569